High-Pressure Shock Compression of Condensed Matter

Editors-in-Chief
Lee Davison
Yasuyuki Horie

Founding Editor
Robert A. Graham

Advisory Board
Roger Chéret, France
Vladimir E. Fortov, Russia
Jing Fuqian, China
Yogendra M. Gupta, USA
James N. Johnson, USA
Akira B. Sawaoka, Japan

High-Pressure Shock Compression of Condensed Matter

L.L. Altgilbers, M.D.J. Brown, I. Grishnaev, B.M. Novac, I.R. Smith, I. Tkach, and Y. Tkach: Magnetocumulative Generators

T. Antoun, D.R. Curran, G.I. Kanel, S.V. Razorenov, and A.V. Utkin: Spall Fracture

J. Asay and M. Shahinpoor (Eds.): High-Pressure Shock Compression of Solids

S.S. Batsanov: Effects of Explosion on Materials: Modification and Synthesis Under High-Pressure Shock Compression

R. Cherét: Detonation of Condensed Explosives

L. Davison, D. Grady, and M. Shahinpoor (Eds.): High-Pressure Shock Compression of Solids II

L. Davison, Y. Horie, and T. Sekine (Eds.): High-Pressure Shock Compression of Solids V

L. Davison, Y. Horie, and M. Shahinpoor (Eds.): High-Pressure Shock Compression of Solids IV

L. Davison and M. Shahinpoor (Eds.): High-Pressure Shock Compression of Solids III

A.N. Dremin: Toward Detonation Theory

Y. Horie, L. Davison, and N.N. Thadhani (Eds.): High-Pressure Shock Compression of Solids VI

R. Graham: Solids Under High-Pressure Shock Compression

J.N. Johnson and R. Cherét (Eds.): Classic Papers in Shock Compression Science

V.F. Nesterenko: Dynamics of Heterogeneous Materials

M. Sućeska: Test Methods of Explosives

J.A. Zukas and W.P. Walters (Eds.): Explosive Effects and Applications

G.I. Kanel, S.V. Razorenov, and V.E. Fortov: Shock-Wave Phenomena and the Properties of Condensed Matter

V.E. Fortov, L.V. Altshuler, R.F. Trunin, and A.I. Funtikov: High-Pressure Shock Compression of Solids VII

L.C. Chhabildas, L. Davison, Y. Horie (Eds.): High-Pressure Shock Compression of Solids VIII

Valery K. Kedrinskii

Hydrodynamics of Explosion

Experiments and Models

Translated by Svetlana Yu. Knyazeva
With 175 Figures

 Springer

Professor Valery K. Kedrinskii
Russian Academy of Sciences
Lavrentyev Inst. Hydrodynamics
Lavrentyev prospect 15
630090 Novosibirsk, Russian Federation
Email: kedr@hydro.nsc.ru

Editors-in-Chief:

Dr. Lee Davison
39 Cañoncito Vista Road
Tijeras, NM 87059, USA
Email: leedavison@aol.com

Dr. Yasuyuki Horie
804 East Lake Drive
Shalimar, FL 32579, USA
Email: horie@eglin.af.mil

ISBN-13: 978-3-642-06130-1   e-ISBN-13: 978-3-540-28563-2

This work is subject to copyright. All rights are reserved, whether the whole or part of the material is concerned, specifically the rights of translation, reprinting, reuse of illustrations, recitation, broadcasting, reproduction on microfilm or in any other way, and storage in data banks. Duplication of this publication or parts thereof is permitted only under the provisions of the German Copyright Law of September 9, 1965, in its current version, and permission for use must always be obtained from Springer. Violations are liable to prosecution under the German Copyright Law.

Springer is a part of Springer Science+Business Media

springeronline.com

© Springer-Verlag Berlin Heidelberg 2010
Printed in Germany

The use of general descriptive names, registered names, trademarks, etc. in this publication does not imply, even in the absence of a specific statement, that such names are exempt from the relevant protective laws and regulations and therefore free for general use.

Cover design: *design & production* GmbH, Heidelberg

Printed on acid-free paper   54/3141/YL   5 4 3 2 1 0

Dedicated to the memory of Academician M.A. Lavrentiev

Preface

In the past century, much progress has been made in the hydrodynamics of explosion, the science dealing with liquid flows under shock-wave loading. This area is concerned primarily with studying underwater and underground explosions, cumulation, and the behavior of metals under the action of the detonation products of high explosives (HE), which produce extreme conditions such as pressures of hundreds of kilobars and temperatures up to several thousand degrees.

This book presents research results of underwater explosions. Included are a detailed analysis of the structure and parameters of the wave fields generated by explosions of cored and spiral charges, a description of the formation mechanisms for a wide range of cumulative flows at underwater explosions near the free surface, and the relevant mathematical models. Shock-wave transformation in bubbly liquids, shock-wave amplification due to collision and focusing, and the formation of bubble detonation waves in reactive bubbly liquids are studied in detail. Particular emphasis is placed on the investigation of wave processes in cavitating liquids, which incorporates the concepts of the strength of real liquids containing natural microinhomogeneities, the relaxation of tensile stress, and the cavitation fracture of a liquid as the inversion of its two-phase state under impulsive (explosive) loading. The problems are classed among essentially nonlinear processes that occur under shock loading of liquid and multiphase media and may be of interest to researchers of physical acoustics, the mechanics of multiphase media, shock-wave processes in condensed media, explosive hydroacoustics, and cumulation.

Obviously, the formulation and solution of problems is often initiated under the influence of our teachers and colleagues. First of all I would like to mention my teachers, Prof. R.I. Soloukhin and Prof. M.A. Lavrentiev. The author is indebted to Prof. L.V. Ovsyannikov for his attention and continuous interest in some problems described in the monograph. The contacts and discussions with my colleagues and friends, Profs. V.M. Titov, V.E. Nakoryakov, R.I. Nigmatullin, E.I. Shemyakin, B.D. Khristoforov, Yu.A. Trishin, Kazujoshi Takayama, Leen van Wijngaarden, Charles Mader, Werner Lauterborn, Brad Sturtevant, David Blackstock, David Crighton, and many others, were very encouraging. Many results were obtained together with researchers in my laboratory: Drs. S.V. Stebnovsky, A.R. Berngardt, A.S. Besov, N.N. Chernobaev, M.N. Davydov, I.G. Gets, V.T. Kuzavov, E.I. Pal'chikov, S.V. Plaksin, and S.P. Taratuta. Recent results obtained in cooperation with Profs. N.F. Morozov, Yu.I. Shokin, Drs. V.A. Vshivkov, G.I. Dudnikova, and G.G. Lazareva are also included in the book. I am grateful to Profs. V.P. Korobeinikov, B.D. Khristoforov, Fumio Higashino, Leif Bjorno, Leen Wijngaarden, Yoichiro Matsumoto, and David Blackstock for valuable comments and recommendations for publication of the English version of the monograph. For this purpose the Russian version of the monograph was considerably reworked and extended.

I would like to express my cordial thanks to Dr. S.Yu. Knyazeva for the high-quality translation of the book. I am also thankful to Dr. M.N. Davydov for technical assistance in the preparation of the camera-ready copy.

The English version of the monograph is based on the 2000 Russian edition, which has been revised and supplemented at the suggestion of the series' editorial board.

Novosibirsk, *V.K. Kedrinskii*
February 2005

Contents

Introduction 1

1 Equations of State, Initial and Boundary Conditions 7
   1.1 Equations of State for Water 7
   1.2 Equations of State for Detonation Products 15
   1.3 Conservation Laws, PU-Diagrams, and Transition Formulas 18
      1.3.1 Transition in a Shock Wave 19
      1.3.2 Transition in a Simple Wave 21
   1.4 Generalized Equation of Pulsations of an Explosive Cavity ... 23
   References 26

2 Underwater Explosions, Shock Tubes,
   and Explosive Sound Sources 29
   2.1 Kirkwood–Bethe Approximation, Cylindrical Symmetry 29
      2.1.1 Basic Assumptions, Initial Conditions 29
      2.1.2 Dynamics of an Explosive Cavity,
             Riemann's Function, Delay Integral 32
      2.1.3 Calculation of the Delay Integral (for $\nu = 1, 2$) 34
      2.1.4 Shock Waves
             (Cylindrical Symmetry, Comparison with Experiment) 35
      2.1.5 Dynamics of Exponent Index, $\theta(r_{\mathrm{fr}})$ 36
      2.1.6 SW Parameters (Trotyl, Calculation and Experiment) 38
      2.1.7 Asymptotic Approximation for Weak Shock Waves
             ($\nu = 1, 2$) 39
   2.2 Hydrodynamic Shock Tubes 42
      2.2.1 Filler's Conical Shock Tube 42
      2.2.2 Glass's One-Diaphragm Shock Tube 43
      2.2.3 Electromagnetic Shock Tube 45
      2.2.4 Two-Diaphragm Hydrodynamic Shock Tubes 47
      2.2.5 Shock Tube Application: High-Rate Reactions
             in Chemical Ssolutions 53
   2.3 Explosive Hydroacoustics 57
      2.3.1 Basic Characteristics of Explosive Sound Sources 57
      2.3.2 Hydrodynamic Sources of Explosive Type 60
      2.3.3 Wave Field, Spectral Characteristics 63

X   Contents

        2.3.4   Array Systems 67
        2.3.5   Explosion of Spiral Charges, Wave Structure 69
  2.4   HE-Nuclear-Tests, Explosive Acoustics and Earthquakes..... 74
  References 78

3   Explosion of Cylindrical and Circular Charges 83
  3.1   Dynamics of a Cylindrical Cavity 83
  3.2   Cylindrical Cavity: Approximated Models
        for Incompressible Liquid 87
        3.2.1   Generalized Equation (for $c \to \infty$) 87
        3.2.2   Cylindrical Cavity under Free Surface 89
        3.2.3   The Model of Liquid Cylindrical Layer 90
        3.2.4   Oscillation Period of a Cylindrical Cavity............ 91
  3.3   Circular Charges 93
  3.4   Dynamics of a Toroidal Cavity, Numerical Models 101
        3.4.1   Ideal Incompressible Liquid 101
        3.4.2   Compressible Liquid 105
        3.4.3   Comparison with Experiments 110
        3.4.4   Basic Characteristics for Toroidal Charges 112
  3.5   Comparative Estimates for Spherical Charges 112
  3.6   Oscillation Parameters 115
  3.7   Explosions of Spatial Charges in Air 116
        3.7.1   Experimental Arrangement 116
        3.7.2   Test Measurement Results 117
  References 123

4   Single Bubble, Cumulative Effects
  and Chemical Reactions 125
  4.1   Passive Gas Phase 125
        4.1.1   Short Shock Waves 125
        4.1.2   Formation of a Cumulative Jet in a Bubble
                (Experiment) 127
        4.1.3   Real State of a Gas 130
        4.1.4   Viscosity and the Effect of Unbound Cumulation 131
        4.1.5   Spherical Cumulation in a Compressible Liquid 134
        4.1.6   Oscillation Parameters 136
  4.2   Chemical Reactions in Gas Phase......................... 137
        4.2.1   Todes' Kinetics, Initiation of Detonation
                by a Refracted Wave............................... 137
        4.2.2   Generalized Kinetics of Detonation in Gas Phase 139
        4.2.3   Dynamics of Bubbles Filled with a Reactive Mixture.. 141
  4.3   Mass Exchange and Chemical Reactions................... 142
        4.3.1   Instant Evaporation of Microdrops 143
        4.3.2   Continuous Evaporation........................... 148
  References 151

5 Shock Waves in Bubbly Media 153
- 5.1 Nonreactive Media, Wave Structure,
       Bubbly Cluster Radiation 153
  - 5.1.1 Shock-Gas Layer Interaction 157
  - 5.1.2 Shock Waves in Bubble Layers 158
  - 5.1.3 Two-Phase Model of a Bubbly Liquid:
           Three Estimates of Wave Effects 170
  - 5.1.4 Amplification, Collision, and Focusing
           of Shock Waves 177
- 5.2 Generation of Radiation by Free Bubble Systems 185
  - 5.2.1 Toroidal Bubble Cloud, Mach Disks 186
  - 5.2.2 Spherical Bubbly Clusters: SW Cumulation
           with a Pressure Gradient Along Front.............. 192
- 5.3 Reactive Bubble Media, Waves of Bubble Detonation 196
  - 5.3.1 Shock Waves in Reactive Bubbly Systems 197
  - 5.3.2 Shock Tube with Changing Cross Sections 210
- References 217

6 Problems of Cavitative Destruction 223
- 6.1 Dynamics of Liquid State in Pulsed Rarefaction Waves 223
  - 6.1.1 Real Liquid State (Nucleation Problems) 224
  - 6.1.2 Formation Mechanism of Bubble Clusters 229
  - 6.1.3 Mathematical Model of Cavitating Liquid 233
  - 6.1.4 Dynamic Strength of Liquid 234
  - 6.1.5 Tensile Stress Relaxation
           (Cavitation in a Vertically Accelerated Tube) 236
  - 6.1.6 Transition to the Fragmentation Stage
           (Experimental Methods) 238
  - 6.1.7 "Frozen" Mass Velocities in a Cavitation Zone 245
  - 6.1.8 Model of "Instantaneous" Fragmentation 248
- 6.2 Disintegration of a Liquid, Spalls 257
  - 6.2.1 Cavitative Destruction of a Liquid Drop............. 257
  - 6.2.2 Cavitative "Explosion" of a Liquid Drop 259
  - 6.2.3 Spall Formation in a Liquid Layer 262
  - 6.2.4 Initial Stages of Disintegration: Solids and Liquids.... 264
- 6.3 Cluster, Cumulative Jets, and Cavitative Erosion 267
  - 6.3.1 Single Cavity and Cumulative Jets:
           Experiment and Models 270
  - 6.3.2 Bubble Cluster Effect 273
- 6.4 Cavitative Clusters and Kidney Stone
       Disintegration Problem.................................... 278
  - 6.4.1 Shock Waves, Bubbles and Biomedical Problems 278
  - 6.4.2 Some Results on Modelling of ESWL Applications.... 283
  - 6.4.3 Hydrodynamic Model of the Disintegration
           in the Cavitation Zone 286

XII   Contents

           6.4.4   Rarefaction Phase Focusing and Cluster Formation ... 290
    References 292

7 Jet Flows at Shallow Underwater Explosions 297
    7.1 State-of-the-Art 297
         7.1.1   Irregular Reflection and Bubbly Cavitation 298
         7.1.2   Directional Throwing Out on the Free Surface
                (Sultans) 299
    7.2 Tensile Stress, Structure of Cavitation Region, Spalls 300
         7.2.1   Development of Cavitation Zone, Spalls
                (Experimental Studies) 301
         7.2.2   Two-Phase Model of Cavitation Region 303
         7.2.3   Parameters of Rarefaction Wave
                in the Cavitation Zone 310
    7.3 Formation of Jet Flows and Their Hydrodynamic Models 313
         7.3.1   Formation of Vertical Jets on the Free Surface
                (Experiment, $H < R_{\max}$) 317
         7.3.2   An Analog Model of a Sultan 325
         7.3.3   Hydrodynamic Model of a Sultan: Pulsed Motion
                of a Solid from Beneath the Free Surface 330
         7.3.4   Abnormal Intensification Mechanism
                of the First Pulsation 333
         7.3.5   Two Models of Formation of Radial Sultans 339
         7.3.6   Basic Parameters of Sultans 343
         7.3.7   Structure of Sultans: Jet Tandem, Analogy
                with a High-Velocity Penetration of a Body
                into Water 344
    7.4 Jet Flows: Shallow Explosions of Circular Charges 346
         7.4.1   Gravity Effect 347
         7.4.2   Flow Structure Produced by Explosions
                of Circular Charges 350
    7.5 Shallow Underwater Explosions, Surface Water Waves 353
    References 356

8 Conclusion: Comments on the Models 359

Introduction

In the past century, much progress has been made in the hydrodynamics of explosion, the science dealing with liquid flows under shock-wave loading. This area is concerned primarily with studying underwater and underground explosions, cumulation, and the behavior of metals under the action of detonation products of high explosives (HE), which produce such extreme conditions as pressures of hundreds of kilobars and temperatures of up to several thousand degrees. Under these conditions, many solid media "forget" about their strength properties and rigid crystal structure and follow the classical laws of hydrodynamics.

A representative example is a shaped charge, i.e., a solid cylinder of HE that has a conical cavity (apex inside) at one end. If the surface of this cavity is lined with a metal layer (liner), the charge gains powerful armor-breaking capabilities. This idea was first patented in 1914. In the 1940s, the shaped-charge theory was developed independently by Birkhoff et al. and by Lavrentiev based on a hydrodynamic model assuming that material properties change radically under high pulse loads. According to this concept, the metal liner behaves as an ideal incompressible fluid, behind the detonation wave initiated at the charge end opposite to the cone, and incident on the cone from its apex.

Under the action of the detonation products, the liner collapses to the axis and the liner walls collide. This results in a jet flow moving in the direction of detonation-wave propagation at a velocity u_{jet} of about 10 km/s. The shaped-charge theory is based on the model of collision of free liquid jets, which is practically identical to that used in penetration theory. The theory suggests that the impact of a shaped-charge jet on armor generates a pressure $p \approx 10^3$ kbar in the deceleration region. This estimate is readily obtained from the formula $p \simeq \rho_{\text{met}} \cdot u_{\text{jet}} \cdot c_{\text{met}}$, if the density of the metal ρ_{met}, the speed of sound in it c_{met}, and the jet velocity u_{jet} are known. Jet penetration into a target is described by the model of jet interaction using reverse velocities. Comparison of results obtained using this model and experimental data confirms the "hydrodynamic origin" of these phenomena.

It is known that hydrodynamic cumulation also arises from hypervelocity (few kilometers per second) impact on a metal specimen with a hollow cavity on the rear surface. It results from the interaction of a plane shock wave generated by an impinging jet in the specimen with the free surface of the cavity, whose points are imparted different initial velocities.

Similar but low-velocity flows are observed in a liquid when its gravity plays a major role, e.g., the fall of a rain drop on a water surface or the impact of a water-filled test tube falling vertically on a table (experiment of G.I. Pokrovsky). In the first case, a hemispherical cumulative cavity on the surface is produced by penetration and spread of the drop. In the second case, the cavity as a meniscus occurs initially due to the capillary effect and wetting. Upon impact the liquid in the test tube "instantaneously" becomes heavy, which leads to cumulative flow in the cavity with jet formation, as in the case with a drop.

These physical effects can be treated as models for the formation of cumulative flows in large-scale underwater explosions. When charge weights are in excess of hundreds of kilograms and typical dimensions of a cavity with explosion products are up to tens or hundreds of meters, the gravity of the liquid can no longer be ignored: the pressure gradient between the upper and lower points of the explosive cavity, which is of the order of $\rho_{\text{liq}} g R$ (ρ_{liq} is the liquid density), becomes comparable with hydrostatic pressure.

Usually, the behavior of various media under explosive loading is studied in the so-called impulsive state. The medium is originally considered as an ideal incompressible liquid, and its flow is considered using the potential: $\mathbf{v} = \nabla \varphi$. Then, using the law of conservation of momentum,

$$\rho_0 \frac{d\mathbf{v}}{dt} = -\nabla p,$$

and replacing v by φ, one can easily obtain

$$\nabla \left(\frac{d\varphi}{dt} + \frac{p}{\rho_0} \right) = 0$$

and

$$\varphi = -\rho_0^{-1} \int_0^\tau p \, dt,$$

where the integral defines the value of the potential on the boundary of the domain under study, τ specifies the duration of explosive loading, and, in essence, determines the initial distribution of φ. This parameter is calculated from the speed of sound in detonation products and from the geometrical characteristics of the HE charge used in each particular case. Then, the Laplace equation is solved for the known domain, and the potential distribution in this domain, its gradients, and, hence, the initial distribution of mass velocities are obtained.

The above approach is general, and the model is too "ideal" to be used without restrictions on the medium strength. For shaped-charge liners, the kinetic energy of the liner elements $\rho u^2/2$ should exceed the dynamic yield point of the liner material σ. For the problems of penetration, hypervelocity impact, or a contact explosion, the situation is not so simple: the targets, as a rule, have large volumes, and the above condition is satisfied only in a very narrow zone near the contact region.

These problems can be solved using the so-called "liquid–solid" model proposed by Lavrentiev, whose idea is extremely simple: the medium is considered an ideal incompressible liquid near the charge or in the impactor contact zone (where the mass velocity is higher than a certain critical value v_*) and in the remaining region, it is treated as an absolutely rigid body. The solution of the problem on deformation of a lead column by explosion of a superimposed charge (V.K. Kedrinskii) illustrates the application of this model in dynamics with changing "phase" boundary as the particle velocity decreases (in the zone of contact with the solid body) and particle "freezing" according to the model.

Explosive magnetic cumulation (collapse of a conducting liner) is among the problems close to the hydrodynamics of explosion. An original solution of the problem of the impact on a planet surface at space velocities (50–100 km/s) was proposed by Lavrentiev late in the 1950s to determine the crater size: the energy lost by the body upon impact is converted into heat, and in the region where this thermal energy exceeds a certain critical value, the solid body is instantaneously gasified.

Finally, we consider the models that are directly concerned with real liquids. A peculiar prototype of the problems of explosion hydrodynamics is the Besant problem (late 19th century) on the flow in an empty spherical cavity, which arises instantaneously in an ideal imponderable incompressible liquid. Now it is recognized as an adequate model for an oscillation stage of the cavity with detonation products during underwater explosion. The problem can be solved using the energy conservation law: the change in the potential energy of the liquid is equal to the increment of its kinetic energy $p_0(V_0 - V) = T_k$. Here $V = (4/3)\pi R^3$ is the current volume of the cavity and the kinetic energy is readily determined from the expression $dT_k = (v^2/2)dm$ and has the form

$$T_k = 2\pi\rho_0 \int_R^\infty v^2 r^2 \, dr \, .$$

In the case of an incompressible liquid, the continuity equation $d\rho/dt + \rho \, \text{div} \, \mathbf{v} = 0$ reduces to $\text{div} \, \mathbf{v} = 0$, whence for arbitrary symmetry, we have $\partial v/\partial r + \nu v/r = 0$ and $vr^\nu = f(t)$. From the condition on the boundary of a spherical cavity, $(f(t) = R^2(dR/dt))$. Hence,

$$T_k = 2\pi\rho_0 R^3 \left(\frac{dR}{dt}\right)^2 \, .$$

The energy conservation law now yields

$$\left(\frac{dR}{dt}\right)^2 = \frac{2p_0}{3\rho_0}\left[\left(\frac{R_0}{R}\right)^3 - 1\right] \, .$$

It is easy to see that as $R \to 0$, the velocity of the cavity wall $dR/dt \to \infty$. This is a typical example of the classical spherical cumulation. We note that in this case, the kinetic energy of the liquid generally approaches its limit – the initial potential energy U_0.

An interesting model of explosion hydrodynamics is the problem of a strong point explosion. The classical model for explosion of a powerful (for example, nuclear) device ignores the dimensions of the device and assumes that all the energy is released at a point (L.I. Sedov). Here we confine ourselves to the case of an incompressible liquid with no counterpressure, i.e., we assume that as $r \to \infty$, the pressure $p \to 0$. The governing parameters of the problem are the explosion energy E_0 and the liquid density ρ_0. Combining these parameters with the independent variable t, one can define the linear dimension of the cavity R as the lower boundary of the perturbation region $R = k(E_0/\rho_0)^{1/5} t^{2/5}$. It is obvious that the flow is self-similar and the self-similarity index is $2/5$. We assume that in our problem the explosion energy is entirely converted into the kinetic energy of the liquid (the expanding cavity is empty):

$$2\pi \rho_0 R^3 \left(\frac{dR}{dt}\right)^2 = E_0 .$$

Substitution of the expressions for the radius and velocity of the explosive cavity yields the coefficient $k = (25/8\pi)^{1/5}$.

It is interesting that the above-mentioned Besant problem is also self-similar in the vicinity of the flow focusing point. Furthermore, the liquid flow near the point $R \to 0$ is completely identical to the case of a strong explosion. Let us assume that $R = a\, t^\alpha$ in the mentioned vicinity. Substitution of this solution into the Besant equation shows that the latter is satisfied if $\alpha = 2/5$ and the coefficient a has the form $a = (25 U_0/8\pi\rho_0)^{1/5}$, where U_0 is the initial potential energy.

The hydrodynamics of explosion involves plenty of interesting models, paradoxes, and unexpected analogies. What do the fracture of a ship's bottom due to underwater explosion, cavitation erosion, and disintegration of kidney stones in lithotriptor facilities have in common? To a certain extent, this is shock-wave loading. But the main feature common to all these phenomena is the impact of cumulative jets on a solid surface under cavity collapse near it.

In the 1950s, the limiting weights of explosive charges W required to damage a ship's bottom in an underwater explosion at various distances h from the ship were studied. One would expect that beginning with a contact explosion, the weight would increase with distance from the bottom. However, experiments revealed a surprising paradox: beginning with a certain distance h_*, the function $W(h)$ "enters" a rather long (within $(2–3)h_*$) horizontal ledge (M.A. Lavrentiev). Thus, the distance increased but fracture was achieved without increase in explosive weight. In addition, the nature of fracture changed: instead of cracks over a large area, the fracture zone was highly localized.

The fracture mechanism was determined by the action of the high-velocity cumulative jet formed at the second stage of an underwater explosion – the collapse of the cavity with detonation products after its first maximum expansion. The proximity of a solid boundary disrupts the one-dimensionality of the flow even if the cavity was spherical at the moment of maximum expansion. Particles on the cavity surface remote from the wall are imparted high velocities, i.e., the classical cumulative effect occurs. The jet is thus directed toward the wall and has a velocity of hundreds of meters per second.

The phenomenon of bubble cavitation in a liquid has long been known. Already in the early 20th century, it was found that bubble cavitation on rotating propeller screws was accompanied by mechanical surface damage. Numerous studies have shown that the damage mechanism is determined by shock-wave generation and the impact of the cumulative microjets produced by the collapse of minute near-surface bubbles in the cavitation zone (bubble cluster). Microstructural analysis of specimens from vibration tests and model experiments in "shock wave–bubble–specimen" systems show that during high-velocity interaction, the microjets penetrate into a specimen at a depth comparable to their length and cause appreciable local damage.

The threshold of the power of hydroacoustic systems is also associated with the development of bubble cavitation near the radiator surface.

In recent years, there have been extensive studies of shock wave focusing in liquids as applied to problems of lithotripsy (breakup of kidney stones). In particular, the mechanism of stone crushing in the focus zone was analyzed. At first glance, the cause of breaking of a stone placed at one of the foci of an ellipse is fairly obvious: the focused wave is refracted into the kidney stone, and its subsequent internal reflection from the interface with the acoustically less rigid media (liquid) results in a rarefaction wave, whose "travel" over the kidney stone leads to the stone breakup (B. Sturtevant). It is argued that the effect of an outer cloud of cavitation bubbles around a kidney stone should be taken into account in this model. Actually, the nature of the load resulting in stone crushing may appear to be much more complex.

Indeed, when analyzing the structure of a shock wave converging to a focus, it is easy to notice that because of diffraction at the edges of a semi-ellipsoid piezoelectric converter, there is a transition in the shock-wave "tail" from the positive postshock pressure to a rarefaction phase with a fairly large amplitude and long duration. Focusing of such a wave should inevitably result in the formation of a bubble cavitation zone in the focus zone (and, naturally, on the surface of the kidney stone). Considering that in practice a series of such shock loads is required for crushing, the periodic action of cavitation cloud (as a whole) on the stone can amplify the Sturtevant effect described above. The question arises of whether in this case, too, the breakup mechanism involves the high-velocity interaction of cumulative microjets (produced by cavitative bubbles in the vicinity of stone surface) with the target.

Shock-loaded liquid still remains a puzzle. And it is difficult to solve these problems by simply writing a complete system of conservation laws in the form of differential equations and closing defining relations, or by developing unique computer programs and codes. Physical models that include all main stages and features of the processes occurring in liquids must remain a key element in research. Otherwise, it would be impossible to explain the hydrodynamic properties of metals during impulsive loading or the brittle fracture of water with fragmentation into flat spall layers due to the action of strong rarefaction waves.

Concluding this short introduction we would like to note that the problems of underwater explosions described in this monograph can be conditionally divided into four groups:

- Shock waves, equations of state, and the dynamics of a cavity with detonation products (for unbounded media)
- Shock waves (transformation and amplification) in multiphase media including a reactive gas phase
- Behavior of a liquid with free boundaries during explosive loading, microinhomogeneities in the liquid and tensile stresses, and inversion of the two-phase state of the liquid and its strength property
- High-velocity jet flows at underwater explosions near the free surface and liquid flows with unknown free boundaries

The above-mentioned lines of research are primarily concerned with understanding the physics of the examined phenomena, searching for the pertinent control mechanisms, developing experimental methods, and constructing adequate mathematical models for describing fast processes and structural changes in liquids. Interest in the problems of explosion hydrodynamics considered here was motivated by the importance of the problems, the apparent illogicality of the phenomena, and the attractiveness of the models proposed by Lavrentiev, which were frequently used as the basis for constructing general concepts.

1 Equations of State, Initial and Boundary Conditions

1.1 Equations of State for Water

Various considerations on the state of continuous media during dynamic compression, shock transition, and unloading are used to solve a broad spectrum of problems on explosions underwater and to analyze the behavior of continuous media under impulsive loading. These approaches are based on certain thermodynamic models and are aimed at providing the most exact description of the state of a medium for a wide range of temperatures and pressures.

Derivation of a unique analytical relation, for example, in the form $p(T, \rho)$, in particular for water, is not a simple task due to the complex character of the dependence of thermodynamic functions, such as pressure, on density ρ, temperature T or internal energy E.

The first known attempt to describe the state of a real medium was made by Van der Waals, who considered the effects of the attraction and repulsion of molecules. Since then the interest in this subject has remained stable not only owing to the endeavor of giving a perfect empirical description of real properties of real media, but also due to the expansion of the assortment of subjects under study, from which we will be concerned with liquids (or "liquid" states) and the detonation products of explosives.

The equations of state are usually presented in one of the following forms:

- As a sum of cooling and heating components (the Mie–Gruneisen equation),

$$p = p_c(v) + G(v) \cdot \frac{c_v T}{v \mu},$$

where v is the specific volume, p_c is the cooling pressure, $G(v)$ is, c_v is the specific heat at constant volume, μ is the specific mass.
- In a form containing a certain function of entropy s

$$p = B(s)F(v)$$

where $B(s)$ is the entropy function, s is the entropy, and $F(v)$ is the function of the specific volume.
- In the most common virial form

$$\frac{pv}{RT} = 1 + \frac{B(T)}{v} + \frac{C(T)}{v^2} + \frac{D(T)}{v^3} + \ldots,$$

where the coefficients $B, C, D \ldots$ are expressed using statistical mechanics in terms of the potential of intermolecular interaction (v is the specific volume).

Let us use the above approaches to consider different equations of state of liquid media. The most well known is the Tait model (see [1, 2]), which was originally designed to describe compressibility of water:

$$\frac{dV}{dp} = -\frac{A}{F(T) + p}.$$

However, the integral of this equation

$$\ln\left(\frac{p + F(T)}{p_0 + F(T)}\right) = \frac{V_0}{A} \frac{V_0 - V}{V_0}$$

results in the contradiction: $\ln \to \infty$, when $p \to \infty$, and the right-hand part of the equation ($V \to 0$) is finite. The second version

$$\ln\left(\frac{p + B(s)}{p_0 + B(s)}\right) = \ln\left(\frac{V(0, s)}{V(p, s)}\right)$$

is more logical and serves as a basis for numerous alternative descriptions of the state of liquid media, both as the shock adiabat and as the equation of state:

$$\frac{p + B(s)}{p_0 + B(s)} = \left(\frac{\rho}{\rho_0}\right)^n.$$

It is assumed that $B(s) = 305$ MPa, while $n = 7.15$ for pressures up to $3 \cdot 10^3$ MPa.

Ridah [3] proposed using the Tait equation with other values of the basic parameters: $B(s) = 321.4$ MPa and $n = 7.00$. Reasoning from the estimates and comparison with larger scale data, some authors believe that $B(s) = $ const and $n = $ const for pressures up to 10^4 MPa. The parameters proposed by Ridah are more convenient and make it possible to obtain simple expressions for the Reimann function σ, which, as shown in [2], is close to mass velocity in the vicinity of the shock wave front in water, as well as for the speeds of sound c and the shock wave velocity U_{sh}:

$$\sigma = \int_{\rho_0}^{\rho} \frac{c}{\rho} d\rho = \frac{c_0}{3}\left[\xi^3 - 1\right],$$

where $\xi = \rho/\rho_0$, $c = c_0\left(1 + \frac{3\sigma}{c_0}\right)$, and $U_{sh} \simeq c_0 + 2\sigma + 2\frac{\sigma^2}{c_0}$. The function can be used to solve the problems of underwater explosion.

Patel and Teja [4] obtained the cubic equation of state for liquids and liquid mixtures as a modification of the Van der Waals equation,

$$p = \frac{RT}{v - b} - \frac{a(T)}{v(v + b) + c(v - b)},$$

where b and c are constants, $a(T)$ is the temperature function, R is gas constant, and $v = \rho^{-1}$ is the specific volume. This equation should satisfy the conditions

$$\frac{\partial p}{\partial v}\Big|_{T_c} = 0, \quad \frac{\partial^2 p}{\partial v^2}\Big|_{T_c} = 0, \quad \text{and} \quad \zeta_c = \frac{p_c v_c}{RT_c}$$

(ζ_c is the critical compressibility factor), which are used to determine the constants b and c and $a(T)$:

$$a(T) = \Omega_a \left(\frac{R^2 T_c^2}{p_c}\right) \beta(T), \quad b = \Omega_b \left(\frac{RT_c}{p_c}\right), \quad c = \Omega_c \left(\frac{RT_c}{p_c}\right).$$

Here

$$\Omega_c = 1 - \zeta_c,$$

and

$$\Omega_a = 3\zeta_c^2 + 3(1 - 2\zeta_c)\Omega_b + \Omega_b^2 + 1 - 3\zeta_c.$$

The function Ω_b is determined as the least positive root of the cubic equation

$$\Omega_b^3 + (2 - 3\zeta_c)\Omega_b^2 + 3\zeta_c^2 \Omega_b - \zeta_c^3 = 0.$$

The function $\beta(T)$ is found from the relations

$$\beta = \left[1 + F(1 - \sqrt{T})\right]^2 \quad \text{or} \quad \beta = \exp\left[C(1 - T^n)\right],$$

where F is an empirical parameter.

Gurtman et al. [5] proposed an analytical form of the equation of state for water (GKH equation), which was consistent with the data available for the densities $50\text{--}3 \cdot 10^4$ MPa. The governing equation was equivalent to the Mie–Gruneisen equation at $c_v = \text{const}$. This condition resulted in the simple relation for the internal energy E. The authors started from the hypothesis that water behaves in a way as a crystalline solid in the vicinity of the Hugoniot adiabat and its equation of state is written analogously to that for melting crystals:

$$p(v, T) = h_1(v) + T h_2(v).$$

Here h_1 and h_2 are arbitrary functions of specific volume and their values can be easily determined from the known thermodynamic equality

$$\left(\frac{\partial E}{\partial v}\right)_T = T \left(\frac{\partial p}{\partial T}\right)_v - p.$$

Comparison yields the obvious relations

$$h_1(v) = -\left(\frac{\partial E}{\partial v}\right)_T \quad \text{and} \quad h_2(v) = \left(\frac{\partial p}{\partial T}\right)_v.$$

Under the above condition $(\partial E/\partial T)_v = c_v = $ const, the internal energy as the function $E(T, v)$ now is determined as

$$E - E_0 = c_v(T - T_0) - \int_{v_0}^{v} h_1(v)\, dv\ .$$

In this case, the GKH equation of state is written as

$$p(v, E) = p_H\left[1 - \frac{G(v)}{2} - \frac{v_0 - v}{v}\right] + \frac{G(v)}{v}E\ ,$$

where $G = h_2 v/c_v$ is the Gruneisen coefficient and p_H is the Hugoniot pressure. In [5], the following quantities were presented as polynomials:

- The Gruneisen coefficient

$$G(v) = a_0 + a_1 v + a_2 v^2 + \ldots + a_7 v^7$$

for $a_0 = 2\,366\,632\,4$, $a_1 = -22\,669\,420$, $a_2 = 91\,259\,368$, $a_3 = -200\,175,85$, $a_4 = 258\,585,11$, $a_5 = -196\,872,84$, $a_6 = 81\,850,023$, and $a_7 = -14\,342,530$.
- The Hugoniot adiabat

$$p_H = b_1 \zeta + b_2 \zeta^2 + \ldots + b_7 \zeta^7\ ,$$

where $\zeta = v_0/v$ is the relative specific volume, $b_1 = 21.953\,4$, $b_2 = 0$, $b_3 = 1\,206.04$, $b_4 = -4\,113.87$, $b_5 = 7\,193.01$, $b_6 = -5\,594.03$, and $b_7 = 1\,566.61$.
- The integral of compression energy

$$I(v) = -\int h_1(v)\, dv = c_0 + c_1 v + c_2 v^2 + \ldots + c_7 v^7\ ,$$

where $c_0 = -3.09 \cdot 10^{12}$, $c_1 = 2.455 \cdot 10^{13}$, $c_2 = -8.255 \cdot 10^{12}$, $c_3 = 1.516 \cdot 10^{14}$, $c_4 = -1.624\,7 \cdot 10^{14}$, $c_5 = 1.006 \cdot 10^{14}$, $c_6 = -3.284 \cdot 10^{13}$, and $c_7 = 4.238 \cdot 10^{12}$. Here G is a dimensionless quantity, p_H is measured in kbar, v, in cm^3/g, and I, in erg/g, and $v_0 = 1,001\,8$ cm^3/g. Following [5], the calculation of heat capacity of water yields $c_p = 0.86$ cal/(g·°C) and $c_v = 0.78$ cal/(g·°C).

Kuznetsov [6] studied a similar version of the equation of state,

$$p = p_c(\rho) + \frac{R \rho T f(\rho)}{\mu}\ ,$$

which presented pressure as cooling $p_c(\rho)$ and heating components, and led to the equation of state of an ideal gas ($\rho \to 0$, $f \to 1$, and $p_0 \to 0$) or the equation of state for solids. The functions $p_c(\rho)$ and $f(\rho)$ were determined

1.1 Equations of State for Water

by extrapolating experimental data. For $0 < \rho < 2.3$ g/cm^3, the following interpolation formula was obtained

$$f(\rho) = \frac{1 + 3.5\rho - 2\rho^2 + 7.27\rho^6}{1 + 1.09\rho^6}.$$

The shock adiabat generalizes the Tait adiabat,

$$p_H = \frac{3050\left(\rho^{7.3} - 1\right)}{1 + 0.7(\rho - 1)^4},$$

with an error of up to $\Delta p/p \simeq 0.05$ with respect to experimental data. In [6], the interpolation formula for temperature in the shock wave front was obtained using the data for f, p_0, and p_H, and the equation of state:

$$T_H \simeq 2.6 \rho_H p_H 10^{-3}.$$

The temperature was measured in degrees Celsius, pressure p_H in bar, and density ρ_H in g/cm^3. The accuracy was $\Delta T/T \simeq 0.1$.

Finally, the equation of state for water becomes

$$p = p_H\left(1 - 0.012\rho^2 f\right) + 4.7\rho f T \quad \text{for} \quad 1 \leq \rho \leq 2.3$$

and

$$p = \zeta^4 - 470\rho f \zeta + 4.7\rho f T \quad \text{for} \quad \rho < 1.$$

Here

$$\zeta = 10(1 - \rho) + 66(1 - \rho)^2 - 270(1 - \rho)^3 \quad \text{for} \quad 0.8 \leq \rho \leq 1$$

and

$$\zeta = 6.6(1 - \rho)^{0.57} \rho^{0.25} \quad \text{for} \quad 0 \leq \rho \leq 0.8$$

For high temperatures ($T \geq 1\,000$ K), Kuznetsov proposed using variable $\tau \simeq \rho\left(T_1/T\right)^3$ instead of ρ in the formula for $f(\rho)$, which considerably reduced the error in energy calculations. Here $T_1 \simeq 900$ K was an estimated value of temperature at which the rotation of water molecules was suggested to be free.

The first items in the right-hand parts of the equation of state (for both ranges) represent the "cooling" component. The contribution of rotationally translational and vibrational degrees of freedom (subscript i) to the water heat capacity c_v was analyzed in [6]. The final result for $T > 273$ K and $\rho \leq 1$ g/cm^3 was presented as follows:

$$c_v = R\left(3 + [f(\rho) - 1]\left[0, 3 + \frac{4R}{(T/273)^2 + 2.7}\right]\right) + \sum_{i=1}^{3} c_i.$$

It was noted that within the range $1 \leq \rho \leq 1.2$, the value of c_v remained invariant.

In solving the problem of underwater explosion of a finite-radius spherical charge in a compressible liquid, Kochina and Melnikova [7] proposed using the three basic equations:

- The equation of state

$$\bar{p} = (49.414\bar{\rho}^{20/3} - 32.62)\left[1 + \frac{\bar{\rho}\bar{T}}{0.00548\bar{\rho}^{20/3} + 0.0799\bar{\rho} + 0.0206}\right]$$

- The shock adiabat

$$p_H = \left(1 + e^{S-S_0}\right)\left(49.414\bar{\rho}^{20/3} - 32.62\right)$$

- The normalized internal energy

$$\varepsilon = 1\,587.564\bar{T} + \bar{\rho}^{-1}\left[280.012\bar{\rho}^{20/3} + 32.62\right]$$

Here the main functions and variables $\bar{\rho} = \rho/\rho_0$, $\bar{T} = T/T_0$, $\varepsilon = E\rho_0/p_0$, $\bar{p} = p/p_0$, and entropy $S = s/c_v$ are dimensionless, while $\rho_0 = 101.865$ kg·c²/m⁴, $p_0 = 61\,926$ kg/m², $T_0 = 288$ K, and $c_v = 3.351\,1 \cdot 10^3$ m²/(c²·K).

To perform calculations for strong underwater explosions (up to $p > 10^2$ MPa), Shurshalov [8] used the equation of state in the form

$$p = A(v) + B(v)\varepsilon \,.$$

In this case,

$$A(v) = -\frac{K' + KS'/c_v}{p_* v_0}, \qquad B(v) = \frac{S'}{c_v v_0},$$

$$K = K_0 \frac{(1-\theta)(0.71 - \theta)}{\theta^{1/3}}\left[1 - \frac{2}{\theta}\exp\left(-\theta^{-2}\right)\right], \quad \text{and} \quad S = S_0 \frac{\theta - 1}{\theta^{3/4}}\,.$$

Here $p_* = 10^4$ kg/m² and $v_* = 10^{-2}$ m³/kg are the characteristic values of pressure and specific volume, respectively; primes mark derivatives with respect to θ, where $\theta = V/V_0$. Other dimensionless variables and parameters are determined as follows: $V = v/v_*$, $V_0 = v_0/v_*$, $S_0 = 22.7 \cdot 10^2$ m²/(c²·K), $c_v = 3.66 \cdot 10^3$ m²/(c²·K), and $v_0 = 1$ cm³/g.

Considering the explosion of a spherical charge at a depth of 60 m for real and instantaneous detonation, Shurshalov used the Kuznetsov equations of state to determine

- The internal energy:

$$\varepsilon = 0.1016 \cdot 10^5 \bar{T} + 0.63 \cdot 10^5 \frac{\left(1 - \bar{\rho}^{-1}\right)\left(0,71 - \bar{\rho}^{-1}\right)}{\bar{\rho}^{-4/3}}$$

$$\times \left[1 - 2\bar{\rho}\exp\left(-\bar{\rho}^2\right)\right] + \text{const}$$

- Entropy:
$$S = c_v \left(\ln \bar{T} + 0,621 \frac{\bar{\rho}^{-1} - 1}{\bar{\rho}^{-3/4}} \right) + \text{const},$$

where ε, $\bar{T}$, and $\bar{\rho}$ are dimensionless quantities, $p_0 = 1$ atm, $\rho = 1$ g/cm^3, and $T_0 = 273$ K.

When calculating the underwater detonation of a pentolite sphere for pressures up to $2.5 \cdot 10^4$ MPa, Sternberg and Walker [9] used the equation of state for water in the virial form,

$$p = f_1 v^{-1} + f_2 v^{-3} + f_3 v^{-5} + f_4 v^{-7},$$

where f_i are polynomials of internal energy E (Mbar·cm^3/g); p and v are measured in Mbar and in cm^3/g, respectively. The initial state of water was considered at $p_0 = 1$ atm, $\rho_0 = 0.998\,21$ g/cm^3, and $T_0 = 20$ °C, so that $E = 0$. In this case, $f_3 = 0.026\,8 - 0.414\,8E$, $f_4 = -0.005 + 0.0741E$, and the other coefficients depended considerably on E subranges [9].

The temperature of water was determined from the equation of state by integrating the following equations along the isentrope:

$$\left(\frac{\partial E}{\partial v}\right)_s = -p \quad \text{and} \quad \left(\frac{\partial T}{\partial v}\right)_s = -T \left(\frac{\partial p}{\partial E}\right)_v.$$

At the moment of incidence of the detonation wave onto the interface with detonation products the temperature of water, T_{sh} turned out to be approximately equal to 1000 °C. When the shock wave front moved away for a distance $r_{\text{sh}} = 1.1 r_{\text{ch}}$ (r_{ch} is the charge radius) in water, the temperature behind the front decreased abruptly to $T_{\text{sh}} \simeq 565$ °C, while the temperature at the interface with detonation products was 640 °C. At a distance $r_{\text{sh}} = 2 r_{\text{ch}}$, the temperature in the shock wave front T_{sh} continued to decrease to about 85°C, while at the interface it was 315°C. The process of heat loss was rather rapid.

Development of experimental approaches to the problem of dynamic compressibility of water for a wide range of parameters is undoubtedly important. Therefore, original, easy, and reliable techniques that provide unambiguous results are classical sources of information. A technique proposed about 40 years ago to study shock adiabat of water by Altshuller et al. [10] is still valid. It suggests preloading an aluminum shield whose shock adiabat is known. A part of the screen volume is filled with the liquid under test. The essence of the experiment is that the same basis is applied for measurements of shock wave velocities: D_{w} in water and D_{sc} in the screen. A pu-diagram method is used to determine the states behind the shock wave front in the screen material and in water (Fig. 1). The state behind the shock wave front (1) in the screen is found from its shock adiabat s and the line D_{sc}, which is easily plotted on the pu-diagram from $p_{\text{sc}} = \rho_{\text{sc}} u_{\text{sc}} D_{\text{sc}}$. The state behind the shock wave front in water (2) is determined from the unloading isentrope r of the

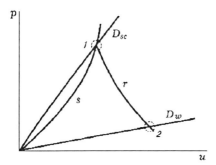

Fig. 1.1. The shock adiabat of water determined from the known shock adiabat of the screen using the pu-diagram method

screen, which is practically symmetrical to the shock adiabat, and line D_w determined by analogy from $p_w = \rho_w u_w D_w$ (Fig. 1).

The screen is made of Al, whose shock adiabat is

$$p = \rho_{0,\mathrm{Al}}(32.5 + 13.9u)u \,,$$

where p is measured in kbar and u is measured in km/s. Analysis of measurement results for the shock adiabat (for pressures up to 0.8 Mbar) showed that as the pressure reached 115 kbar, a phase transition started in water within the zone of which the shock front velocity was "frozen" at $D_w \simeq 5.44$ km/s.

Using this technique, Podurets et al. [11] extended the registration range of pressure in water to 14 Mbar. The following values of basic characteristics were registered in the screen (Al) and in water: $D_{\mathrm{Al}} = 36.4$ km/s, $u_{\mathrm{Al}} = 25.55$ km/s, $p_{\mathrm{Al}} = 25.2$ Mbar, $D_w = 43.95$ km/s, $u_w = 32.42$ km/s, $p_w = 14.25$ Mbar, and $\rho_w = 3.815$ g/cm^3 (the relative error of density estimates was approximately 0.03). Good correlation between the data of [11] extrapolated to pressures on the order of tens of megabars and calculations following the Thomas–Fermi model (statistical description of electrons of strongly compressed atoms) was noted in the survey [12]. The latter provided detailed information on equations of state for water and detonation products.

Comprehensive analysis of the state of water under strong underwater explosions was given by Kot [13], who presented the results graphically for thermal (p, ρ, T) and caloric (p, ρ, E) equations of state. Butkovich [14] analyzed the Bjork equation of state and identified three shock compression ranges of water in terms of its physical conditions: at $p > 700$ kbar water evaporated completely during unloading, at $p < 50$ kbar no vapor resulted from unloading, and at 50–700 kbar water evaporated partially with the formation of a liquid–vapor mixture.

Analyzing the equations of state for water [15], Baum et al. noted a tendency to retaining a rather convenient presentation of shock adiabat in the form analogous to the Tait equation. In particular, they presented possible variants:

$$p = 4.3\left[(\rho/\rho_0)^{6.4} - 1\right]$$

for the range 30–115 kbar and

$$p = \left[140\,(\rho/\rho_0)^2 - 284\right]$$

for the range 115–450 kbar.

Despite some uncertainties and limitations noted in [10, 17], the paper of Rice and Walsh [16] became classical and formed a basis for many studies of the shock adiabat of water. Its results can be presented as the dependence of the shock wave front velocity D on the mass velocity u behind the front:

$$D = 1.483 + 25.306\,\lg(1 + u/5.19)\,,$$

where D and u are measured in km/s.

Of interest is the analysis performed by Dharmadurai in [18], where the correlation between the density of a substance at the critical point ρ_c and that in the solid state ρ_s was tested. This has to do with an interesting experimental fact: for many substances, $\rho_s/3\rho_c \simeq 1$, i.e., it is a constant close to unity (in the experiment, the coefficient in the denominator is equal to 3.07). Using this feature as a new criterion for comparison of various models of the Van der Waals equation for liquids, Dharmadurai proposed a modification of the equation of state

$$p = \frac{RT}{V(1 - b/4V)^4} - \frac{a}{V^2}\,.$$

Although a reference to the equations of state for liquids might seem meaningless, while speaking about solid state, whose density depends on the crystalline structure, one should pay attention to the fact that, when the matter is cooled below its critical point, it condenses. In this case, due to cohesion forces there is a tendency to "correct" arrangement of molecules analogously to the structure of the solid state. As was noted by Eyring and John [19], the structure of a liquid is close to that of a solid and can be considered as a "mixture" of the degrees of freedom in solid and gaseous states. Actually, for many liquids, the ratio of density ρ_{liq} at the point of normal boiling to the critical density ρ_c is close to 2.7, as, for example, for liquid hydrogen, argon, and oxygen. Adding 10 % for the density increase during solidification, we get a value that is close to the above constant.

1.2 Equations of State for Detonation Products

Determination of the state of detonation products is no less complex than the problem considered in the first section, because it is necessary to adequately describe thermodynamic properties of gases for wide ranges of temperatures and pressures, and consider the change of molecular structure, as well as the features of chemical reactions at high pressures [1]. Nevertheless, since the "relaxation" of the main parameters of the detonation products in the expanding cavity is rather rapid, simplified models are preferable. Following [1], the application of the models "owes a great deal of its success to the fact that the description of detonation phenomena depends slightly on the form of the equation of state". Some of these models, as, for example, the Hulford–Kistiakowsky–Wilson equation [1],

$$\frac{p\tilde{V}}{RT} = 1 + KT^{-1/4} \cdot e^{0.3KT^{-1/4}},$$

became a classic. Here $K = $ const and is determined from the relation $K = \sum n_i K_i$, where n_i is the number of moles of the ith component of gas in 1 cm^3 of the mixture and K_i are empirical constants that characterize each chemical component ($K_i = 108$ cm^3/mol for water) [1].

Using more precise experimental data, Kuznetsov and Shvedov [20, 21] proposed the following system of transcendental relations that govern the state of detonation products of *hexogen* for the range of pressures $0 < \rho_* < 2.3$ g/cm^3:

$$p - p_0 = \frac{R\rho\varphi\left(E - \Delta - E_0\right)}{c_v^* + 0.3R(\varphi - 1)},$$

$$p_0 = 15.4\rho^3 - 12.6\rho^2,$$

$$E_0 = 0.77\rho^2 - 1.26\rho,$$

$$\varphi = 1 + \frac{2.7\tau + 15\tau^4}{1 + 2.6\tau^4}, \qquad \tau = 0.72\rho(T^*)^{-0.3}, \qquad T^* = 0.3\frac{p - p_0}{\rho\varphi(\tau)},$$

$$c_v^* = \frac{1.76R\Theta^2 \exp\left(\Theta\right)}{(\exp\left(\Theta\right) - 1)^2} + 2.624R,$$

and

$$\Delta = \frac{1.76RT\Theta}{\mu(\exp\left(\Theta\right) - 1)}\left[1 - \frac{\Theta\exp\left(\Theta\right)}{\exp\left(\Theta\right) - 1}\right].$$

Here $\Theta = 3{,}200/T$, $\mu = 25$, $T = 10^3 T^*$ K, p_0 and E_0 are elastic components depending only on density (p_0 in kbar and E_0 in kJ/g), μ is the molecular mass, and R is the gas constant. Sought characteristics are determined from the given values of ρ and E (or p and ρ), while the total error does not exceed 10 %. Using the equations obtained in [21], pressure, its thermal component, the work of adiabatic expansion of the products, the speed of sound in them, and the isentrope index were calculated versus density (the data were presented in tables for real and instantaneous detonation).

To describe the state of detonation products, Fonarev and Chernyavskii [22] used the relation for internal energy

$$E = \frac{pv}{\gamma - 1},$$

1.2 Equations of State for Detonation Products

assuming that the adiabatic index γ depended on specific volume V. In this case, the range of γ was determined as follows:

$$\gamma = 2.63 - 0.96\varepsilon \qquad \text{at } \varepsilon < 1,$$
$$\gamma = 1.16 + 0.12/(\varepsilon - 0.63) \qquad \text{at } 1 \leq \varepsilon \leq 4.7,$$
$$\gamma = 1.18 + 0.061\,6/(\varepsilon - 2.4) \qquad \text{at } 4.7 \leq \varepsilon \leq 24.3,$$
$$\gamma = 1.18 + 0.081(0.23 + 0.001\varepsilon)^2 \qquad \text{at } 24.3 \leq \varepsilon \leq 815,$$
$$\gamma = 1.27 \qquad \text{at } \varepsilon \geq 815.$$

Here $\varepsilon = V/V_2$, $V_2 = 1$ cm^3/g, $v = V\rho_1$, $p = P/P_1$, P_1 and ρ_1 are the pressure and the density of the liquid, respectively. The above equation of state was used in [18, 22] to calculate the explosion of a *trotyl* charge.

Jacobs [23] proposed an equation of the Mie–Gruneisen type

$$p = p_x(\rho) + G(\rho)\rho(E - E_*) ,$$

where

$$G(\rho) = \frac{1}{3} + \frac{7.425[\exp(-4.95/\rho)]}{\rho} ,$$

and

$$E_* = a\exp(-k/\rho) - b\rho^m + A\left[(\rho - \rho_1)^3 + B\right]\exp\left[-\alpha(\rho - \rho_1)\right] .$$

The constants were obtained from processed experimental data and calculations. Here, ρ_1 is the density at the Jouguet point, $a = 2.3466 \cdot 10^{12}$, $k = 11.12$, $b = 4.2077 \cdot 10^8$, $A = -4.8557 \cdot 10^9$, $B = -1.5574 \cdot 10^{10}$, $m = 2$, and $\alpha = 2.1051$.

Using the same approach, Lee and Hornig [24] calculated the state of products of *trinitrotoluene* (*TNT*). They considered the equation of the isentrope to be known, while the Gruneisen coefficient was determined from the dependence of the detonation velocity on the initial density of the explosive.

A simpler form was used by Sternberg and Walker [9] to calculate the state of detonation products of *pentolite*:

$$p = A\rho E + B\rho^4 + C\exp(-K/\rho)$$

for $A = 0.35$, $B = 0.002\,164$, $C = 2.0755$, and $K = 6$. Here, E is measured in Mbar·cm^3/g, p, in Mbar, and ρ, in g/cm^3, while B, C, and K have the appropriate dimensions. They assumed that the energy released during explosion of pentolite was equal to 0.0536 Mbar·cm^3/g, which was equivalent to 1.280 kcal/g, and $\rho_0 = 1.65$ gr·cm^{-3} was the density of the explosive prior to detonation. The constants were determined in terms of the following detonation parameters at the Chapman–Jouguet point (index j): $D = 0.765\,5$, $E_j = 0.077\,5$, $p_j = 0.245\,2$, $c_j = 0.571\,4$, and $\rho_j = 2.21$.

18   1 Equations of State, Initial and Boundary Conditions

Finally I would like to mention two monographs of Ch. Mader [25, 26], where one can find the detail information on the state equations.

The present survey of the equations of state for liquids and detonation products does not pretend to be comprehensive and represents mainly the most typical models. For each specific problem, the appropriate model is selected depending on the parameter range and with regard to presentation convenience. Sometimes an experiment prompts unexpected solutions, which makes it possible to simplify the statement, as, for example, the solution proposed for the problem on underwater explosion of a spherical charge by Kirkwood and Bethe [2].

Considering the typical exponential shape (in the vicinity of the wave front) of a shock wave generated by explosion in a liquid, they put forward the so-called "peak" approximation. They proposed considering the dynamics of enthalpy on the wall of the explosive cavity as exponential attenuation and determining parameters of this function from boundary and initial conditions. It is obvious that this "physical" approach eliminated the necessity of dealing with the equation of state of detonation products, as if the latter were not available.

1.3 Conservation Laws, PU-Diagrams, and Transition Formulas

Modern computing facilities make it possible to solve the problem on generation and propagation of shock waves during explosions underwater within the complete statement. However, it is obvious that calculations of the wave field even in the nearest zone (of the order of $10^3 \cdot R_{\mathrm{ch}}$) involve a number of difficulties caused primarily by different scales of characteristic time of the processes occurring in the detonation products and those associated with the interface dynamics of the explosive cavity and formation of the shock wave. The problem on shock wave parameters in the far zone, the so-called asymptotics, can turn out to be unsolvable. In this respect, of most interest is the Kirkwood–Bethe model that splits the process into two parts: the dynamics of explosive cavity, whose equation is derived and analyzed independently of the external wave field, and calculations of the formation of a shock wave on the basis of disturbances generated by the cavity at each moment of time.

As a rule, the initial conditions are formulated and determined using the pu-diagram method within the problems on decay of an arbitrary discontinuity. To solve these problems, one needs information on transition conditions in shock and simple waves (see, for example, [27, 28]). In this case, first the type of detonation which should be selected, it can be initiated in the charge center (allowance should be made for the propagation of the detonation wave up to the interface with the liquid), otherwise the process should be considered as instantaneous detonation with constant volume. It is obvious that

1.3 Conservation Laws

in each of the cases the problem of the decay of an arbitrary discontinuity will yield different initial conditions. Proper solution of the problem on the formation and propagation of a shock wave will be considered in the chapter devoted to generation of shock waves, now we shall dwell on transition conditions only.

1.3.1 Transition in a Shock Wave

If we use the following equation of the state for water

$$\bar{p}/\rho^n = \text{const},$$

where $\bar{p} = p + B$, (B and n are the parameters of one of the above models of the Tait equation), as well as the appropriate expression for specific internal energy $\bar{\varepsilon}$

$$\bar{\varepsilon} = \frac{\bar{p}}{\rho(n-1)},$$

the laws of conservation of mass, momentum, and energy for a strong discontinuity are written in the known gas-dynamic form

$$\frac{v_2}{\tau_2} = \frac{v_1}{\tau_1}, \tag{1.1}$$

$$\bar{p}_2 + \frac{v_2^2}{\tau_2} = \bar{p}_1 + \frac{v_1^2}{\tau_1}, \tag{1.2}$$

and

$$\bar{p}_2 v_2 + \frac{v_2}{\tau_2} \cdot \left(\bar{\varepsilon}_2 + \frac{v_2^2}{2}\right) = \bar{p}_1 v_1 + \frac{v_1}{\tau_1} \cdot \left(\bar{\varepsilon}_1 + \frac{v_1^2}{2}\right). \tag{1.3}$$

Here $\tau = \rho^{-1}$, $v_{1,2} = u_{1,2} - U$ is the relative mass velocity, U is the shock wave velocity, and subscripts 1 and 2 denote the conditions ahead and behind the shock wave, respectively. It follows from Eq. (1.2) that

$$\bar{p}_2 - \bar{p}_1 = \rho_1 v_1^2 - \rho_2 v_2^2 = \rho_1 v_1 (v_1 - v_2).$$

Assuming that the mass velocity ahead of the shock wave front is $u_1 = 0$ and using Eq. (1.1), we easily obtain the pressure jump as a function of the mass velocity u_2 and the shock wave velocity U:

$$\bar{p}_2 - \bar{p}_1 = \rho_1 u_2 U.$$

Simple combinations of Eqs. (1.1) and (1.2) result in the following equations for relative velocities:

$$v_1^2 = \tau_1 \frac{\bar{p}_2 - \bar{p}_1}{1 - \frac{\tau_2}{\tau_1}} \quad \text{and} \quad v_2^2 = \tau_2 \frac{\bar{p}_2 - \bar{p}_1}{\frac{\tau_1}{\tau_2} - 1}.$$

From the law of energy conservation (1.3) on the jump and from the expressions obtained for v_1^2 and v_2^2, it follows that

$$\bar{\varepsilon}_2 - \bar{\varepsilon}_1 = \frac{1}{2}(\bar{p}_1 + \bar{p}_2)(\tau_1 - \tau_2) .$$

Substituting the expressions for internal energy, one can readily obtain the shock adiabat for water

$$\frac{\bar{p}_2}{\bar{p}_1} = \frac{\tau_1 - \mu^2 \tau_2}{\tau_2 - \mu^2 \tau_1} ,$$

where $\mu^2 = (n+1)/(n-1)$.

In the equation for shock adiabat, we replace $\bar{p}_2$ and τ_2 by u_2 on the basis of Eqs. (1.2) and (1.1) and obtain

$$(\mu^2 v_2 - v_1) v_1 = (1 + \mu^2) \tau_1 \bar{p}_1 . \tag{1.4}$$

Using the condition $u_1 = 0$ and replacing v by u and U, we obtain the equation for the shock front velocity

$$U^2 - 2\beta u_2 U - c_1^2 = 0 ,$$

where $\beta = (n+1)/4$. Introducing the Mach number $M_2 = u_2/c_1$, we get the following solution:

$$U = c_1 \left(\beta M_2 + \sqrt{1 + (\beta M_2)^2} \right) . \tag{1.5}$$

In the first approximation, the solution (1.5) has the form

$$U^{(1)} \simeq c_1 + \beta u_2 ,$$

or, for the Ridah equation of state, ($n = 7$ and $\beta = 2$),

$$U^{(1)} \simeq c_1 + 2u_2 .$$

In the second approximation ($\beta M < 1$),

$$U^{(2)} \simeq c_1 \left(1 + \beta M_2 + (\beta M_2)^2 / 2 \right) .$$

Transforming Eq. (1.4) (substituting $v_2 - v_1$ with u and U), we get the relation on the jump for mass velocities

$$u_2 - u_1 = \frac{2c_1}{n+1} \left(\frac{1}{M_1} - M_1 \right) . \tag{1.6}$$

Here $M_1 = v_1/c_1 = -U/c_1$. For the pressure jump, we use Eqs. (1.2) and (1.6):
$$\bar{p}_2 - \bar{p}_1 = \frac{2n}{n+1}\bar{p}_1\left(M_1^2 - 1\right). \tag{1.7}$$

The pressure jump in the shock wave front is determined from Eqs. (1.1) and (1.6):
$$\rho_2 = \rho_1 \frac{(n+1)M_1^2}{2+(n+1)M_1^2}. \tag{1.8}$$

The relations (1.6)–(1.8) determine the transition in the shock wave.

1.3.2 Transition in a Simple Wave

Let us consider an isentropic flow that is described by the laws of conservation of

- Mass,
$$\frac{\partial \rho}{\partial t} + \mathbf{u}\nabla\rho + \rho\nabla\mathbf{u} = 0;$$

and

- Momentum,
$$\frac{\partial \mathbf{u}}{\partial t} + (\mathbf{u}\nabla)\mathbf{u} + \rho^{-1}\nabla\bar{p} = 0.$$

This system is closed by the equation of state
$$\frac{\bar{p}}{\bar{p}_1} = \left(\frac{\rho}{\rho_1}\right)^n.$$

Using the Tait equation of state, we can derive the relation of density and pressure to the sound velocity c
$$d\rho = \rho_1 \frac{2}{n-1}\left(\frac{c}{c_1}\right)^{(3-n)/(n-1)} \frac{dc}{c_1},$$

$$d\bar{p} = \bar{p}_1 \frac{2n}{n-1}\left(\frac{c}{c_1}\right)^{(n+1)/(n-1)} \frac{dc}{c_1}$$

and transform the first two equations of the above system (for one-dimensional case) to the form
$$\frac{\partial c}{\partial t} + u\frac{\partial c}{\partial x} + \frac{n-1}{2}c\frac{\partial u}{\partial x} = 0, \quad \text{and} \quad \frac{\partial u}{\partial t} + u\frac{\partial u}{\partial x} + \frac{2}{n-1}c\frac{\partial c}{\partial x} = 0.$$

Multiplying the first equation by $\pm 2/(n-1)$ and summing up the result with the second equation, we obtain after simple transformations the known gas-dynamics equation
$$\frac{\partial}{\partial t}\left(u \pm \frac{2}{n-1}c\right) + (u \pm c)\frac{\partial}{\partial x}\left(u \pm \frac{2}{n-1}c\right) = 0,$$

from which it follows that the Riemann invariants $r = u + 2c/(n-1)$ and $s = u - 2c/(n-1)$ remain unchanged for characteristics propagating at velocities $u + c$ ($dx/dt = u + c$) and $u - c$ ($dx/dt = u - c$), respectively. The invariance means that under transition from the state p_* (Fig. 2 b) governed, for example, by the relevant parameters of detonation products, to the state 2 along the r-wave the changes in

- Mass velocity
$$u_2 - u_* = \frac{2c_*}{n-1}(1 - \zeta),$$

- Density
$$\rho_2 = \rho_* \zeta^{2/(n-1)},$$

and

- Pressure
$$\bar{p}_2 = \bar{p}_* \zeta^{2n/(n-1)}$$

determine the transition to a simple wave, where $\zeta = c_2/c_*$.

From the relations and the equality of mass velocities and pressures at the contact discontinuity, one can easily define the initial conditions in the problems on the generation of shock waves in a liquid. Figure 2 illustrates two classical cases of this problem by means of pu-diagrams:

- Determination of pressure p_2 and mass velocity of the contact boundary u_2 in an originally resting liquid and in the material of a piston moving at u_p after their collision, when shock waves propagate both in the piston and in water (Fig. 1.2a, discontinuity decay in a two-diaphragm hydrodynamic shock tube; s_w and s_p are the shock adiabats of the liquid and piston material, respectively).
- Determination of initial conditions in the problem of underwater explosion (or in the Glass hydrodynamic shock tube): discontinuity decay at the detonation products–liquid interface; a transition from pressure p_* to

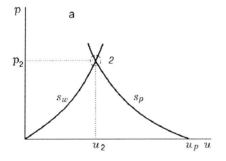
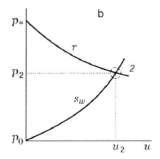

Fig. 1.2a,b. Conditions on the contact boundary determined using the pu-diagrams: the impact of a piston moving at velocity u_p on a resting liquid (**a**) and the rupture of a diaphragm at the compressed gas-liquid interface (**b**)

a final state (p_2, u_2) occurs in the detonation products in the rarefaction wave along r and in the liquid, along the shock adiabat s_w (Fig. 1.2b).

1.4 Generalized Equation of Pulsations of an Explosive Cavity

It is known that the energy released during the detonation of an explosive charge underwater is distributed between detonation products and the shock wave. The internal energy of detonation products during expansion of the cavity to $R = R_\mathrm{max}$ converts into the potential energy of the liquid ($U_* = p_\infty 4\pi R_\mathrm{max}^3/3$), which, in particular enables experimental assessment (based on the oscillation period) of the share of α in the heat of explosion Q (kcal/g) retained in the detonation products

$$\alpha Q \rho_\mathrm{ch} V_\mathrm{ch} = p_\infty V_\mathrm{max}$$

For standard cast or pressed explosives, the value of α is close to 1/2 in spherical charges, but it changes considerably depending on the flow geometry and the density of specific explosive.

The energy distribution is a dynamic process, while the explosive cavity serves as a piston in the formation of a shock-wave field. Obviously, the more exact the law of motion of the cavity, the closer to reality the calculation results for shock wave parameters. In the Kirkwood–Bethe model, this equation is derived independently to describe the dynamic boundary at which the parameters of disturbances governing the wave field are permanently calculated.

The first publications on cavity dynamics in compressible liquids appeared about half a century ago. One should first of all mention the results of Herring [29], Kirkwood and Bethe [2], and Gilmore [30], as well as calculations of Trilling [31]. Some generalized analysis of these models was reported in [32], similar information is also available in the review made by Flinn [33].

Of major interest is the Kirkwood–Bethe approach to the problems of underwater explosion, which allows one to calculate the shock wave parameters in the near zone, as well as the wave asymptotics, and to derive the equation of pulsation for various types of symmetry.

Let us consider a one-dimensional isentropic liquid flow,

$$\frac{d\rho}{dt} + \rho\left(\frac{\partial v}{\partial r} + \frac{\nu v}{r}\right) = 0,$$

$$\frac{dV}{dt} + \rho_0^{-1}\frac{\partial p}{\partial r} = 0,$$

and

$$p = p(\rho),$$

which for the potential φ, $\mathbf{v} = -\nabla\varphi$, and the enthalpy h is presented as the system of equations:

$$c^{-2}\frac{dh}{dt} - \frac{\partial^2\varphi}{\partial r^2} - \frac{\nu}{r}\frac{\partial\varphi}{\partial r} = 0, \tag{1.9}$$

and

$$-\frac{\partial^2\varphi}{\partial r \partial t} + \frac{\partial}{\partial r}\left(v^2/2 + h\right) = 0. \tag{1.10}$$

Here $\nu = 0, 1, 2$ governs the type of symmetry.

Let us introduce the so-called kinetic enthalpy $\Omega = h + v^2/2$ as a new unknown function. Since the potential is determined to the accuracy of an arbitrary function of t, the law of conservation of momentum takes the simple form

$$\frac{\partial \varphi}{\partial t} = \Omega.$$

Substitution of the function h for Ω and then for φ transforms Eq. (1.9) to the form

$$c^{-2}\frac{\partial^2\varphi}{\partial t^2} - \left(1 - \frac{v^2}{c^2}\right)\frac{\partial^2\varphi}{\partial r^2} - \nu r^{-1}\frac{\partial\varphi}{\partial r} = c^{-2}\frac{\partial v^2}{\partial t}$$

or, in the acoustic approximation,

$$c_0^{-2}\frac{\partial^2\varphi}{\partial t^2} - \frac{\partial^2\varphi}{\partial r^2} - \nu r^{-1}\frac{\partial\varphi}{\partial r} = 0.$$

For the new function $\Phi = r^{\nu/2}\varphi$, the governing system of conservation laws becomes more convenient for analysis:

$$c_0^{-2}\frac{\partial^2\Phi}{\partial t^2} - \frac{\partial^2\Phi}{\partial r^2} - \frac{\nu(2-\nu)}{4r^2}\Phi = 0, \tag{1.11}$$

and

$$\frac{\partial \Phi}{\partial t} = r^{\nu/2}\Omega.$$

For plane ($\nu = 0$) or spherical ($\nu = 2$) symmetries, Eq. (1.11) is reduced to the form

$$c_0^{-2}\frac{\partial^2\Phi}{\partial t^2} - \frac{\partial^2\Phi}{\partial r^2} = 0.$$

The solution of the equation has the form $\Phi = \Phi(\xi)$, where $\xi = t - r/c_0$. We assume that in the asymptotic approximation (to the accuracy of one member $\Phi/4r^2$) it also describes a cylindrically symmetrical flow ($\nu = 1$).

Introducing the function $G = r^{\nu/2}\Omega$, we see that

$$\frac{\partial \Phi}{\partial t} = \frac{\partial \Phi}{\partial \xi} = G, \quad \frac{\partial \Phi}{\partial r} = -c_0^{-1}\frac{\partial \Phi}{\partial t} = -c_0^{-1}G.$$

1.4 Generalized Equation of Pulsations

Upon substitution, the wave equation transforms to the form

$$\frac{\partial G}{\partial t} + c_0 \frac{\partial G}{\partial r} = 0, \qquad (1.12)$$

which points to invariance of the function G that remains unchanged at the characteristics diverging at velocity c_0. Following the Kirkwood–Bethe model [2], this result is generalized to the case when the disturbance propagates at velocity $c + v$, yielding

$$\frac{\partial G}{\partial t} + (c+v) \frac{\partial G}{\partial r} = 0, \qquad (1.13)$$

where c is the local speed of sound. Undoubtedly, validity of this generalization should be proved by experimental data. It is noteworthy that this result is of fundamental character; it allows one to determine the shock wave parameters in the near and far zones and investigate the dynamics of an explosive cavity. These are the most important problems in underwater explosion studies.

From Eq. (1.13) one can derive the generalized equation of one-dimensional oscillation of a cavity in a compressible liquid [34, 35]. Substituting the expression $G = r^{\nu/2}(h + v^2/2)$, we obtain

$$\left(\frac{\partial}{\partial t} + (c+v)\frac{\partial}{\partial r}\right) \left[r^{\nu/2}\left(h + v^2/2\right)\right] = 0,$$

or upon easy transformations,

$$r\frac{dh}{dt} + rv\frac{dv}{dt} + \frac{\nu}{2}h(c+v) + \frac{\nu}{2}(c+v)\frac{v^2}{2} + rc\frac{\partial h}{\partial r} + rcv\frac{\partial v}{\partial r} = 0.$$

Substitution of partial derivatives $\partial h/\partial r$ and $\partial v/\partial r$ for complete derivatives with regard to the laws of conservation of momentum and mass,

$$\frac{\partial h}{\partial r} = -\frac{dv}{dt}, \quad \frac{\partial v}{\partial r} = -c^{-2}\frac{dh}{dt} - \nu\frac{v}{r},$$

yields

$$r\left(1 - \frac{v}{c}\right)\frac{dv}{dt} + \frac{3}{4}\nu\left(1 - \frac{v}{3c}\right)v^2 = \frac{\nu}{2}\left(1 + \frac{v}{c}\right)h + \frac{r}{c}\left(1 - \frac{v}{c}\right)\frac{dh}{dt}. \qquad (1.14)$$

Equation (1.14) is valid for any point of the liquid, including that on the surface of the cavity containing detonation products. For the cavity, assuming $r = R$, $v = dR/dt = \dot{R}$, and $h = H$ (H is the enthalpy on the cavity wall from the side of the liquid) and keeping the notation c for the appropriate speed of sound [35], Eq. (1.14) becomes:

$$R\left(1 - \frac{\dot{R}}{c}\right)\ddot{R} + \frac{3}{4}\nu\left(1 - \frac{\dot{R}}{3c}\right)\dot{R}^2 = \frac{\nu}{2}\left(1 + \frac{\dot{R}}{c}\right)H + \frac{R}{c}\left(1 - \frac{\dot{R}}{c}\right)\frac{dH}{dt}. \qquad (1.15)$$

It should be noted that Eq. (1.15) is common for all the three types of one-dimensional flows and does not result in the known peculiarities at infinity for plane and cylindrical geometries even in the case of an incompressible liquid. We obtain fairly real equations of pulsations [34, 35] of

- A plane cavity,

$$\frac{d^2 R}{dt^2} = c^{-1} \frac{dH}{dt} \tag{1.16}$$

- A cylindrical cavity,

$$R\left(1 - \frac{\dot{R}}{c}\right) \ddot{R} + \frac{3}{4}\left(1 - \frac{\dot{R}}{3c}\right)\dot{R}^2 = \frac{1}{2}\left(1 + \frac{\dot{R}}{c}\right) H + \frac{R}{c}\left(1 - \frac{\dot{R}}{c}\right) \frac{dH}{dt}, \tag{1.17}$$

- A spherical cavity,

$$R\left(1 - \frac{\dot{R}}{c}\right) \ddot{R} + \frac{3}{2}\left(1 - \frac{\dot{R}}{3c}\right)\dot{R}^2 = \left(1 + \frac{\dot{R}}{c}\right) H + \frac{R}{c}\left(1 - \frac{\dot{R}}{c}\right) \frac{dH}{dt}. \tag{1.18}$$

For plane geometry, R means half of the cavity width. Equation (1.18), which, as we see, is a particular case of Eq. (1.15) [35], is sometimes called the Gilmore equation [30].

The expression for the local sound speed c and enthalpy H on the cavity wall can be obtained if the equation of state for the liquid is known. For example, for the most frequent form of the Tait equation $(p+B)/(p_\infty+B) = (\rho/\rho_\infty)^n$, we have

$$c^2 = \frac{n(p+B)}{\rho} = c_\infty^2 \left[1 + \frac{p(R) - p_\infty}{B}\right]^{(n-1)/n},$$

and

$$H = \frac{c^2 - c_\infty^2}{n-1}, \quad \text{where} \quad c_\infty^2 = \frac{n(p_\infty + B)}{\rho_\infty}.$$

Here the subscript ∞ marks the appropriate parameters of undisturbed state, $p(R)$ is the pressure in detonation products, which can be calculated, for example, using one of the above equations of state. The easiest way is to use the equation of the adiabat type with the variable γ.

References

1. J.O. Hirschfelder, C.F. Curtiss, R.B. Bird: *The Molecular Theory of Gases and Liquids* (John Wiley, New York, 1954)
2. R. Cole: *Underwater Explosions* (Dover, New York, 1965)
3. S. Ridah: *Shock Waves in Water*, J. Appl. Physics **64**, 1 (1988)

4. N.C. Patel, A.S. Teja: *A new Cubic Equation of State for Fluids and Fluid Mixture*, Chem. Eng. Science **37**, 3 (1982)
5. G.A. Gurtman, J.W. Kirsch, C.R. Hastings: *Analitical Equation of State for Water Compressied to 300 kbar*, J. Appl. Physics **42**, 2 (1971)
6. N.M. Kuznetsov: *State Equation and Thermocapacity of Water for Wide Spectrum of Thermodynamic Parameters*, Zh. Prikl. Mekh. i Tekhn. Fiz. **2**, 1 (1961)
7. N.N. Kochina, N.S. Melnikova: On explosion in water with regard to compressibility. In: Unstable Motions of Compressed Media with Explosive Waves. Transactions of the Steklov Mathematical Institute (1966) pp. 35–65
8. L.V. Shurshalov: *Calculation of Powerful Underwater Explosions*, Izv. Akad. Nauk SSSR, Gidromekh. **5** (1971)
9. H.M. Sternberg, W.A. Walker: *Calculated Flow and Energy Distribution Following Underwater Detonation of a Pentolite Sphere*, Phys. Fluids **14**, 9 (1971)
10. L.V. Altshuller, A.A. Bakanova, R.F. Trunin: *Phase Transformations Under Compression of Water by Strong Shock Waves*, Dokl. Akad. Nauk SSSR **121**, 1 (1958)
11. M.A. Podurets, G.V. Simakov, R.F. Trunin et al.: *Water Compression by Strong Shock Waves*, Zh. Exp. Teor. Fiz. **62**, 2 (1972)
12. V.P. Korobeinikov, B.D. Khristoforov: *Underwater Explosion*, Itogi Nauki Tekh. Gidromekh. **9**, (1976)
13. C.A. Kot: Intense underwater explosions: *IIT Research Institute, Chicago, Illinoise, USA* preprint, 1970
14. T.R. Butkovich: *Influence of Water in Rocks on Effects of Underground Nuclear Explosion*, J. Geophys Research. 1971 **76**, 8 (1971)
15. F.A. Baum, L.P. Orlenko, K.P. Stanyukovich et al.: *Explosion Physics* (Nauka, Moscow, 1975)
16. M.H. Rice, J.M. Walsh: *Equation State of Water to 250 kbar*, J. Chem. Phys. **26**, 4 (1957)
17. R.A. Papeti, M. Fijisaki: *The Rice and Walsh Equation of State for Water: Discussion, Limitation and Extension*, J. Appl. Physics **39**, 12 (1968)
18. G. Dharmadurai: *Solid State Density in Equations of State for Liquids*, J. Physics III France, 6 (1996)
19. H. Eyring, M. John: *Significant Liquid Structures* (John Wiley, New York, 1969)
20. N.M. Kuznetsov, K.K. Shvedov: *State Equation of Detonation Products of Hexogen*, Fiz. Goren. Vzryva **2**, 4 (1966)
21. N.M. Kuznetsov, K.K. Shvedov: *Isentopic Expansion of Detonation Products of Hexogen*, Fiz. Goren. Vzryva **3**, 2 (1967)
22. A.S. Fonarev, S.Yu. Chernyavskii: *Calculation of Shock Waves Under Exlosion of Spherical Charges of Explosives in Air*, Izv. Akad. Nauk SSSR, Gidromekh. **5**, (1968)
23. S.J. Jacobs: On the equation of state for detonation products of high density. In: 12th Intern. Symp. Combustion, Poitiers, 1968, Abstracts papers (Combust. Inst., S.A., Pittsburgh PA, 1968) pp. 88–89
24. E.L. Lee, H.C. Hornig: Equation of state of detonation products gases. In: 12th Intern. Symp. Combustion, Poitiers, 1968, Abstracts papers (Combust. Inst., S.A., Pittsburgh PA, 1968) pp. 87–88
25. Ch.L. Mader: *Numerical Modeling of Detonations* (Los Alamos Series in Basic and Applied Sciences, 1979) University of California Press, p. 485
26. Ch.L. Mader: *Numerical Modeling of Explosives and Propellants* (CRC Press, 1998), p. 439

27. L.V. Ovsyannikov: *Lectures on Fundamentals of Gas Dynamics* (Novosibirsk State University, Novosibirsk, 1967)
28. B.L. Rogdestvesky, N.N. Yanenko: *Systems of Quasi-Linear Equations* (Nauka, Moscow, 1968)
29. C. Herring: *Theory of Pulsation of the Gas Bubble Produced by an Underwater Explosion*, OSRD Rep. 236 (1941)
30. F.R. Gilmore: *The Collaps and Growth of a Spherical Bubbl in a Viscous Compressible Liquid*, Californ. Tech. Univ. Hydrodynamics Lab. 1952. Rep. No. 26-4.
31. L. Trilling: *The Collaps and Rebound of a Gas Bubble* J. Appl. Physics **23**, 14 (1952)
32. R.T. Knapp, J.W. Daily, F.G. Hammitt: *Cavitation* (McGraw-Hill, New York, 1970)
33. G. Flinn: Physics of acoustic cavitation in liquids. In: Physical Acoustics, vol. 1, W. Mason (ed.), part B (Academic Press, New York and London, 1966) pp. 7–138
34. V.K. Kedrinskii: On pulsation of a cylindrical gas cavity in an unbounded liquid: In: Continuum Dynamics, vol. 8 (Institute of Hydrodynamics, Novosibirsk, 1971) pp. 163–168
35. V.K. Kedrinskii: *Hydrodynamics of Explosion*, Zh. Prikl. Mekh. i Tekhn. Fiz. **28**, 4 (1987)

2 Underwater Explosions, Shock Tubes, and Explosive Sound Sources

2.1 Kirkwood–Bethe Approximation, Cylindrical Symmetry

Research of explosive processes in liquid media both for practical purposes and for laboratory modeling is concerned with the estimate of shock wave parameters in the near and far zones, the assessment of the distribution of explosion energy between detonation products and the shock wave, as well as with the analysis of basic features and characteristics of the dynamics of an explosive cavity with detonation products.

Some of these problems were surveyed by Cole, Korobeinikov and Khristoforov, and Zamyshlyaev and Yakovlev [1–3]. This chapter reports the results obtained for the problem of shock wave generation in a liquid using the Kirkwood–Bethe (KB) model [1] initially developed for spherical flows and generalized to cylindrical charges [4]. As for any approximation, the results obtained within the framework of the KB model should be compared with experimental data or with calculations carried out using other techniques. The KB model enables derivation of the equation of oscillations of a cylindrical cavity in a compressed liquid and, on this basis, solution of the problem on explosion of cord charges (cylindrical symmetry). Without this equation, the resolution of many practical problems, in particular on the use of elongated charges for underground explosions or for directed emissions at underwater explosions, is complicated. The research results presented in this chapter are based on the data of [4–7].

2.1.1 Basic Assumptions, Initial Conditions

A charge of radius R_{ch} (of infinite length for the case of cylindrical symmetry) is placed in a boundless ideal liquid. We consider an isentropic potential liquid flow whose state is described by the Tait equation.

Initial conditions for detonation products and at the boundary of a gas cavity from the side of the liquid are defined assuming an arbitrary discontinuity decay for instantaneous detonation at constant volume and adiabatic character of the process, with the adiabatic index $\gamma(\rho)$ of detonation products.

When defining the behavior of the cavity boundary, we ignore the internal reflections of rarefaction waves propagating in detonation products after the decay.

The problem is considered in the so-called "peak" approximation, which implies that, first, the shock wave parameters are determined only in the near front area and, second, the time variation of pressure p and enthalpy h on the "cavity-detonation products" interface is specified by the exponential law

$$h|_{r=R_{ch}} = h(0) \cdot e^{-t/\theta},$$

where the index of the exponent decay θ is determined from the parity condition of total time derivatives with respect to pressure (p, p_g) and velocity (u, v) on both sides of the interface:

$$\frac{dp}{dt} = \frac{dp_g}{dt}, \quad \text{and} \quad \frac{du}{dt} = \frac{dv}{dt}. \tag{2.1}$$

We determine $\theta(0)$ from the generalized equation of oscillations of an explosive cavity,

$$R\left(1 - \frac{\dot{R}}{c}\right)\ddot{R} + \frac{3}{4}\nu\left(1 - \frac{\dot{R}}{3c}\right)\dot{R}^2 = \frac{\nu}{2}\left(1 + \frac{\dot{R}}{c}\right)H + \frac{R}{c}\left(1 - \frac{\dot{R}}{c}\right)\frac{dH}{dt}, \tag{2.2}$$

by separating an enthalpy derivative and, upon replacement of $dH = \rho^{-1}dp$ and $\dot{R} = u$, reducing Eq. (2.2) to the form

$$\frac{1}{\rho c}\frac{dp}{dt} = \frac{du}{dt} - \frac{\nu/2\left[(c+u)\left(H + u^2/2\right) - 2cu^2\right]}{R(c-u)}. \tag{2.3}$$

An analogous equation should be derived for detonation products. To this end, we use the solution of the wave equation for converging waves: $\Phi = \Phi(t + r/c_{g,0}) = \Phi(\zeta)$. Upon the known transformations similar to those performed in Chap. 1, $\Phi_{tt} = G_t$, $\Phi_r = \Phi'/c_{g,0} = \Phi_t/c_{g,0} = G/c_{g,0}$ ($c_{g,0}$ is the speed of sound in undisturbed gas, subscripts t and r denote the appropriate derivatives, and the prime marks ζ derivative), we obtain the equation for converging characteristics,

$$\left(\frac{\partial}{\partial t} - c_{g,0}\frac{\partial}{\partial r}\right)G = 0,$$

which, upon generalization, yields the final form

$$\left(\frac{\partial}{\partial t} - (c_g - v)\frac{\partial}{\partial r}\right)G = 0.$$

Here G is the invariant on characteristics converging at velocity $(c_g - v)$ in the detonation products.

2.1 Underwater Explosion

Expanding this equation by the known analogy with the replacement of partial derivatives by total derivatives, we obtain for detonation products the following:

$$\frac{1}{\rho_g c_g} \frac{dp_g}{dt} = -\frac{dv}{dt} + \frac{\nu/2\left[(c_g - v)\left(H_g + v^2/2\right) - 2c_g v^2\right]}{R(c_g + v)}. \tag{2.4}$$

Summing up the right- and left-hand parts of Eqs. (2.3) and (2.4), with regard to the conditions of Eq. (2.1) at the contact boundary, we have

$$\left.\frac{dp}{dt}\right|_{t=0} = \frac{\nu \rho \rho_g c c_g}{2R_{\text{ch}}\left(\rho_g c_g + \rho c\right)} (\alpha_1 - \alpha_2), \tag{2.5}$$

where in view of the condition $u = v$,

$$\alpha_1 = \frac{1}{c_g + u}\left[(c_g - u)\left(H_g + \frac{u^2}{2}\right) - 2c_g u^2\right]$$

and

$$\alpha_2 = \frac{1}{c - u}\left[(c + u)\left(H + \frac{u^2}{2}\right) - 2cu^2\right].$$

On the basis of the "peak" approximation we have

$$\frac{dH}{dt} = \frac{1}{\rho}\frac{dp}{dt} = -\frac{1}{\theta}H(0)e^{-t/\theta}$$

or

$$\frac{1}{\rho(0)}\left.\frac{dp}{dt}\right|_{t=0} = -\frac{1}{\theta(0)}H(0),$$

which, in terms of Eq. (2.5), yields the expression for the initial value of the exponent decay constant:

$$\theta(0) = \frac{2H(0)R_{\text{ch}}\left[\rho_g(0)c_g(0) + \rho(0)c(0)\right]}{\nu \rho_g(0)c(0)c_g(0)(\alpha_1 - \alpha_2)}. \tag{2.6}$$

In addition, the initial conditions include the data on $p(0)$ and $u(0)$ from discontinuity decay (Chap. 1), as well as basic characteristics at the explosive cavity–liquid interface

$$\frac{\rho(0)}{\rho_0} = \left(\frac{p(0) + B}{p_0 + B}\right)^{1/n}, \quad c(0) = c_0\left(\frac{\rho(0)}{\rho_0}\right)^{(n-1)/2},$$

and

$$H(0) = \frac{c_0^2}{n-1}\left[\left(\frac{\rho(0)}{\rho_0}\right)^{n-1} - 1\right]$$

and in detonation products

$$\rho_g(0) = \rho_* \left(\frac{p(0)}{p_*}\right)^{1/\gamma_*}, \quad c(0)^2 = c_*^2 \left(\frac{p(0)}{p_*}\right)^{(\gamma_*-1)/\gamma_*},$$

and

$$H_g(0) = \frac{1}{\gamma_* - 1} \frac{p(0)}{\rho_g(0)} + \frac{p(0)}{\rho_g(0)},$$

where the initial parameters of instantaneous detonation are determined from the known relations $p_* = \rho_*(\gamma_* - 1)Q$, $c_*^2 = D^2\gamma_*/2(\gamma_* + 1)$. Here the asterisk denotes detonation products.

2.1.2 Dynamics of an Explosive Cavity, Riemann's Function, Delay Integral

The assumption on the invariance of the function G on the characteristics $(c + v)$ is of principal importance, because it simplifies considerably the solution of the problem on an isentropic flow of an ideal compressible liquid. In view of the invariance of G, to determine the function at any point of the liquid with due regard for the delay time, it suffices to know its value at every moment of time on the surface of the explosive cavity. Thus, if the assumption is accepted, the problem is reduced to the equation of dynamics of the explosive cavity (2) with one of the equations of state of detonation products presented, for example, in Chap. 1.

To determine the shock wave parameters in the vicinity of the front, we use the original Kirkwood–Bethe model. We apply the peak approximation not only to determine the initial value of $\theta(0)$ (exponent index in the pressure profile), but also to simulate the enthalpy at the contact boundary.

The equation of motion of the cavity boundary in the "peak" approximation ($H(t) = H(0)\exp(-t/\theta(0))$, $dR/dt = u$, and $M_b = u/c$) becomes

$$\frac{du}{dt} = H(t)\left[\frac{\nu}{2}(1 + M_b) - \frac{R}{c\theta(0)}(1 - M_b)\right]$$
$$- \frac{3}{4}u^2\nu\left(1 - \frac{1}{3}M_b\right)[R(1 - M_b)]^{-1}, \qquad (2.7)$$

which makes it possible to avoid the equation of state of expanding detonation products when considering the near zone.

The next fundamental step in the statement of the problem on underwater explosion is replacing the mass velocity u in the vicinity of the shock wave front by the Riemann function

$$\sigma = \int_{\rho_0}^{\rho} c \frac{d\rho}{\rho},$$

2.1 Underwater Explosion

which for the Tait equation of state for liquid becomes

$$\sigma = \frac{2c_0}{n-1}\left[\left(\frac{\rho}{\rho_0}\right)^{(n-1)/2} - 1\right].$$

This substitution makes it possible to express all basic characteristics and functions of the problem in terms of σ:

$$c = c_0\left(1 + \frac{n-1}{2}\frac{\sigma}{c_0}\right), \quad c + u = c_0\left(1 + \frac{n+1}{2}\frac{\sigma}{c_0}\right),$$

$$h = c_0\sigma\left(1 + \frac{n-1}{4}\frac{\sigma}{c_0}\right), \quad \Omega = c_0\sigma\left(1 + \frac{n+1}{4}\frac{\sigma}{c_0}\right),$$

and

$$U = c_0\left\{1 + \frac{n+1}{4}\frac{\sigma}{c_0} + \frac{1}{2}\left(\frac{n+1}{4}\right)^2\left(\frac{\sigma}{c_0}\right)^2 - \frac{1}{8}\left(\frac{n+1}{4}\right)^4\left(\frac{\sigma}{c_0}\right)^4 + \ldots\right\}$$

$$\simeq c_0\left(1 + \frac{n+1}{4}\frac{\sigma}{c_0}\right),$$

where, in this case

$$G(r, t_{\text{fr}}) = r^{\nu/2}c_0\sigma\left(1 + \frac{n+1}{4c_0}\sigma\right) = G(R, t) = R^{\nu/2}\Omega(R, t).$$

We introduce $\beta = (n+1)/4c_0$. In the last expression, the following time notation was introduced intentionally: t is the time "rigidly" associated with the explosive cavity dynamics and t_{fr} is the time "rigidly" associated with the shock wave front. The latter is defined by the evident condition

$$t_{\text{fr}} = \int_{R_{\text{ch}}}^{r_{\text{fr}}} \frac{dr}{U} = t + \int_{R(t)}^{r_{\text{fr}}} \frac{dr}{c+v}. \tag{2.8}$$

The integral in the right-hand part of Eq. (2.8), which defines the propagation time of the disturbance from its "inception" on the cavity surface at the time t until the moment when the disturbance overtakes the shock wave front, is the so-called delay integral. In calculating the integral, the function G determined for the time t on the explosive cavity boundary is an invariant.

The problem on underwater explosion is solved following the scheme:

1. Equation (2.7) for the dynamics of an explosive cavity with detonation products (within the peak approximation) is solved regardless of the surrounding pressure field. Thus, it can be considered as a certain law of motion of a flat, cylindrical or spherical piston generating a shock wave in the liquid.
   Solution of this equation for each moment of time t yields $R(t)$ and $G = R^{\nu/2}[H(R) + \dot{R}^2/2]$.

2. Since G is invariant on the characteristics propagating at velocity $(c+v)$ or $(c+\sigma)$, the following equality is valid

$$G = R^{\nu/2}\left[H(R) + \dot{R}^2/2\right] = r^{\nu/2}\left(h + \sigma^2/2\right) = \text{const}. \qquad (2.9)$$

3. The disturbance $G(R)$ "arising" on the surface of the explosive cavity at a time t overtakes the shock front in the delay time

$$t_{\text{del}} = \int_{R(t)}^{r_{\text{fr}}} \frac{dr}{c+\sigma}. \qquad (2.10)$$

We would like to remind readers that $c + \sigma = c_0(1 + 2\beta\sigma)$, while $U \simeq c_0(1 + \beta\sigma)$, i.e., $c + \sigma > U$. The position of the front r_{fr} is not known in advance and the front continues to move away from the charge as the disturbance approaches it.

4. Pressure is determined using the relation $\Omega(\sigma)$:

$$\beta\sigma = (1/2)\left(\sqrt{1 + 4\beta\Omega/c_0} - 1\right),$$

which with regard to the equation $\Omega = G/r^{\nu/2}$ can be rewritten as

$$\beta\sigma = \frac{1}{2}\left(\sqrt{1 + 4\beta G(R)/c_0 r^{\nu/2}} - 1\right).$$

In the expression for the Riemann function, ρ is replaced by p:

$$\frac{\bar{p}}{\bar{p}_0} = \left[1 + \frac{n-1}{2c_0}\sigma\right]^{2n/(n-1)}.$$

Upon substitution of $\beta\sigma$ and simple transformations we finally obtain

$$\frac{\bar{p}}{\bar{p}_0} = \left\{1 + \frac{n-1}{n+1}\left[\sqrt{1 + \frac{4\beta G(R)}{c_0 r^{\nu/2}}} - 1\right]\right\}^{2n/(n-1)} \qquad (2.11)$$

To determine the coordinate of the shock wave front r_{fr}, we have to find the delay integral (2.8).

2.1.3 Calculation of the Delay Integral (for $\nu = 1, 2$)

To transform the integral we use the invariance of the function G ($dG = 0$), which is preserved over the whole integration interval for each of its value at fixed $R(t)$:

$$\frac{\nu}{2}\sigma(1+\beta\sigma)dr + r(1+\beta\sigma)d\sigma + r\beta\sigma d\sigma = 0.$$

2.1 Underwater Explosion

This enables the change of variables

$$dr = -\frac{2r}{\nu}\frac{(1+2\beta\sigma)}{\sigma(1+\beta\sigma)}d\sigma$$

and the reduction of the integral to the intermediate form

$$t_{\text{del}} = -\frac{2}{\nu c_0}\int \frac{rd(\beta\sigma)}{\beta\sigma(1+\beta\sigma)}.$$

It follows from the invariance condition that $r = [G(R)/\Omega(r)]^{2/\nu}$ and, consequently,

$$t_{\text{del}} = -\frac{2G(R)^{2/\nu}}{\nu c_0}\int \frac{d(\beta\sigma)}{\beta\sigma(1+\beta\sigma)\Omega^{2/\nu}}.$$

From the definition of Ω, $\beta\sigma(1+\beta\sigma) = \beta\Omega/c_0$, it follows that $d\beta\Omega/c_0 = (1+2\beta\sigma)d(\beta\sigma)$ and $(1+2\beta\sigma) = \sqrt{1+4\beta\Omega/c_0}$. Their substitution in the integral and the change of variables $y = 4\beta\Omega/c_0$ reduce the integral to the convenient form

$$t_{\text{del}} = -\frac{1}{2\nu}\left(\frac{4\beta(R)}{c_0}\right)^{2/\nu}\frac{4}{c_0}\int_{4\beta\Omega(R)/c_0}^{4\beta\Omega(r)/c_0}\frac{dy}{(1+y)^{1/2}y^{2/\nu+1}}. \qquad (2.12)$$

The calculation scheme for pressure profile in this statement involves the following steps:

1. The spatial coordinate r of the point is specified.
2. Solving Eq. (2.7) for the cavity dynamics, we obtain the values of R and $G(R)$ for each time t "rigidly" associated with the explosive cavity.
3. From the integral (2.12) for a fixed r we determine the delay times t_{del} corresponding to the range of R and $G(R)$ values and, consequently, the time interval $t_r = (t+t_{\text{del}})$, the time of arrival of disturbances at a point with coordinate r (the disturbances being "generated" on the cavity surface during expansion at different times t and covered the distance $(R-r)$ over the time t_{del} at velocities $(c+\sigma)$).
4. Based on Eq. (2.11) we construct the distribution $p(t_r)$ at a given point of space.
5. The function $p(t_r)$ is ambiguous, and it allows the front position t_{fr} and the amplitude p_{fr} to be found from the geometrical condition of "equal areas."

2.1.4 Shock Waves (Cylindrical Symmetry, Comparison with Experiment)

As an example, we calculated the shock wave parameters near the front as a function of distance for detonating cord charges made of different high explosives (HE): pentaerythritol tetranitrate (PETN) ($\rho_* = 1.42$ g/cm^3, $Q = 1.4$ kcal/g, $W = 10$ g/m), trotyl ($\rho_* = 1.59$ g/cm^3, $Q = 1.0$ kcal/g, $W = 11,2$ g/m) and hexogen ($\rho_* = 1.32$ g/cm^3, $Q = 1.29$ kcal/g, $W = 15$ g/m). The calculation results were compared with experimental data [8]. For all HE types, the detonation velocity was $D = 7$ km/s. According to the statement, the problem was considered in the context of the condition of "instantaneous detonation at constant volume", which suggested the obvious equality $R(0) = R_{\text{ch}}$.

To compare and to process experimental data, for pressures up to $p_{\text{fr}} > 200$ MPa the profile of a conical shock wave was built in [8]. This wave's front velocity N was sought by considering the finiteness of the detonation velocity of the detonating cord (method of independent sections) from the formula $N = U[1 + (U/D)^2]^{-1/2}$, then the front pressure was determined from N.

In Fig. 2.1, the calculation results (curve 1) for pressure in the shock wave front ($G \propto r^{0.5}$, (1)) are compared with experimental data of [8] (dots) for the given distances $r_{\text{fr}}^0 = r_{\text{fr}}/\sqrt{q}$, m$^{3/2}$/kcal$^{1/2}$; $q = QW$ is the explosion heat per length unit of the detonating cord. There was satisfactory consistency between the data. The calculation results of Rice and Ginell for trotyl ($G \propto r^{0.4}$) (curve 2) are compared in Fig. 1 with experimental data of [1] (circles). The curve 3 corresponds to the self-similar dependence $p_{\text{fr}} = 0.29(r_{\text{fr}}^0)^{-2}$.

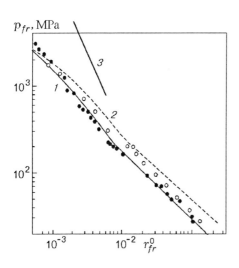

Fig. 2.1. Calculation of amplitude of cylindrical shock wave at underwater explosion; circles and dots show the experimental data of [1] and [7], respectively

2.1.5 Dynamics of Exponent Index, $\theta(r_{\text{fr}})$

It is obvious that the problem on shock wave propagation is solved, if in addition to the wave front pressure at each point of the space, the dynamics of the exponent index is known. This dependence can be determined from the relation between the time scales of the gas cavity t and those of the front t_{fr} [1].

Let us fix the cavity boundary at a certain instant t_0 that will correspond to the time $t_{fr,0}$ of arrival of a shock wave at the point r. Restricting ourselves to the first-order term, we disintegrate t in the Taylor series in the vicinity of t_0:

$$t \simeq t_0 + \left.\frac{\partial t}{\partial t_{fr}}\right|_r (t_{fr} - t_{fr,0}) = t_0 + \frac{1}{\eta}(t_{fr} - t_{fr,0}) \ .$$

Here η is the coefficient governing the dynamics of the exponent index $\theta(r_{fr})$.

From (2.8) one can easily determine

$$\eta = \left.\frac{\partial t_{fr}}{\partial t}\right|_r = 1 + \frac{\partial}{\partial t} \int_{R(t)}^{r_{fr}} \frac{dr}{c_0(1+2\beta\sigma)} \ .$$

If the variable $g = G(R,t)/G(0)$ is introduced, using the above relation for $\beta\sigma$ and Ω and substituting the variables (dr for $d\sigma$), upon simple transformations we get

$$\eta = 1 - \frac{\beta\sigma_*}{\beta c_0(1+2\beta\sigma_*)} + \frac{4}{\bar{\theta}(0)}\left(\frac{\beta g \Omega(0)}{c_0}\right)^2 \left[\frac{1+2\beta\sigma}{\beta\sigma(1+\beta\sigma)} - \frac{1+2\beta\sigma_*}{\beta\sigma_*(1+\beta\sigma_*)}\right.$$

$$\left. +8\left(\frac{1}{1+2\beta\sigma} - \frac{1}{1+2\beta\sigma_*}\right) - 6\ln\frac{\beta\sigma_*(1+\beta\sigma)}{\beta\sigma(1+\beta\sigma_*)}\right] \ .$$

Here $\bar{\theta}(0) = \theta(0)c_0/R(0)$ and the asterisk denotes parameters on the cavity boundary. If the relation between t and r_{fr} is known, one can easily determine η.

Calculation results for the dependence of the exponent index behind the front $\theta^0 = \theta/\sqrt{\bar{q}}$ (s·m$^{1/2}$/kcal$^{1/2}$) on the reduced distance r_{fr}^0 are compared with experimental data of [8] in Fig. 2.2.

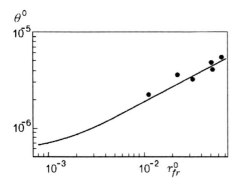

Fig. 2.2. Calculations for the pressure decay index behind the front of cylindrical shock wave vs. distance for underwater explosions: dots show experimental data for trotyl [8]

2.1.6 SW Parameters (Trotyl, Calculation and Experiment)

The pressure in the front of cylindrical shock wave and the exponent index can be presented as a function of the dimensionless coordinate $\bar{r} = r/R_{\mathrm{ch}}$:

$$p_{\mathrm{fr}} = 13\,660\bar{r}^{-1.92}\ \mathrm{MPa} \quad \text{at} \quad 1 \leq \bar{r} \leq 2.5,$$
$$\theta/R_{\mathrm{ch}} = 0.61\bar{r}^{1.64}10^{-4}\ \mathrm{s/m} \quad \text{at} \quad 1.6 \leq \bar{r} \leq 8.5;$$
$$p_{\mathrm{fr}} = 10\,200\bar{r}^{-1.60}\ \mathrm{MPa} \quad \text{at} \quad 2.5 \leq \bar{r} \leq 6,$$
$$\theta/R_{\mathrm{ch}} = 5.50\bar{r}^{0.61} \cdot 10^{-4}\ \mathrm{s/m} \quad \text{at} \quad 8.5 \leq \bar{r} \leq 50;$$
$$p_{\mathrm{fr}} = 4\,000\bar{r}^{-1.08}\ \mathrm{MPa} \quad \text{at} \quad 6 \leq \bar{r} \leq 15;$$
$$\theta/R_{\mathrm{ch}} = 10.4\bar{r}^{0.45} \cdot 10^{-4}\ \mathrm{s/m} \quad \text{at} \quad 50 \leq \bar{r} \leq 3\,200;$$
$$p_{\mathrm{fr}} = 1\,540\bar{r}^{-0.72}\ \mathrm{MPa} \quad \text{at} \quad 15 \leq \bar{r} \leq 3\,200.$$

Figure 2.3 shows experimental data for the front velocity U_{fr}, the lowermost curve (triangles) is for a PETN charge of bulk density. The values of U_{fr} as a function of the front coordinate $\bar{r}_{\mathrm{fr}}$ of a one-dimensional cylindrical shock wave obtained using the Kirkwood–Bethe model (the index fr is omitted below) can be presented as the following power approximations:

$$U = 3.67\bar{r}^{-0.28} \quad \text{at} \quad 1 \leq \bar{r} \leq 10,$$
$$U/c_0 = 1 + 0.276\,5[1 + (0.163 - 1.483 \cdot 10^{-3}\bar{r} +$$
$$6.23 \cdot 10^{-6}\bar{r}^2 + 6.54 \cdot 10^{-8}\bar{r}^3)(\bar{r} - 10)]^{-1} \quad \text{at} \quad \bar{r} > 10,$$
$$\bar{r} = (1 + 3.135\tau)^{0.78} \quad \text{at} \quad 0 \leq \tau \leq 5.75,$$
$$\bar{r} = 2.52\tau^{0.787\,5} \quad \text{at} \quad 5.75 \leq \tau \leq 26.3,$$

and

$$\bar{r} = 1.86\tau^{0.88} \quad \text{at} \quad 26.3 \leq \tau \leq 263.$$

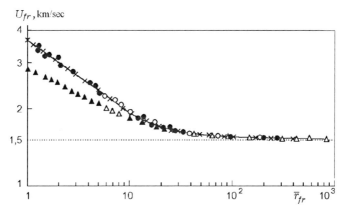

Fig. 2.3. Front velocity of a cylindrical shock wave at underwater explosion of a cord charge (experimental data for PETN, TNT, and RDX; the lowermost curve is for a bulk PETN charge)

2.1.7 Asymptotic Approximation for Weak Shock Waves ($\nu = 1, 2$)

Asymptotic dependences for pressure and the exponent index behind the front were determined using the known relations of the Kirkwood–Bethe theory [5], which are substantially simplified for $r \gg R_{\mathrm{ch}}$, when the dependence of the Riemann function σ on pressure becomes linear. Let us transform σ for $p \ll B$:

$$\sigma = \frac{2c_0}{n-1}\left[\left(\frac{p+B}{p_0+B}\right)^{(n-1)/2n} - 1\right] \simeq \frac{p-p_0}{\rho_0 c_0}.$$

Now

$$G \simeq r^{\nu/2}\frac{p-p_0}{\rho_0}, \quad c+\sigma \simeq c_0 + 2\beta\frac{p-p_0}{\rho_0}, \quad \text{and} \quad U_{\mathrm{fr}} \simeq c_0 + \beta\frac{p-p_0}{\rho_0}.$$

Let us assign a certain surface r_1 as an initial surface whose parameters govern the subsequent wave pattern. We shall assume that pressure on the surface is known as a function of time, $p_1(t_1)$, where $t_1 = 0$ corresponds to the moment of arrival of the shock wave front

$$p_1 = p_{\max}\exp(-t_1/\theta_1)$$

at the point r_1. The basic equation for time has the form

$$t = t_1 + \int_{r_1}^{r}\frac{dr}{c+\sigma}.$$

For the given dependence $p_1(t_1)$, we determine the function $G(t_1)$, which upon "inception" at the moment t_1 on the radius surface r_1 remains unchanged under further propagation of the wave at velocity $c+\sigma$, which enables the change of the variables. Since

$$G(t_1) = r^{\nu/2}\frac{p-p_0}{\rho_0} = r_1^{\nu/2}\frac{p_1-p_0}{\rho_0},$$

we have

$$dr = -\frac{2r}{\nu}\frac{dp}{p-p_0},$$

where $r = (\rho_0 G(t_1)/(p-p_0))^{2/\nu}$. The substitution and simple transformations within the accepted approximations make it possible to determine the delay time for the asymptotics

$$t_{\mathrm{del}} = -\frac{2}{\nu c_0}(\rho_0 G(t_1))^{2/\nu}\int_{p_1-p_0}^{p-p_0}\left(1 - \frac{2\beta}{\rho_0 c_0}p\right)p^{-2/\nu - 1}\,dp.$$

This integral can be easily taken and, if the invariance of the function G is taken into account, eventually leads to the following results:

- *For spherical symmetry,*

$$t_{\text{del}}^{\text{sph}} = \frac{r - r_1}{c_0} - \frac{2\beta G(t_1)}{c_0^2} \ln \frac{r}{r_1}$$

- *For cylindrical symmetry,*

$$t_{\text{del}}^{\text{cyl}} = \frac{r - r_1}{c_0} - \frac{4\beta G(t_1)}{c_0^2} \left(r^{1/2} - r_1^{1/2} \right) \qquad (2.13)$$

Let's confine ourselves to the cylindrical asymptotics for which

$$t_{\text{fr}}^{\text{cyl}} = t_1 + \frac{r_{\text{fr}} - r_1}{c_0} - \frac{4\beta G(t_1)}{c_0^2} \left(r_{\text{fr}}^{1/2} - r_1^{1/2} \right).$$

In this case t_1 can be written as

$$t_1 = \theta_1 \ln \frac{p_{\max} - p_0}{p_1 - p_0}$$

or, using the invariance of the function G,

$$t_1 = \theta_1 \ln \frac{(p_{\max} - p_0)\, r_1}{(p_{\text{fr}} - p_0)\, r_{\text{fr}}},$$

which makes it possible to write t_{fr} as

$$t_{\text{fr}}^{\text{cyl}} = \theta_1 \ln \left(\frac{p_{\max} - p_0}{p_1 - p_0} \right) + \frac{r_{\text{fr}} - r_1}{c_0} - \frac{4\beta G(t_1)}{c_0^2} \left(r_{\text{fr}}^{1/2} - r_1^{1/2} \right).$$

If we take the differential of both parts and, reasoning from the definition $U_{\text{fr}} = dr_{\text{fr}}/dt_{\text{fr}} \simeq c_0 + \beta(p-p_0)/\rho_0$, substitute dt_{fr} for $dr_{\text{fr}}/U_{\text{fr}}$, we easily obtain the following differential equation for the asymptotics of pressure distribution in the shock wave front:

$$\frac{dp_{\text{fr}}}{dr_{\text{fr}}} = -\frac{\frac{\theta_1}{2r_{\text{fr}}} - \frac{n+1}{2c_0}\frac{p_{\text{fr}} - p_0}{\rho_0 c_0^2}\left(\frac{r_1}{r_{\text{fr}}}\right)^{1/2} + \frac{3}{4}\frac{n+1}{c_0}\frac{p_{\text{fr}} - p_0}{\rho_0 c_0^2}}{\frac{r_{\text{fr}}}{p_{\text{fr}} - p_0}\left[\frac{\theta_1}{r_{\text{fr}}} - \frac{n+1}{c_0}\frac{p_{\text{fr}} - p_0}{\rho_0 c_0^2}\left(\frac{r_1}{r_{\text{fr}}}\right)^{1/2} + \frac{n+1}{c_0}\frac{p_{\text{fr}} - p_0}{\rho_0 c_0^2}\right]}. \qquad (2.14)$$

Numerical analysis of this equation shows that within the range $3\,200 < \bar{r}_{\text{fr}} < 10^6$, where $\bar{r}_{\text{fr}} = r_{\text{fr}}/R_{\text{ch}}$, the amplitude distribution (in MPa) can be approximated by the following simple dependence:

$$p_{\text{fr}} \simeq \frac{1580}{\bar{r}_{\text{fr}}^{0.72}}. \qquad (2.15)$$

Here we used the following boundary conditions for r_1: $r_1 = 3\,200$, $p_1 = 4.73$ MPa and $\theta_1 = 59 R_{\text{ch}}/c_0$.

2.1 Underwater Explosion

For $r_{fr} \to \infty$, Eq. (2.14) is considerably simplified: only the last terms remain in the numerator and in the denominator (in brackets). Thus, the function $p_{fr}(r_{fr})$ is defined by the following differential equation

$$\frac{dp_{fr}}{dr_{fr}} = -\frac{3}{4}\frac{p_{fr}-p_0}{r_{fr}},$$

whence it follows that the asymptotic dependence of pressure in the shock wave front is determined from the law

$$\frac{p_{fr}-p_0}{p_{in}-p_0} = \left(\frac{r_{fr}}{r_{in}}\right)^{-3/4},$$

where the subscript "in" denotes the shock wave parameters determined from Eq. (2.15) at a distance at the limit of the range mentioned for this dependence. This solution coincides with the known Landau–Khristianovich asymptotics [8, 9].

The asymptotics for θ can be determined as follows: using the relation (2.15), we obtain the equation

$$t_{fr}\frac{c_0}{r_1} = \int_1^{\bar{r}_{fr}} \frac{c_0}{U_{fr}}d\bar{r}$$

for the propagation time of the shock wave front to the point r_{fr} and defining the front velocity $U \simeq c_0 \cdot [1+(n+1)p_{fr}/4nB]$, we obtain

$$\tau_{fr} = \bar{r}_{fr} - 1 - \frac{n+1}{1.18n}\frac{p}{B}\bar{r}_{fr}^{0.75}\left(\bar{r}_{fr}^{0.295}-1\right),$$

where $\tau_{fr} = t_{fr}c_0/r_1$.

The time τ, over which the pressure at the point $\bar{r}_{fr}$ becomes p, is determined with regard to the delay time of arrival of the appropriate disturbance to this point, the invariance of the function G, and the change of t_1 through p_{max}, and θ_1:

$$\tau = \bar{\theta}_1\left(\ln\frac{p_{max}}{p} - \frac{1}{2}\ln \bar{r}_{fr}\right) + (\bar{r}_{fr}-1) - \frac{n+1}{n}\frac{p}{B}\bar{r}_{fr}^{1/2}\left(\bar{r}_{fr}^{1/2}-1\right).$$

Subtracting τ_{fr} from τ and assuming $\tau - \tau_{fr} = \bar{\theta}$ and $p = p_{fr}/e$, one can easily obtain the distribution for the decay constant,

$$\bar{\theta} = \bar{\theta}_1\left(1+0.205\ln \bar{r}_{fr}\right) + 5.12\bar{r}_{fr}^{0.295}\left[0.53 - 0.848\bar{r}_{fr}^{-0.295} + 0.3185\bar{r}_{fr}^{-0.5}\right].$$

For distances up to $\bar{r}_{fr} \simeq 10^6$ at least, the calculation results can be approximated by the simple dependence $\bar{\theta} \simeq 8.4\bar{r}_{fr}^{0.244}$.

The asymptotic solution has another power: $\bar{\theta} \propto \bar{r}_{fr}^{0.25}$, which is also analogous to the Landau–Khristianovich asymptotics [9, 10].

2.2 Hydrodynamic Shock Tubes

Some problems of high-velocity hydrodynamics require laboratory simulation of rather strong shock-wave loads with controlled parameters. For this purpose, a variety of hydrodynamic shock tubes was designed to apply different generation techniques selected depending on specific research objectives. The four basic designs are a conical shock tube, a hydrodynamic one-diaphragm tube, an electromagnetic tube, and a hydrodynamic pulse two-diaphragm tube (Fig. 2.4).

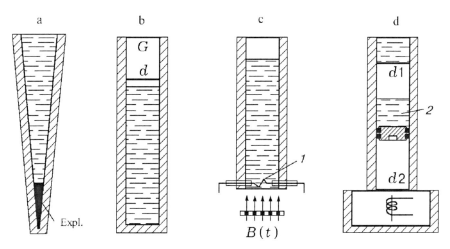

Fig. 2.4a–d. Hydrodynamic shock tubes: (a) Filler's conical tube; (b) Glass's tube; (c) electromagnetic tube; (d) two-diaphragm tube

2.2.1 Filler's Conical Shock Tube

The tube is designed for direct modeling of underwater explosions. The scheme (Fig. 2.4a) suggests that the parameters of a shock wave generated by explosion of a spherical change in an unbound liquid can be realized, if a certain spatial angle is separated in the space and a certain portion of the charge is put at its vertex.

It is clear that a system of cones cannot completely occupy the entire spherical space; however, on the basis of experimental results, Filler [11] formulated the simulation condition for spherical explosion from the charge mass within the scheme:

$$Q \simeq k \frac{q}{\sin^2(\theta/2)} \ .$$

Here Q is the simulated charge mass, q is the charge mass used in the setup, $k \simeq 0.21$ is the experimental loss coefficient, and 2θ is the full flat angle of the cone. The shock tube with $2\theta = 7°$ was used in the experiments. The explosion of a 0.5-*g trotyl charge* proved to be equivalent to that of a *113-g charge* in an unbounded liquid.

Within this scheme, the amplification factor is determined by the relation of the complete spatial angle 4π to the spatial angle separated by the shock tube cone $2\pi(1 - \cos\theta)$ or as $\sin^{-2}(\theta/2)$. The ideal amplification factor (without losses) is equal to 1,070, its actual value is approximately equal to 230. Since the laws of attenuation of the amplitude and of the pressure decay constant with distance behind the front proved to be similar to the known expressions for an unbound liquid, main losses in the scheme (the coefficient k) are governed by the conditions of detonation initiation in a microcharge in the shock wave "breech".

2.2.2 Glass's One-Diaphragm Shock Tube

The tube diagram is shown in Fig. 2.4b: a diaphragm (d) separates a high-pressure chamber with gas (G) from a low-pressure chamber filled with a liquid [12]. When the diaphragm is broken, a shock wave (subscript "sh") propagates in the liquid and a rarefaction wave (subscript "rf") propagates in gas. The subscripts "g" and "l" denote gas (adiabatic index γ) and liquid (adiabatic index n), respectively.

The equation for the hydrodynamic shock tube [12],

$$\frac{c_l(\gamma-1)}{2c_g}\left\{\left[1-\left(\frac{p_l+B}{p_{sh}+B}\right)^{1/n}\right]\frac{p_{sh}-p_l}{n(p_l+B)}\right\}^{1/2} = 1 - \left(\frac{p_{sh}}{p_g}\right)^{(\gamma-1)/2\gamma},$$

can be obtained by combining the conditions of the parity of mass velocities $u_{sh} = u_2$ and pressures $p_{sh} = p_2$ on the contact discontinuity with the following relations:

- For mass velocity in the rarefaction wave,

$$u_2 = \frac{2c_g}{\gamma-1}\left[1-\left(\frac{p_2}{p_g}\right)^{(\gamma-1)/2\gamma}\right]$$

- For laws of conservation of mass and momentum on the jump,

$$\rho_l V_l = \rho_{sh} V_{sh} = m, \quad p_l + m V_l = p_{sh} + m V_{sh}$$

- For relations between densities and the speed of sound in liquids, using the state equation for liquids (Tait or Ridah equations, see Chap. 1),

$$\frac{\rho_{sh}}{\rho_l} = \left(\frac{p_{sh}+B}{p_l+B}\right)^{1/n} \quad \text{and} \quad c_l = \left(\frac{n(p_l+B)}{\rho_l}\right)^{1/2}$$

Two interesting conclusions follow from the equation. First, this generation scheme produces a maximum shock wave amplitude $p_{\text{sh,max}}$ in the liquid, which cannot be exceeded, since it appears when $p_g \to \infty$:

$$\frac{p_{\text{sh,max}}}{B}\left[1 - \left(\frac{p_{\text{sh,max}}}{B}\right)^{-1/n}\right] = 4n\left(\frac{c_g}{c_l(\gamma-1)}\right)^2.$$

This proves that the ultimate value $p_{\text{sh,max}}$ (assuming $p_l \ll p_{\text{sh}}$ and B, whereas $B \ll p_{\text{sh,max}}$) depends significantly on the type of gas used in the high-pressure chamber. Table 2.1 presents asymptotic values of shock wave parameters obtained in the one-diaphragm hydrodynamic shock tube for different gases, assuming that the Tait equation of state is valid for the entire range.

Second, an interesting implication for practical use of this type of shock tubes is obtained when relatively weak shock waves are generated in the liquid, when p_g is of the order of several tens of megapascals. In this case, since $p_{\text{sh}} < p_g$, one can assume that $p_{\text{sh}} \simeq \rho_l u c_l$, where u is the mass velocity of the liquid behind the shock wave front. Then, the condition on the contact discontinuity is noticeably simplified to

$$\frac{2c_g}{\gamma - 1}\left[1 - \left(\frac{p_{\text{sh}}}{p_g}\right)^{(\gamma-1)/2\gamma}\right] = \frac{p_{\text{sh}}}{\rho_l c_l},$$

and after simple transformations is reduced to the form

$$p_{\text{sh}}/p_g \simeq 1 - \gamma p_g / c_l c_g \rho_l$$

whence it follows that pressure in the shock wave p_{sh} upon the rupture of the diaphragm is practically equal to the initial gas pressure p_g. Thus, pressure gauges can be easily calibrated in the shock tube.

The results of [12] also suggest applicability of the Tait adiabat as a shock adiabat for water for dynamic pressures of up to 45 000 MPa, which is confirmed by experimental data of Rice and Walsh [13] and the Kirkwood–Richardson theory [14] on the shock wave velocity, mass velocities, and the speed of sound. Analogous results were noted in [15].

Table 2.1. Asymptotic values of shock wave parameters obtained in the one-diaphragm hydrodynamic shock tube for different gases. The Tait equation of state is assumed to be valid here

Gas	$p_{\text{sh,max}}$, MPa	U, m/s	u_{sh}, m/s	$\bar{\rho}_{\text{sh}}$	$\bar{c}_{\text{sh}}$	M
Water	$1.165 \cdot 10^3$	$2.4 \cdot 10^3$	480	1.25	1.76	1.6
Air	$8.05 \cdot 10^3$	$4.65 \cdot 10^3$	1 720	1.59	4.19	3.1
Helium	$2.0 \cdot 10^4$	$6.7 \cdot 10^3$	2 980	1.80	6.09	4.46
Hydrogen	$7.4 \cdot 10^4$	$1.17 \cdot 10^4$	6 300	2.165	10.8	7.8

Note: Here $\bar{\rho}_{\text{sh}} = \rho_{\text{sh}}/\rho_l$, $\bar{c}_{\text{sh}} = c_{\text{sh}}/c_l$, and $M = U/c_l$.

2.2.3 Electromagnetic Shock Tube

This scheme is used in two variants, both are based on the same principle of energy storage in a capacitor bank. As a load, the first version applies either a discharge gap in the working medium or an exploding wire that stabilizes the discharge channel. In the second version, the bank is loaded on the conductor as a flat spiral coil, wherein a pulsed magnetic field is generated by an alternating current. Let us consider the second scheme shown in Fig. 2.5a [16].

An electromagnetic hydrodynamic shock tube designed for the generation of ultrashort (microsecond duration) shock waves in a liquid consists of a transparent working section (1) 80 mm in diameter filled with distilled water. The central part of the section bottom is a conducting duralumin membrane (2) that is 30 mm in diameter and 0.8 mm thick. The control section of the shock tube (electromagnetic source) is composed of a high-voltage capacitor bank (CB) and a flat spiral coil (3) placed between the membrane and the massive copper disc (5). The bank capacity is 2 µF, its inductance

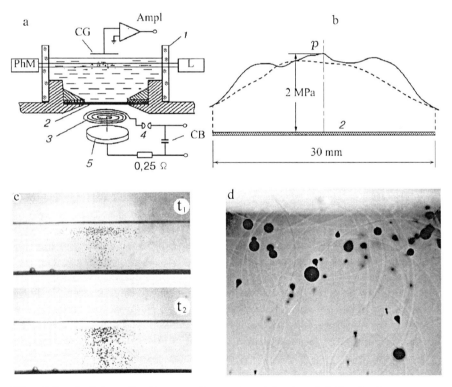

Fig. 2.5a–d. Schematic diagram of experimental setup (**a**) for shock wave generation in liquids by magnetic field, wave front profile (**b**) and cavitation zones, (**c**) 1 µs exposure time, (**d**) 3 ns exposure time

is 25 nH. The bank accumulates energy up to 100 J at a voltage of 10 kV. In the discharge chain, the capacitor bank is connected with the spiral by a spark discharge gap (4) which operates as a cut-off switch.

Ignition of the discharge gap by an external high-voltage pulse results in the discharge of the capacitor bank on the flat coil that generates powerful pulsed magnetic field. This magnetic pulse puts into motion the membrane, which excites a flat shock wave in the liquid specimen. If the inductance of the discharge chain is low, the "battery–spiral" contour can be closed very rapidly. The contour parameters are adjusted to generate shock waves with amplitudes of the order of 10 MPa, the rise time of about 1 μs, and the duration of 3–4 μs.

The minimal magnetic field losses in the gap between the membrane and the disc and practically aperiodic discharge is achieved, when

$$h > \delta = \sqrt{2\varepsilon c^2/\sigma\omega}, \quad R = 2\sqrt{L/C},$$

where h is the membrane thickness, δ is the skinlayer depth, ε is the dielectric constant, c is the speed of light, σ is the material conductivity, ω is the angular frequency of the discharge, R is the wave resistance, and L and C are the contour inductance and capacity, respectively.

If the liquid column in the working section is 30 mm high, due to diffraction effects the linear dimension of the wave zone generated by the membrane reduces from 30 mm near the membrane to 20 mm near the free surface of the liquid. One can state that a free-wave zone arises in the specimen. Thus, within a reasonable registration time span one can neglect the effect of the working chamber walls on the process, for example, when studying the initial cavitation stage near the free surface under the reflection of a shock wave from it (Fig. 2.5c,d). Figure 2.5c presents two sequential frames of cavitation zone development between free surface and membrane (1 μs exposure time).

The thin structure of cavitation zone and wave field was resolved using practically instantaneous photograph with 3 ns exposure time (Fig. 2.5d). Here one can see cavitative bubbles, cumulative jets and a system of shock waves radiated by bubbles. One can note that bubbles in cavitation zone have different sizes and hence they are in different stages of pulsations. The latter gives the base to conclude that the effect of bubble collapse with cumulative jet formation is a result of interaction of shock waves generated by bubble groups which are characterized by relative high frequency of pulsation and more "slow" bubbles. Comparison of Figs. 2.5c and 2.5d shows that obviously 1 μs exposure time is enough in order that a general image was smoothed and a cavitation zone became practically uniform one.

Experimental measurements of the mass velocity by an electromagnetic transducer proved that over the observation time, only an incident wave and a wave reflected from free surface as a rarefaction wave were recorded in the region under study [16].

The electromagnetic transducer was made of a narrow band of aluminum-backed lavsan film (15 μm thick and 20 mm long) loosely connected with the contacts. It was placed in a constant magnetic field with the induction $B = 0.06$ T. As the band moved at the flow velocity u behind the front of the incident and the reflected waves, the voltage corresponding to the known relation $U = -Blu$ was induced in the band. Here l is the band lenght.

The experiments showed that the parameters of the shock wave generated by the electromagnetic source are stable and reproducible. Figure 2.5b presents experimental data for the shock wave front measured by the pressure gauge on the polyvinylidene fluoride-(PVDF)-base in the region under study at distances of 5 and 23 mm from the membrane (dashed line). One can see that the wave profile was stabilized when the gauge moved from the membrane to the free surface near which the profile of the shock wave front was smoothened.

A He–Ne laser (L) and photomultipliers (PhM) were used to study the initial stage of bubble cavitation, while a variable-capacitance transducer (CG) recorded the displacement of the free surface (see Fig. 2.5a). The transducer was made of two surfaces placed at a distance of 0.5 mm. The internal and the external surfaces were 15 mm and 30 mm in diameter, respectively. The external surface made of 15-μm thick backed aluminum film was placed immediately on the liquid free surface. The voltage of about 100 V was maintained between the plates. Displacement sensitivity of the transducer was of the order of 1 μm.

2.2.4 Two-Diaphragm Hydrodynamic Shock Tubes

As is obvious from the foregoing, despite the positive results achieved for HE microcharges in conical shock tubes, this method does not enable laboratory adjustment of the amplitudes and durations of generated waves in a fairly wide parameter range due to design difficulties (large charges) and the critical size restrictions (light-weight charges). The gas-dynamic one-diaphragm tube requires high pressures in the gas chamber and, for technical reasons, fails to provide a constant wave amplitude and the adjustment of the wave duration. To apply a wide range of parameters in the studies of the propagation of shock waves in liquids and multiphase media, the shock tubes with exploding wires require rather large storage capacitor banks, complicated block, and framing prevention systems.

However, high amplitudes and durations can be obtained without using the sources with even higher pressures of compressed detonation products or large capacitor banks. High-velocity liquid flows can be used instead. Indeed, for example, for pressures of about 100–150 MPa and lower, the dependence between the amplitude p of the stationary shock wave and the mass velocity u of particles behind the front $p \simeq \rho_l c_l u$ is linear, it follows that the mass velocity of only 100 m/s corresponds to pressure of 150 MPa. A one-dimensional liquid flow can easily achieve these velocities, its instantaneous deceleration

48   2 Shock Waves in Liquids

on a solid or liquid obstacle produces the appropriate pressures. This method was first proposed in [17] and then studied in [18]. Let us consider it in more details.

2.2.4.1 Experimental Setup

The method was based on instantaneous deceleration of a liquid (solid) piston thrown by compressed gas. Schematic diagram of the setup is shown in Fig. 2.4d.

The shock tube was positioned vertically and, as a rule, consisted of three sections: high-pressure gas receiver (designed for static pressure up to 20 MPa), acceleration channel with a liquid or solid piston 2 (the channel was partially or completely vacuumed depending in the piston type), and the working section with the liquid separated from the acceleration channel by a diaphragm (d1).

A setup can consist of only two sections, if the liquid under study is used as a piston. The liquid piston was separated from the receiver diaphragm (d2) by a light hermetically packed transmitting piston (Fig. 2.4d) that prevented outbreak of gas in the liquid when the diaphragm d2 opens. In this scheme, the shock wave was generated when the liquid piston impinged on the rigid closed butt end of the acceleration channel. The two-section system with a liquid piston was used to generate a powerful step-like shock wave with an amplitude on the order of hundreds of megapascals and a duration on the order of hundreds of microseconds.

The diaphragm d2 was made of annealed sheet copper 0.5 mm thick. The depth of special cuts made in the diaphragm was calibrated depending on the required breaking pressure (load on the piston). The length of the acceleration channel varied from one to two meters. For visual observation, a part of the section near the butt end made of plexiglas was placed in a steel yoke with narrow longitudinal slits. To synchronize the recording with the illumination flashes, the pressure gauges and the contact pickups (light guides) were inserted in the lateral and front channel walls. The final stage of motion of the liquid piston before the impact was recorded by a high-speed photorecorder in the regime of continuous scan.

The experiments proved the importance of wall smoothness, because even a slight unevenness led to disturbance in the thrown liquid column, which resulted in its discontinuity before the impact. Since the volume of the gas receiver exceeded that of the acceleration channel considerably, the piston motion could be considered as uniformly accelerated, its velocity could be estimated from the elementary relation $u_r \simeq \sqrt{2p_r S x m^{-1}}$, where p_r is the pressure in the receiver, S is the area of the channel cross-section (here, $\simeq 7$ cm^2), x is the path traversed by the piston, and m is the piston mass. Thus, for example, if the piston was 0.4 m long and its mass was 370 g with the initial pressure of 4 MPa in the receiver, the piston was accelerated to 140 m/s over the section of about 1.25 m. Under the impact on the tube butt, a step-like shock wave was excited with an amplitude of over 200 MPa and a duration of about 600 μs. If the piston length was reduced by a factor of four, other conditions being equal, the wave amplitude increased to 500 MPa and its duration was reduced to approximately 130 μs.

For relatively weak waves with amplitude of the order of 10 MPa [19], both diaphragms (in the two-diaphragm scheme) were made of thin film. The process was triggered by an electromagnet with a needle in the gas receiver designed for forced rupture of the diaphragm d2. In the experiments with weak waves, a fluorocarbon polymer piston had the form of a disc with a circular stabilizer used to stabilize piston motion along the channel. Generally, four piston types of length (including the stabilizer) 33, 18.5, 11.5, and 8 mm and disc thickness 8.5, 11, 7, and 4 mm, respectively, were used. The piston velocity was recorded by two fiber-optic transducers, whose operation principle was based on light reflection from the piston wall. A bundle of optical fibers with total diameter of 2 mm was flush mounted on the shock tube wall in the end part of the acceleration channel (in front of d1). The other end of the bundle was split into two parts, with a light source attached to one end and a photoreceiver FD-27K, to the other end. Along one part of fibers, the light from the source was emitted in the acceleration channel. When the piston obstructed the transducer, the light was reflected from its wall and passed along the other part of fibers to a photodiode. The signal from the photodiode was fed to the frequency counter Ch3-34a through an amplifier.

The piston velocity was measured from the time when it passed by the transducer on the base of 20 mm. One transducer was placed closer to the diaphragm, at a distance of 20 mm. The shock wave amplitude was determined from the known relation $p \simeq \rho_l u c_l$, where u is the velocity of the contact surface after the impact. The velocity was determined from the data on the decay of an arbitrary discontinuity, in the case of impact of a solid piston on the diaphragm d1, or was equal either to the velocity u (under the impact of a liquid piston on the rigid butt end) or to $u/2$ ("liquid–liquid" scheme).

The measurement accuracy depended on the distance between the transducers, their diameter, and the distance from the diaphragm. The system amplifying the signals from the transducers ensured their actuation when the piston edge passed the axis of each transducer to the accuracy of 0.25 mm, which yielded a relative error of velocity measurements of about 2.5%. One can easily estimate that the measured velocity differed from the velocity at the moment of impact by less than 5%, which was equal to the systematic error.

The diaphragm d1 made of 0.05 mm thick lavsan film was deflected for several millimeters under the vacuumization of the channel, which could distort the wave pattern due to the impact of the piston on the convex surface. To form a flat shock wave front, a metallic disc 0.8 mm in diameter was glued on the lavsan diaphragm using an epoxy adhesive. To eliminate the damping effect of air remaining after vacuumization, the upper part of the acceleration channel had a narrow circular slit and a special cavity.

2.2.4.2 Measurement Techniques

The pulse profile of weak shock waves ($\approx$ 10 MPa) in a liquid was recorded by piezotransducers, whose working element was flush mounted on the wall of the working section of the shock tube. Since the low mechanical strength of piezoelectric elements restricted their use for measurement of pressures up to hundreds of megapascals, germanium sensors were used in [17, 20]. The measurement approach was based on the known resistance of germanium as function of static pressure [21]. Physically, this effect was caused by a change in the width of the forbidden zone E_G (the width of the energy gap between the filled valent and vacant zones, that was the conducting zone) with pressure p, which, according to experimental data [21], could be determined from the dependence

$$\left(\frac{\partial E_G}{\partial p}\right)_T = \alpha$$

($\alpha = 5 \cdot 10^{-6}$ eV/atm), that was valid up to 1.5 GPa (15 kbar). The variation of the width of the forbidden zone eventually affected the value of the squared concentration of proper carriers [21], determined as

$$n_i^2 = 9.3 \cdot 10^{31} T^3 \exp\left(-\frac{E_G^{00} + \alpha p}{kT}\right),$$

where p is the pressure in atm; T is the temperature, in K, and k is the Boltzmann constant. For germanium at room temperature and atmospheric pressure $E_G^{00} = 0.75$ eV, while at $p = 10$ kbar, $E_G = 0.8$ eV with n_i^2 changing almost an order of magnitude.

The scheme of pressure recording is shown in Fig. 2.6 [20]. A germanium element 1 (1 × 1 × 0.5 mm) was covered by epoxy resin 3 with an extender (corundum powder) polymerized at room temperature. One of the electrodes was closed directly on the metallic casing, in which it was sealed by the resin, while the second electrode 2 was isolated. In the electric circuit of the transducer (SpresG) shown in the figure, R is the load resistance (of the same order as that of the transducer), $C = 0.6$ μF is a standard storage bank with voltage on the order of several tens of volts. The gauge casing was mounted in the steel chamber 4, so that the semiconductor element became the part of the butt end of the shock tube in order to record a shock wave generated under the impact on the butt end, or an incident shock wave generated under the impact on a resting liquid 5 and reflected from the butt end. The chamber 4 was the working section of the shock tube that served as a part of the acceleration channel under the impact on the wall or as the third section with a liquid under study.

Semiconductor gauges were used to register shock waves within 30–500 MPa [17, 20]. The lower limit of the range was defined by the sensitivity of the amplifier used (0.25 cm/mV). Recording sensitivity was not lower than 1/300 mV/atm. The rise time of the front pressure was about 2 μs due to the

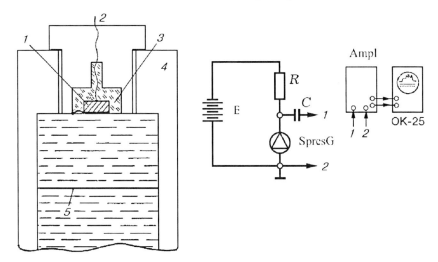

Fig. 2.6. Pressure recording by a semiconductor gauge under the impact of a liquid piston on the tube end: (1) germanium element; (2) electrode; (3) epoxy resin; (4) chamber; (5) shock wave

availability of the epoxy layer on the surface of the semiconductor element, while its own frequency was approximately 30 MHz and the speed of sound in germanium was 4,930 km/s. The sensitivity of the recording scheme could be enhanced by increasing the bank voltage. However, the experience showed that for each specific germanium element there was an optimal experimentally adjusted voltage; thermal noise occurred when it was exceeded.

The semiconductor gauges were calibrated by optical (depending on the time between two neighboring density gradients and using the pressure oscillogram) and static (oil, $T = 18\,°C$) techniques [20]. In the latter case, the gauges were calibrated to the resistance of elements 471 Ω (1) and 747 Ω with p–n transition (2). The relative resistance of the element depended linearly on pressure and remained as such up to 20 kbar. Availability of the p–n transition (gauge 2) enhanced considerably the sensitivity of the recording system, but reduced its resistance to mechanical loads.

Figure 2.7 shows a typical pressure oscillogram recorded at the butt end of the two-diaphragm shock tube for the impact of a liquid piston on a resting liquid in the working section. Upon the impact, shock waves propagate in resting and braked media. The butt end gauge records the first reflected wave 1 with practically constant pressure behind the front. The second jump 2 is a result of the interaction between the waves reflected from the piston and the tube end. Behind the wave front, there is a pressure decline due to unloading at the lower end of the piston. The third jump 3 is a result of repeated interaction between the waves. Comparison of the oscillogram with the wave interaction scheme in the shock tube proves that the sensor

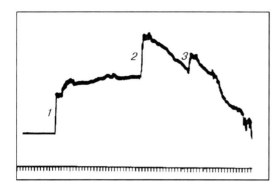

Fig. 2.7. Pressure oscillogram of shock waves at the butt end of a two-diaphragm shock tube (germanium sensor): $p_1 = 200$ MPa, time scale is 20 μs/div

reproduces precisely all the wave processes occurring upon the collision of the two liquid columns.

Pressure gauges can be reliably calibrated in the hydrodynamic shock tube [17]. The pressure wave amplitude is calculated from the velocity of the liquid column u during deceleration and from the equation of state (for example, the Tait equation):

$$p_2 - p_0 = \frac{u^2}{4}\left[\frac{1}{\rho_0} - \frac{1}{\rho}\right]^{-1}, \quad \bar{p} = \left(\frac{\rho}{\rho_0}\right)^n$$

where ρ and p_2 are the density and pressure in the shock wave front.

2.2.4.3 PU-Diagrams for Calculating the Parameters of Hydrodynamic Shock Tubes for Different Impact Schemes

A pu-diagram of the impact scheme is presented in Fig. 1.2a, where the states (w) and (p) are attributed to immovable and moving media, respectively. One can easily determine that during collision, shock waves s_w and s_p propagate in both media. In this case, pressure p_2 in the shock wave generated in the resting liquid as a function of the piston velocity (u_p) using the transition formulas is defined by the dependence

$$p_2 - p_1 = \frac{u_p^2}{\left[1 - ((p_1+B)/(p_2+B))^{1/n}\right]\left[\rho_1^{-1/2} + \rho_{1,p}^{-1/2}\right]^2},$$

which proves that the estimate of pressure in an acoustic approximation ($p_2 \simeq \rho_1 u_p c_1$) for pressures over 10^3 atm differs noticeably from the exact solution. The above estimates show that in the above cases one can obtain step-like shock waves with pressure on the order of 10^4 atm. To this end, a liquid piston should be accelerated to several hundreds of meters per second. Changing the piston length, one can assign the required duration of the compressed state of the liquid prior to the arrival of a wave reflected from the piston's free end.

2.2 Hydrodynamic Shock Tubes

In conclusion, we note that, following [18], at $u_p/c_p \leq 1.2$ (where u_p is the mass velocity behind the shock wave front) the front velocity can be approximated by the relation $c_p + ku_p$, where $k \simeq 2$ for water, $k \simeq 1.56$ for gold, $k \simeq 1.36$ for steel, and $k \simeq 1.28$ for tungsten. These values were obtained in solving the problem on the collision of particles with an obstacle at $u_p \leq 3.6$ km/s.

If the target is rigid, the following relations are valid for pressure and velocity of the shock wave front in a liquid piston:

$$p_p = \rho_p u_p (c_p + k u_p) \quad \text{and} \quad U = c_p + k u_p, \quad \text{at} \quad k \simeq 2.$$

Here the subscript p denotes the characteristics of the medium interacting with the target (liquid, thrown particles, etc.). If the target is elastic, the approximation $p_2 \simeq \rho_1 u_2 c_1$ is valid, where $\rho_1 c_1$ is the acoustic impedance of the target. For liquid targets, we have

$$p_2 = \rho_p u_1 (c_p + k u_1) \quad \text{and} \quad U = c_p + k u_1,$$

where $u_1 = u_p - u_2$ is determined from the equation

$$\frac{u_1}{u_p} = \left[\left(\frac{1+\rho_1 c_1/\rho_p c_p}{2kM_p}\right)^2 + \frac{\rho_1 c_1/\rho_p c_p}{kM_p}\right]^{1/2} - \frac{1+\rho_1 c_1/\rho_p c_p}{2kM_p}$$

($M_p = u_p/c_p$), or approximately

$$\frac{u_1}{u_p} \simeq \frac{\rho_1 c_1}{\rho_1 c_1 + \rho_p c_p},$$

which is valid, if the impedances of the liquid and the target differ significantly.

2.2.5 Shock Tube Application: High-Rate Reactions in Chemical Ssolutions

As is known, the method of shock waves is widely used for experimental studies of relaxation processes in gases at high temperatures and sometimes for physico-chemical studies of condensed media.

The above method of generating strong one-dimensional shock waves in a liquid was proposed in [17]. It can be used to produce a temperature jump of 3–30 °C in a liquid [22].

Following the estimate [23], the temperature jump is determined from the pressure p_{sh} behind the shock wave front,

$$T - T_0 \simeq 2.6 \rho_{sh} p_{sh} \cdot 10^{-2},$$

where T_0 is the initial temperature in °C, ρ_{sh} is the density in kg/m^3, and p_{sh} is the pressure behind the shock wave front in MPa. One can easily determine that, if the amplitude is 100 MPa, the temperature jump behind the front does not exceed 3 °C.

The method of temperature jump is rather efficient for studying high-rate chemical reactions in solutions with the relaxation time of $\geq 10^{-6}$ s. The temperature jump is arranged in different ways:

a) High-voltage discharge (a liquid is heated for about 1 µs; however, this method requires highly conductive solutions, i.e., only the solutions with high ion concentration can be used)

b) Super-high frequencies (high SHF absorption factor is required, which provides a temperature increase only within 0.1–1 °C)

c) Optical heating (absorption of visible light, ultra-violet flash lamp and laser heating systems) [24]

All the methods have some advantages and shortcomings and, as a rule, apply strong restrictions on the working volume of the solution under study. In addition, these methods are characterized by a *short-term* pulsed equilibrium shift, when the temperature jump is of negligible duration, at least much shorter than the relaxation time.

High-velocity collision, which is used to determine the velocity and the shift of thermodynamic equilibrium for reversible chemical transformations in solutions, allows one to study the relaxation process for the case of temperature or pressure conservation (by analogy with [25]) after the jump and can be applied in a complex study of liquid-phase reactions with simultaneous pressure and temperature changes.

The left part of Fig. 2.8 shows the scheme of an experimental setup: (1) pressure gauge; (5) condenser; (6) diaphragm; (7) Jupiter-9 lens; (8) light filter; (10) conical Plexiglas windows ($d_{\min} = 3$ mm, $d_{\max} = 8$ mm, $l = 42$ mm); (11) 0.05-mm thick brass membrane; (12) piston. A solution under study (2) was placed in a stainless steel chamber (3). An incandescent lamp (4) (12 V, 100 W) was used as a light source. The shift of equilibrium behind the shock wave front was recorded from the absorption on the appropriate wavelength using a photomultiplier FEU-18A (9). Interference light filters (8) or a monochromator PM-2M were used to separate the required wavelength. An incident shock wave produced a pressure of 110 MPa, the temperature jump was 3 °C, and the heating time of the solution was less than 2 µs.

The efficiency of the technique was tested for two chemical reactions [22] studied before using other relaxation techniques [26, 27]. The reactions with pronounced temperature effect were selected. Following [28], the system parameters p and T were selected to eliminate the effect of the pressure jump on reaction rate constants. The system allowed one to study the kinetics of high-rate reactions with half-reaction time varying from several microseconds to 600 µs.

As the shock wave passed through an equilibrium solution, the temperature jumped behind the front produces an equilibrium shift, which was determined from the change of concentration of the component with the highest

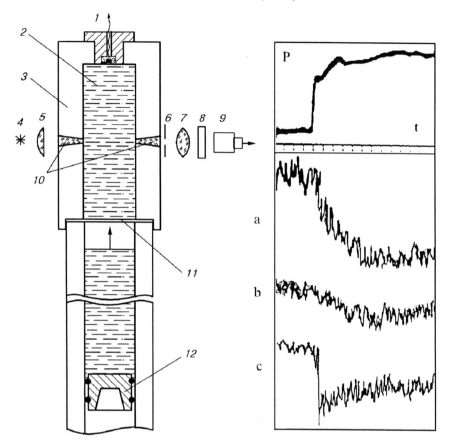

Fig. 2.8. Schematic diagram of an experimental setup (left) for studying fast reactions and relaxation processes in the fast reactions in solutions (right)

absorption factor. The wavelengths that provided maximum light absorption for each reagent were measured preliminarily using a spectrophotometer.

The rate constants of direct k_1 and reverse k_2 reactions were determined from the known relations (for the same initial concentration) $1/\tau = 2c_0 k_2 \sqrt{k} = 2c_0 k_1/\sqrt{k}$, where τ is the experimentally determined time constant, c_0 is the initial concentration of reagents, and $k = k_1/k_2$ is the equilibrium constant (determined by an SF-10 spectrophotometer using the static method).

The reaction of tropeolin-0 (weak acid) with water made it possible to test the sensitivity of the method and determine the relaxation time, whose value was compared with that determined in [26]. The shift behind the shock wave front in water solution of tropeolin in neutral medium was experimentally recorded as the tropeolin concentration changed from $0.78 \cdot 10^{-5}$ to $6 \cdot 10^{-5}$ mol/l. The maximum absorption of tropeolin was achieved at 4250 Å.

A typical oscillogram is shown in curve b on the right side of Fig. 2.8 (the time scale is 10 μs, the pressure jump amplitude in the shock wave front $\simeq 10^2$ MPa). Experimental results for $T = 22\,°C$ and the time constant τ of the first relaxation stage are presented below:

$c_0 \cdot 10^5$ mol/l	0.78	1.08	1.86	2.90	3.97	6.0
$\tau \cdot 10^6$ s	30.5	20.0	16.6	13.5	10.1	6.0

Two stages of the relaxation process are clearly seen in the oscillogram: an increase in the concentration c after the temperature jump (the solution darkened in about 50 μs) and then a decrease to a new equilibrium state. Each value of τ was determined in several sets of measurements. The dispersion of values did not exceed ±10%. The experiments showed that with increasing concentration of tropeolin, the relaxation time decreased considerably.

For tropeolin in an alkaline medium, the relaxation process (Fig. 2.8, curve c, rightside), whose character was completely consistent with the experimental data of [26], was observed. The relaxation time constants were close as well.

The measurements showed that the relaxation time constant depended slightly on concentration in the region with low concentration of tropeolin (at $T = 20°C$ and $k = 0.714 \cdot 10^{-2}$):

$c_0 \cdot 10^5$ mol/l	0.78	1.56	1.92	2.25
$\tau \cdot 10^6$ s	25	23.7	21.7	21.7
$k_1 \cdot 10^{-4}$ 1/(mol·s)	1.2	1.30	1.40	1.40
$k_2 \cdot 10^{-6}$ 1/(mol·s)	1.7	1.83	2.0	2.0

The constants of electron-transfer rates were measured using the described technique for the reaction

$$Fe(DMBPy)_3^{2+} + Ir\,Cl_6^{2-} \underset{k_2}{\overset{k_1}{\rightleftharpoons}} Fe(DMBPy)_3^{3+} + Ir\,Cl_6^{3-}.$$

The optical characteristics of the reagents [22] are presented in Table 2.2.

Table 2.2. Optical characteristics of reagents

Complex	λ_{max}, Å	ε_{max}, mol/cm
$Ir\,Cl_6^{2-}$	4 900	3 920
$Fe\,(DMBPy)_3^{2+}$	5 290	8 470
$Fe\,(DMBPy)_3^{3+}$	5 430	350
$Ir\,Cl_6^{3-}$	4 150	76

Note: λ_{max} is the wavelength at which maximum absorption occurs (ε_{max} is the absorption factor).

The complex of bivalent iron had the greatest absorption factor, thus, the equilibrium shift was established on the basis of changing concentration. A typical oscillogram of the relaxation process is shown in curve a on the right side of Fig. 2.8. For equal initial concentrations of reagents $c_0 = 4.35 \cdot 10^{-5}$ mol/l and the ionic force of 0.1 mol, a new equilibrium state had the relaxation time $\tau = 20.7$ μs at $T = 22.1\,°C$. The calculated rate constants of direct and reverse reactions (for equilibrium constant $k = 1.95$) were equal to $\sim 8 \cdot 10^8$ and $\sim 4 \cdot 10^8$ l/(mol·s), respectively, and were close to the values obtained in [27].

The above experimental results proved the applicability of this technique to studying nonequilibrium processes induced by the temperature jump behind the front of a strong shock wave in solutions.

2.3 Explosive Hydroacoustics

High-rate changes in the physical and chemical states of a medium accompanied by abrupt heat release and pressure and temperature increase are usually defined as explosive processes. Chemical media that are capable of such transformations are called explosives. The processes proceeding with velocities on the order of 1 km/s and higher are called explosive processes. Radiation sources or systems based on such processes are called *explosive sound sources*. They generate shock and compression waves, which can be defined as high-pressure zones with high energy density.

Such waves can be generated not only by HE and explosive gas mixtures, by high-voltage electric discharges and different types of hydrodynamic shock tubes, but also by the so-called *explosive-type sources*: airguns, waterguns, watershocks, and powerful laser beams focused into a liquid. The main advantage of these sources over traditional sources is the high level of transformation of explosive energy into an acoustic energy in the form of a strong wide-band signal that is noise resistant and capable of propagating over hundreds of kilometers.

Beyond the different types of sources and their characteristics, detonation parameters as well as shock waves in near and far fields of standard HE, and their spectral characteristics are considered along with non-traditional sources designed for solving the problems of directivity and control over radiation energy [29].

2.3.1 Basic Characteristics of Explosive Sound Sources

An explosion can be considered as a two-stage process. During the first stage, high energy is accumulated within a limited volume. For example, in the case of HE, this energy is released at the detonation front due to chemical transformation (~ 1 kcal/g). Under electric discharge the energy accumulated in the capacitor bank is "pumped" into the discharge channel. The radiation process itself may be defined as the second stage, although there is often no exact time threshold and it may start before the energy accumulation is over. To analyze the possibilities of the sound sources let us estimate the energy characteristics of the sources.

2.3.1.1 Detonation Parameters

Combustion and detonation are the two main types of explosive chemical transformations. Combustion, as a rule, occurs place in gaseous mixtures. Detonation is typical of high-energy explosives and is understood as the process of shock wave propagation in HE, with a stable zone of chemical reaction accompanied by heat release behind the front. The rate of this process, $D = $ const for a given HE, along with the density ρ_{HE}, explosion heat Q, heat capacity of the reaction products c_{v*}, and isentropic index γ, defines the main physical characteristics of the reaction products (pressure p_*, density ρ_*, and temperature T_*) [30]:

$$p_* = \frac{\rho_{HE} D^2}{\gamma + 1}, \quad \rho_* = \frac{\gamma + 1}{\gamma} \rho_{HE} \quad \text{and} \quad T_* = \frac{2\gamma}{\gamma + 1} \frac{Q}{c_{v,*}}.$$

Detonation parameters of gas mixtures and condensed explosives are presented in Table 2.3.

One can see that the values of pressure in detonation waves of condensed explosives ($\gamma \simeq 3$) and in gas mixtures ($\gamma \simeq 1.25$) differ by several orders of magnitude. The orders of magnitudes of the same parameters at underwater nuclear explosions are estimated as $p_{\text{nucl}} \approx 10^7$ kbar and $T_{\text{nucl}} \approx 10^7$ K. In most cases, the physical model of instantaneous detonation is quite applicable to HE. Following this model, detonation of the whole charge occurs instantaneously. In this case, the charge volume remains unchanged, pressure $\bar{p}_*$ and temperature $\bar{T}_*$ are the same throughout the volume, with density of detonation products $\bar{\rho}_*$ equal to the initial density of the charge ρ_{HE}. Following the model of instantaneous detonation, $\bar{p}_*$ is found from the condition that the the internal energy of the detonation products fully depends on the explosion heat:

$$\bar{p}_* = (\gamma - 1)\rho_{HE} Q = p_*/2.$$

Table 2.3. Detonation parameters of gas mixtures and condensed explosives

HE	ρ_{HE}, g/cm^3	D, m/s	p_{*a}, kbar	T_*, K
$2H + O_2$	$1.17 \cdot 10^{-3}$	2630	0.038	3 960
$CH_4 + 2O_2$	$1.17 \cdot 10^{-3}$	2220	0.027	4 080
$2C_2H_2 + 5O_2$	$1.17 \cdot 10^{-3}$	3090	0.051	5 570
TNT	1.62	7050	215	2 350
RDX	1.8	8600	360	3 750
PETN	1.77	8400	340	4 150
TETRYL	1.7	7850	265	2 940

2.3.1.2 Electric Explosion

A scheme of the generation of a hydroacoustic signal using this method is shown in Fig. 2.9 [31]. A capacitor bank with capacitance C is charged through the resistance R from a high-voltage source up to voltage U. The circuit comprises an electric contour (1) with the gaseous gap G and the working gap W (of length l) submerged in the water, and the inductance L that enables control of the discharge regimes in the contour. As a rule, the gap G is ionized by an external high-voltage pulse and the capacitor bank is closed in the load, being a column of weakly conducting liquid between the electrodes of the gap W. The two main stages are distinguished in this complicated process:

1) Pre-discharge stage, in which a conducting channel is created in the dielectric due to the dissociation and ionization of molecules
2) Breakdown stage, when the characteristics of the explosive gap W as a source of high-density energy are formed

When the discharge is aperiodic (i.e., there is no oscillation process in the circuit 1), the hydrodynamic and physical characteristics of the discharge in the liquid are similar to those typical of an explosion. The energy stored in the capacitor bank is determined from the known relation

$$E_0 = \frac{1}{2}CU^2 \ .$$

Experience shows that the total coefficient of energy transformation (into the explosive cavity η_{cav} and into the shock wave η_{sh}), $\eta = \eta_{cav} + \eta_{sh}$, depends on the capacitance C and length l of the discharge gap W, the value of E_0 being fixed. However, the value of η_{sh}, as an efficiency parameter, is used more often.

The data on η_{sh} and the discharge time τ_d are presented in Table 2.4 as a function of contour parameters for different types of discharges.

To estimate the energy retained in the explosion products after shock wave generation, it suffices to take into account the oscillation period of the explosive cavity, which governs the maximum cavity radius and, consequently, the internal energy, if the hydrostatics is known. In the experiments for $C =$

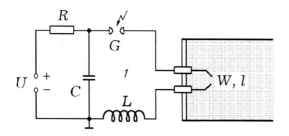

Fig. 2.9. Schematic diagram of an experimental setup for spark discharge in water

Table 2.4. Explosion data as a function of contour parameters for different types of discharge

U, kV	C, µF	l, cm	τ_d, µs	η_{sh}, %	Discharge type
30	0.1	5.0	7.5	32	Aperiodic
21.2	0.2	2.5	17.5	25	Aperiodic
15	0.4	1.2	25.0	13	Periodic
10	0.9	0.5	140.0	3	Periodic

0.1 µF, $U = 30$ kV, and $l = 4$ cm, the period is within 5–6 ms. The internal energy of a (cylindrical) cavity found from the theoretical estimate is

$$E_* \simeq \pi \rho l^3 \left(\frac{p_0 T_b}{\rho l}\right)^2,$$

and is about 0.1 E_0 at the ambient pressure $p_0 = 0.1$ MPa. Consequently, $\eta_{cav} \simeq 10\%$. Thus, taking into account the estimate η_{sh} (see Table 2.4), the full performance factor is $\eta \simeq 40\,\%$, which is twice as low as the appropriate value for cylindrical HE charges. An energy of several kilojoules close to the heat released by a 1-g HE charge, may in principle be stored in the compact capacitor bank due to the increase in the capacitance C and voltage U. However, as is seen from Table 2.4, as C increases η_{sh} sharply decreases. To stabilize the discharge, a thin manganine wire can be mounted in the gap W (shock wave generation by exploding wires).

2.3.2 Hydrodynamic Sources of Explosive Type

As stated above, airguns, waterguns and watershocks are explosive-type sources. The principle of their operation is based on the storage of compressed gas energy and the transfer of this energy directly into the liquid in the form of explosively expanding cavity in airguns, by means of transformation of potential energy of ambient liquid in waterguns or through formation of compression wave applying the principle of a two-section hydrodynamic shock tube.

2.3.2.1 Airguns

In the simplest case, an airgun comprises two elements (Fig. 2.10): a pressure chamber (PC) of constant volume and a source as a gas cavity (GC) in the ambient liquid [32]. The compressed air is supplied from the pressure chamber through the system of ports equipped with a controlling piston into the ambient liquid. Ideally, the losses in the port and the remaining portion of gas in the pressure chamber are disregarded, and the internal energy of compressed gas is converted into the internal energy of the emerging gas cavity. The cavity is assumed to be spherical or cylindrical and the gas inside

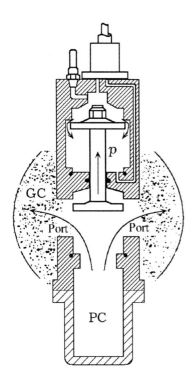

Fig. 2.10. Schematic diagram of the airgun

it to obey the adiabatic law. The dynamics of a cavity of radius R and volume V can be described by the law of energy conservation

$$\frac{p_{01}V_{01}}{\gamma - 1} - \frac{p_1 V_1}{\gamma - 1} = \frac{pV}{\gamma - 1} + p_0 V + \frac{\rho \dot{V}^2}{8\pi R} + E_{\text{rad}} \ .$$

Here the internal energy of the compressed gas (the left-hand part of the equation) is spent for producing the internal gas energy in the cavity $pV/(\gamma - 1)$, the potential energy $p_0 V$, and the kinetic energy $\rho \dot{V}^2/8\pi R$ of the liquid surrounding the cavity, as well as for the generation of the acoustic wave energy E_{rad}. It should be noted that $V_{01} = V_1$, since the volume of the pressure chamber is constant.

The values of E_0 for airguns can be compared with the explosion energy of a "classical" HE: for $V_{01} \simeq 2 \cdot 10^4$ cm^3 and $p_{01} = 34.5$ MPa, E_0 is approximately 1.7 MJ. This is equivalent to 0.4 kg of TNT. However, the efficiency of energy transfer into acoustic energy in the case under consideration is by far lower because the formation of the "explosive" cavity is relatively slow. In the case of explosives, it is formed at the detonation rate; in the case of airguns, this is the velocity of gas mass transfer from the pressure chamber [32]

$$\frac{dm}{dt} = k_e A \sqrt{2\rho (p_{\text{u}} - p_{\text{d}})} \ ,$$

where k_e is the coefficient of losses, which depends on the system design, A is the port area, p_u and p_d are the upstream and downstream pressures, respectively, in the orifice. Such systems are often used in seismic studies to design high-power pulsed source of low-frequency sound in the frequency range 10–200 Hz. Their efficiency is likely to be dependent on depth.

2.3.2.2 Watershocks and Waterguns

Figure 2.11 illustrates the operation principles of hydrodynamic shock tubes that result in a powerful sound source, the so-called "watershock" [33]. A piston (2), placed inside a pneumatic cylinder (1), is accelerated under the effect of pressure supply p in a vacuum and stores a kinetic energy, which, through a transmitting piston (3) is partially transformed into the energy of a shock wave generated by the acceleration of a liquid column in the horn.

When stopped abruptly at the rest (5), a cavitation zone and rarefaction waves occur on the surface of the transmitting piston. They suppress possible oscillations and define the profile of a resulting wave. At the horn exit, the shock wave profile is triangular with a subsequent rarefaction phase. The wave amplitude in the initial part of the horn is estimated by the simple expression $p_{sh} = \rho u_p c$, where u_p is the velocity of the transmitting piston at the moment of impact on the liquid column. Pressure variations in the wave as the latter propagates over a cross-sectional area of the horn are associated with the decrease in the energy flux density.

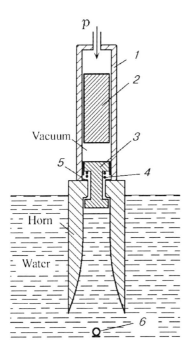

Fig. 2.11. Experimental setup for a watershock: (1) pneumatic cylinder; (2) and (3) plungers; (4) spring; (5) rest; (6) pressure gauge

The waterguns are pulsed pneumatic systems, designed for the generation of powerful single pulses under flow cumulation, comprise the main chamber that is partially filled with seawater and compressed gas. The latter ejects water through the special ports with the help of shuttles. After the shuttle has stopped, the liquid moves by inertia and separates from the system, and a vapor cavity with potential energy $p_0 V_{max}$ and a very low pressure is formed in the surrounding space. The subsequent collapse of this cavity, being almost empty, generates an implosion pulse in the surrounding space.

2.3.3 Wave Field, Spectral Characteristics

For most of the above-mentioned sources, the wave pattern and the relation with the source dynamics are analogous to an underwater explosion (Fig. 2.12). After the detonation wave has refracted at the charge–water interface, a shock wave is formed in the liquid, and the explosion products form a gas cavity with high internal pressure. The cavity makes decaying radial oscillations that produce acoustic waves (oscillations) in a sea. The number of oscillations is limited, as a rule, by the shape instability of the collapsing cavity and by the energy loss under radiation.

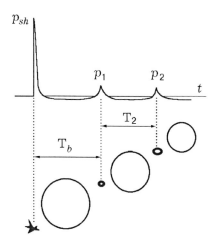

Fig. 2.12. Wave field structure at underwater explosion

2.3.3.1 Parameters of Underwater Explosion

The general features of the structure and parameters of the wave produced by such sources are well known [1, 34–37]. Presented in Fig. 2.12 are classical wave structure (shock wave and two successive oscillations) and synchronous dynamics of the explosion cavity, including the explosion moment itself, three maxima and two minima. Basic explosion characteristics in the field close to the charge can be calculated using the following relations [34]:

$$p_{sh} = 2.08 \cdot 10^4 \left(W^{1/3}/r\right)^{1.13}, \quad p_1 = 3\,300 W^{1/3}/r,$$

$$p_2 = 0.22 p_1, \quad \tau_0 = 0.555 W^{1/3}/z_a^{5/6}, \quad \tau_1 = 1.1 W^{1/3}/z_a^{5/6},$$

$$T_b = 4.34 W^{1/3}/z_a^{5/6}, \quad \text{and} \quad T_3 = 7.4 W^{1/3}/z_a^{5/6}.$$

Here W is the explosive weight (in pounds, 1 pound = 0.453 kg), $z_a = h + 33$ is the full "depth," h and r are the explosion depth and distance (in feet, 1 ft = 0.3048 m), p is the pressure (in psi, 1 psi = $6.8948 \cdot 10^3$ Pa), and T is the time (in s). The shock wave has the form of an exponent near the front with amplitude p_{sh} and positive phase duration τ_0. Oscillations are in the form of double exponents with amplitudes p_1 and p_2 and positive phase durations τ_1 and τ_2. The pressure wave profile (Fig. 2.12) is characterized by the rarefaction phase [35] $\Delta t = 3.1 W^{1/3}/z_a^{5/6}$, which determines the time interval over which the pressure of detonation products in the cavity is lower than hydrostatic pressure.

2.3.3.2 Characteristic Energy Spectrum of Explosive Sources

The energy distribution of an explosive source is shown in Fig. 2.13. It consists of three components [36–38]. The first component defines the spectral characteristics of the shock wave (SW) and describes a high-frequency part of the spectrum where the energy level falls off with the increase of frequency at 6 dB/octave ($\sim f^{-2}$). The second defines a spectral composition of the first and the second oscillations. The third component is associated with the determination of a low-frequency interval of the spectrum (LF) via the combination of pulses I_0, I_1, and I_2 of the shock wave and oscillations.

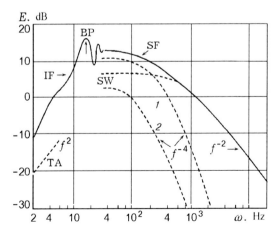

Fig. 2.13. Theoretical energy distribution for a 453-g charge exploded at a depth of 56.6 m as measured from a distance of 91.4 m

Its characteristic times are T_b and T_2, and the duration of the rarefaction phase between I_0 and I_2: $T_3 \simeq T_b + T_2$. Figure 2.13 also presents a theoretical asymptote (TA), the summation result (SF), and bubble pulse frequency (BP, 15 Hz).

For a fundamental oscillation frequency of the cavity filled with explosion products,

$$f_b = \frac{1}{T_b} = 0.23 \frac{z_a^{5/6}}{W^{1/3}},$$

the spectrum (Fig. 2.14, [37]) has a sharp maximum that decreases with increasing explosion depth [39]

$$10 \log \left(\frac{E_{\max}}{W^{4/3}} \right) = -6.2 \log z_a + 33.5$$

and in the region $f < f_b$ it is described by the relation

$$\frac{E}{E_{\max}} = \left(\frac{f}{f_b} \right)^{2.67}.$$

The experimental data for relatively small charges (less than 4.5 kg) at large depths (6708 m and less) recorded above the charge near a free surface demonstrated that the energy distribution between a shock wave and oscillations changed with depth. This was due to the transformations of a wave profile and oscillations. The explosion depth became one of the most significant parameters and was determined from the characteristic period of the spectrum T_s [34]:

$$h = 5.82 W^{0.4} (T_s + \theta_e)^{-1.2} - 33.$$

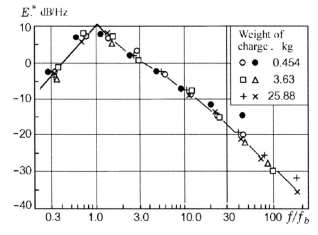

Fig. 2.14. Spectral distribution of the energy flux under large-scale explosion at a depth of 1372 m

The spectrum became scalloped in form and remained the same when propagating over large distances, even if the signal was noticeably distorted. Here $\theta_e = 2.3$ ms was the empirical constant obtained from the data recorded at distances of 460–570 km and 1 150–1 300 km from the explosion site of a 817-g charge at depths of $\sim$ 18.3 m, the receiver being at a depth of about 3 km.

It should be noted that, for the reason mentioned above, the spectrum transformation should be estimated with great care, since upon explosion of large charges at large depths, the explosive cavity could migrate (emerge). The kinetic energy of the radial oscillations of the cavity was converted into the energy of vertical motion under collapse, thereby the minimum cavity size and, consequently, the oscillation amplitude might sharply vary.

2.3.3.3 Peculiarities of Pneumatic Systems

As stated above, the operation principle of waterguns [40] consists in formation of a low-pressure vapor cavity, which irradiates a single-implosion pulse when collapsing under ambient pressure p_0 (Fig. 2.15). The radiation process is adequate here to the phase of the first oscillation of the explosive cavity, and its characteristics can be used (with appropriate coefficient corrections) to estimate the performance of watergun systems.

The profile of the wave generated by an airgun is similar to that shown in Fig. 2.12 for HE, but its amplitude is significantly less and the oscillation decay is weaker. The latter is explained by the negligible radiating capability of the system. In the frames of a spherical or cylindrical model their oscillation frequency, being the fundamental one in the spectrum for the airgun, is defined by the relations [41]:

$$f_{\text{sph}} = 0.027 p_0^{0.93} E_0^{-0.27} \quad \text{and} \quad f_{\text{cyl}} = 2.86 p_0^{0.93} E_{0,\text{l}}^{-0.5},$$

where $E_{0,\text{l}}$ is the initial energy of the system per unit length. On the basis of the above law of energy conservation, the wave pressure may be estimated by

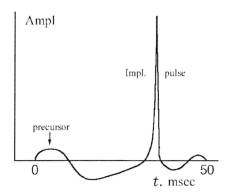

Fig. 2.15. An ideal near-field wave signal of a watergun

2.3 Explosive Hydroacoustics

Table 2.5. Efficiency of energy conversion in different types of explosives

No	Type of radiator, energy, mass	Transformation efficiency, %			
		Total	0–100 Hz	20–50 Hz	100–1000 Hz
1	TNT, 1.5 g	24	0.3	0.09	2.77
2	TNT, 44 g	24	1.07	0.317	5.94
3	TNT, 255 g	24	1.96	0.595	10.1
4	TNT, 1274 g	24	3.24	1.03	12.8
5	Combustion of gaseous mixture, 45 kcal	1.95	1.2	0.415	0.656
6	Detonation of mixture, 39 kcal	1.45	0.7	0.233	0.659
7	Airgun*, 35 kcal	0.75–1.5	0.74–1.47	0.31–0.61	0.03

* For different rates of port opening.

$$p_r(t) = \frac{\rho}{4\pi r}\dddot{V}.$$

It has been confirmed experimentally that airgun efficiency is maximized at low frequencies (on the order of 10 Hz) and sharply decreases as the explosion depth increases.

2.3.3.4 Comparative Characteristics of Spectra

Spectral efficiency of the conversion of accumulated energy into acoustic energy can be illustrated by experimental data on the low-frequency region of the energy spectrum of acoustic signals emitted by different sources [42] (a stoichiometric mixture of propane–butane with oxygen was used in the experiments in the airgun chamber of volume 1 dm^3 at pressure 138 atm (2000 psi)).

As is seen from Table 2.5, the most amount of the energy is released in the band up to 1 kHz. However, for airguns it is already released in the range 0–100 Hz. With almost the same energies of radiators, from 35 to 45 kcal (cases 2 and 5–7), the efficiency of conversion into acoustic energy over the range 20–50 Hz is of the same order.

The use of low-energy explosives as an acoustic source is less effective as compared to high explosives due to the low detonation velocities. Their spectrum features are closer to gaseous mixtures.

2.3.4 Array Systems

2.3.4.1 Airgun–Watergun Systems.

The aim of airgun–watergun arrays is to cumulate the wave energy in a given direction [43,44]. The system comprises n individual sources (Fig. 2.16) with

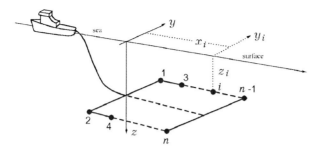

Fig. 2.16. Sketch of experiments with an n-airgun array

fixed distance d between them. A maximum energy $E_{\max}$ of the signal generated by the system depends on the array parameters and the wavelength. The configuration of the sources and time intervals between detonations determines the direction of radiation, whose index is defined by the relation [41]

$$D_1 = 10 \log \left(\frac{E_{\max}}{nE_0} \right) .$$

The problem of synchronization in firing each airgun in the array is, certainly, of great importance. If all the sources are initiated simultaneously, the released energy for seven sources will be 18.6 ± 2.3 kJ, which is much higher than 7.5 ± 0.9 kJ in the case of individual initiation. The total efficiency of airgun systems is somewhat lower than 2%. To describe the behavior of arbitrary arrays of airguns, numerical methods have been successfully used in [32].

2.3.4.2 Explosive Sources of Array Type

As stated above, explosive sound sources as a main element of various sonars designed for generation of quasitonal, directional, strong, long-duration pulses of high acoustic power have been the focus of research for many years. Their range includes spark discharge generators [45], condensed liquid [46], and solid [47–49] explosives, gaseous explosive mixtures [49–51], as well as devices in which shock waves are generated by collapsing cavities [38, 53, 54]. The integral energy parameters of radiation of some explosive sources in water are compared in Table 2.5 [42]. Their spectral characteristics were experimentally studied by Christian, Urick, and Turner and Scrimger [38, 54, 55].

It is known that explosive sources, which are often applied in attempts to solve a wide spectrum of problems of geophysics, navigation and propagation in the ocean, possess high power and their radiation is registered at large distances. However, these acoustic sources must provide directivity and low-frequency radiation that is a nontrivial task for pointwise explosive sources. Nevertheless, a special distribution of array type or charge geometry prove to be appropriate for this purpose.

Some approaches to solving these problems are based on the well-known concepts of classical acoustics on directed radiation by specially arranged distributed sources. Linear cord charge ensures primary radiation of a shock wave in the plane perpendicular to its axis [49]. A vertically arranged chain of HE charges initiated individually and successively with a given time interval between initiations [47] provides the radiation of shock wave packet with given duration and tonal "coloration." Explosion of a charge on the basis of a specially profiled cone is used to arrange directed radiation [48]. These problems were considered in [57–60].

Continuous systems of the type of plane or spatial spiral charges designed for high explosive cord are of particular interest [61–63]. Their advantages over other types are controllable frequency characteristics of the spectrum and a wave shape that is prescribed by the parameters of a spiral. In such a system, the detonation front moves along the spiral coils rotating relative to its axis and provides the cyclic radiation of the shock waves in the given direction.

It should be noted that the detonation rate of a standard detonating cord is approximately five times as high as the speed of sound in liquids. This fact allows one to apply the model of instantaneous explosion of a coil to preliminary estimates and to investigate experimentally the peculiarities of wave formation process by the example of an exploding circular conductor. Experiments in this statement and research results for the explosion of real circular charges will be considered in the next chapter.

2.3.5 Explosion of Spiral Charges, Wave Structure

Figures 2.17 and 2.18 present streak-camera records of the formation of a periodic succession of shock waves [29,63] in the ambient space due to continuous rotation of the detonation front around the spiral axis.

Figure 2.17a,b illustrates the formation of a succession of three shock waves radiated in a liquid at underwater explosion of a three-coil spatial spiral charge. The succession period of waves in this packet was exactly equal to the time it took for the detonation front to pass one coil of the spiral. Figure 2.18 presents the results of the full-scale experiment with a flat spiral charge (spiral of Archimedes) made of detonating cord (about 35-m long) turned in ten coils with a pitch of about 10 cm. Detonation was initiated from the external coil. One can see that at a distance of 20 m from the spiral plane a packet of shock waves was registered on its geometrical axis (Fig. 2.18a). Figure 2.18b shows the oscillogram of the shock wave registered from the detonating cord lined horizontally.

It is quite obvious that depending on the detonation velocity and linear dimensions of the spirals (pitch and diameter), the signal shape might be either a prescribed succession of shock waves or a single positive signal with modulated amplitude. In the first case, over one period of the detonation front turn along the coil, the front of radiated wave could pass the distance

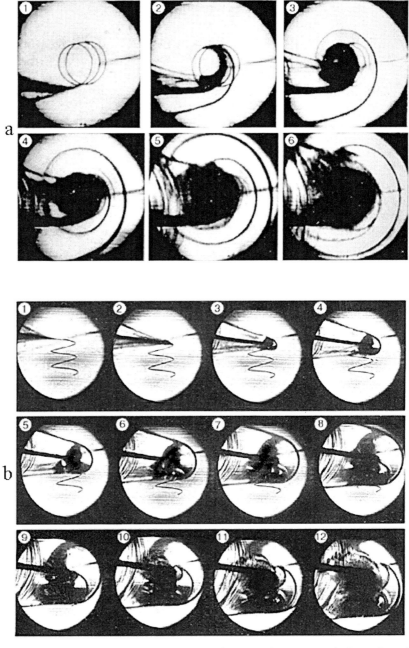

Fig. 2.17a,b. Formation of a shock wave packet at underwater explosions of spatial spiral charges

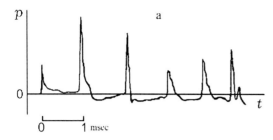

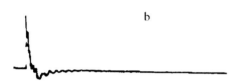

Fig. 2.18a,b. Hydroacoustic signal recorded at underwater explosion of a flat spiral charge at a distance of 20 m along its axis (**a**) and short single signal from a cord charge drawn in a line (**b**)

that was much greater than that covered by detonation along the spiral axis. This corresponded to the condition $D_{ax} \ll U_{sh}$, where D_{ax} is the axial component of the detonation velocity and U_{sh} is the velocity of the shock wave in the liquid. In this case, each next coil would explode in the region far behind the front of the shock wave radiated by the previous coil. Thus, if the spiral parameters were constant, the radiation would be determined by the succession of shock waves of practically equal amplitude for spirals with uniform parameters.

In the second case, the shock front velocity and the axial component of detonation velocity coincided ($D_{ax} \approx U_{sh}$). Figure 2.17b shows this variant of the wave formation for vertical arrangement of the spiral axis. In this case, the pressure wave was amplified and the general character of radiation was changed: the succession of shock waves was transformed into a long solitary wave with modulated amplitude according to the frequency of rotation of the detonation front around the spiral charge axis.

Naturally, these frequencies could vary if circular elements of the charge had different linear characteristics. In any case, the wave duration was defined by the length of the detonating cord l and the detonation velocity D. Shots 6, 7 and 10, 11 in Fig. 2.17b show the explosion of the second and the third coils of the spiral, respectively, immediately behind the shock wave front in the liquid. In addition, it is clearly seen in photo 10 that the interaction of the shock wave from the second coil with the explosive cavity from the first coil led to the formation of an additional wave, which significantly complicated the overall radiation pattern.

Figure 2.19a,b presents the results of modeling the shock wave formation during underwater explosion of a flat three-coil spiral charge made of detonating cord (frame-by-frame scanning). The formation of a series of shock waves during the explosion of a flat five-coil spiral initiated from the outer side was

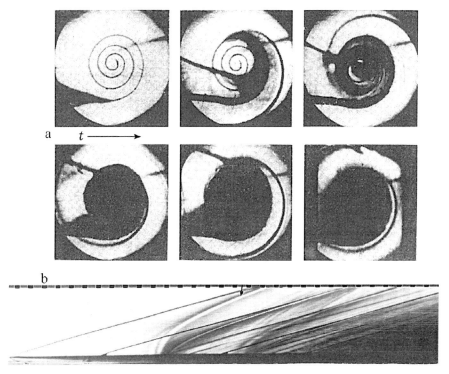

Fig. 2.19a,b. Formation of a shock wave package under the explosion of flat spirals (a) and the wave pattern ahead of the charge plane for the charges of the type of the spiral of Archimedes (b)

visually demonstrated by continuous scanning (Fig. 2.19b, the spiral plane was positioned horizontally and was perpendicular to the figure plane). One can expect predominant radiation of the acoustic signal toward the symmetry axis or in close direction. Initiation of the charge from the external section led to converging detonation fronts and to complicated interaction between radiated waves, which evidently resulted in a periodic amplitude modulation due to different radii of circular elements of the flat spiral.

The parameters of radiation induced by the explosion of charges of complicated geometry can be estimated in a rather simple way. The wave packet radiated by the charge consists of the succession of shock waves whose amplitudes are determined from the data for concentrated charges with equivalent HE weight of the coils. The frequency of the succession of shock waves in the packet is determined by the coil length and the detonation velocity of the HE, while the duration of the whole packet (or of a solitary wave with modulated amplitude) depends on the full travel time of the detonation front along the spiral made of detonating cord. This points to the unique possibility to control the duration of radiation.

The results obtained for the dynamics of cylindrical and toroidal cavities containing detonation products (see details in Chap. 3) can be used to analyze the wave field structure in the circular source approximation, assuming that the source is "switched" instantaneously or approximated by a set of pointwise sources "switched" in series with the detonation velocity. The time variation of the source power at each point of the circle is assumed identical and is determined from Eq. (1.17) (Chap. 1) with regard to the phase shift of oscillations due to finite velocity of switching of the source elements (detonation velocity). Thus, the wave structure radiated during the explosion of a circular charge is calculated with regard to the delay time for the detonation velocity D and the velocity of the acoustic wave c_0. In accordance with the above statement, the acoustic pressure can be determined as

$$p \simeq \rho \frac{a}{4\pi} \int_0^{2\pi} \frac{\ddot{S}(t-\tau)}{f} d\alpha,$$

where $f = \sqrt{z^2 + r^2 + a^2 - 2a\,r\cos\alpha}$; r, α, and z are the coordinates of the cylindrical system, the angle α is counted off from the "circle center–initiation point" direction, and S is the section area of the toroidal (or cylindrical) cavity. The calculations can be substantially simplified using the "peak" approximation [1,4]. Figure 2.20 shows the calculation results for the wave structure at the point lying at a distance $r = 2a$ in the circular plane on the "initiation point–center" line. The calculations were performed for $R_0 = 0.15$ cm, $a = 15$ cm, and $D = 7.7$ km/s. It illustrates the dynamics of shock wave formation for relative pressure $\bar{p} = 4\pi p/ap(0)$, where $p(0)$ is the initial pressure in the detonation products calculated using the model of instantaneous detonation for discontinuity decay. The time τ_1 was taken with respect to the constant $\theta_0 = 2.7 R_{\text{ch}}/c_0$.

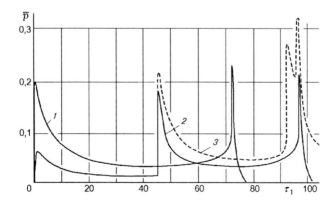

Fig. 2.20a,b. Shock wave structures under instantaneous explosion of a circular charge in a liquid (1), under explosion with finite detonation velocity (2) and at the first coil under continuous rotation of the detonation front (3) (calculation results)

The curve (1) corresponded to the instantaneous explosion at which only two (direct and reflected from the circle axis) waves were generated; the profile (2) is obtained for finite detonation velocity. The three waves arose in complete accordance with the experiment. The first and the second waves were shock waves of initial and final sections of the circle, respectively, while the third wave was the compression wave reflected upon focusing. Curve (3) described the initial unsteady wave profile for the case of continuous rotation of the detonation front about the circle of radius a (the initial section coincided with the profile (2)).

2.4 HE-Nuclear-Tests, Explosive Acoustics and Earthquakes

An analysis of the investigation results published during the last 10–15 years has shown that more 70 papers use the keywords "underwater explosions", though most of them are not interesting in view of the problems of underwater explosions considered in this monograph. Nevertheless, two types of papers can be mentioned as containing useful information. The first one is devoted to the classical problems of explosive hydroacoustics, to numerical modelling underwater explosions and shock wave interactions with targets as well as to some methods of shock wave parameter measuring. The other group of interesting papers is the class of problems the solution of which is determined by the interests of monitoring the Comprehensive Nuclear-Test-Ban-Treaty (CTBT) and by interest to an interdisciplinary problem of the discrimination between earthquakes and underwater explosions. The short survey of mentioned results is presented in this section.

Two presumed underwater explosions, detonated at a site located approximately 215 km southwest of Oahu, Hawaii, were tested by Reymond et al. [64] on several criteria. In particular, these observations indicate a strong frequency dispersion in the spectrograms and the identification of a bubble period (0.45 s) in the spectra of the signals, which translates into a yield of 275 kg of equivalent TNT for a depth of 50 m. This test provides a perfect opportunity to assess the capabilities of T-phase stations and hydrophones for detection, location, identification, and quantification of mentioned explosive sources.

A collaborative research program for the purpose of monitoring CTBT, and examining and analyzing hydroacoustic data from underwater explosions was carried out by Eneva, Stevens, Khristoforov et al. [65]. These data were used as constraints on modeling the hydroacoustic source as a function of depth below the water surface. This is of interest to the CTBT because explosions in the ocean are more difficult to identify if they are on or near the ocean surface. Unique historical Russian data sets have now become available from test explosions of 100-kg TNT cast spherical charges in a shallow reservoir with a low-velocity air-saturated layer of sand on the bottom. In these measurements a reduction of peak pressure by about 60–70% was observed for half-immersed charges as compared with deeper explosions. Besides, peak-pressures were measured during 1957 underwater nuclear explosion (10 kT and depth 30 m) in the Bay of Chernaya (Novaya Zemlya). The 100-kg TNT data were compared with model predictions. Shockwave modeling was based on spherical wave propagation and finite element calculations, constrained by empirical data from US underwater chemical and nuclear tests. Peak-pressure measurements and pressure-time histories were simulated at a 10 km distance from hypothetical 1-kT and 10-kT nuclear explosions conducted at various depths in the ocean. The ocean water was characterized by a realistic sound velocity profile featuring a velocity minimum at 700-m depth. Simulated measurements at that same depth predict at least a tenfold increase in peak pressures from explosions in the SOFAR channel as compared with very shallow explosions (e.g., similar to 3-m depth).

The results of using of the Israel Seismic Network (ISN) for discrimination between low magnitude earthquakes and explosions were described by Gitterman, Pinsky, Shapira [66], Gitterman, Ben-Avraham, Ginzburg [67] and Gitterman, Shapira in [68]. As noted in [66] the results for the Middle East region are important for CTBT monitoring, especially when considering small nuclear tests which may be conducted under evasive conditions. The performance of efficient discriminants based on spectral features of seismograms using waveforms of 50 earthquakes and 114 quarry and underwater blasts with magnitudes 1.0–2.8, recorded by ISN short-period stations at distances up to 200 km was explored. The single-station spectral ratio of the low and high-frequency seismic energy has shown an overlap between explosions and earthquakes. Different frequency bands were tested; the (1–3 Hz)/(6–8 Hz) ratio provided the best discriminant performance.

In [67], ISN as a spatially distributed multichannel system was utilized for the discrimination of earthquakes and underwater explosions in the Dead Sea based on different spectral features of the seismic radiation from underwater explosions and earthquakes, i.e. spectral semblance statistics. The analysis of 28 single-shot underwater explosions (UWEs) and 16 earthquakes in the magnitude range M–L = 1.6–2.8, within distances of 10–150 km, recorded by the ISN, has found that the classical discriminant of the seismic energy ratio between the relatively low-frequency (1–6 Hz) and high-frequency (6–11 Hz) bands, averaged over an ISN subnetwork and shown an overlap between UWEs. The method demonstrates a distinct azimuth-invariant coherency of spectral shapes in the low-frequency range (1–12 Hz) of short-period seismometer systems. The modified statistics provided an almost complete separation between earthquakes and underwater explosions.

Spectral analysis of two events (probable earthquakes or probable UWE) that occured off the coast of the Levant and were recorded by the ISN helped to characterize these events. The observed spectral modulation of the seismic waves was compared with the expected spectra due to source mechanism of UWE and possible effects of the liquid layer (i.e. bubbling, reverberation and channelling). The results of the study demonstrated that spectral analysis of seismograms can be used to discriminate between off-coast earthquakes and underwater explosions recorded in Israel [68].

Wardlaw and Luton [69] studied the interaction structure of shock wave, explosive bubble and target. Shock reflects off the bubble as an expansion that reduces the pressure level between the bubble and the target, inducing cavitation and its subsequent collapse that reloads the target. By comparing deformable and rigid body simulations, it is shown that cavitation collapse can occur solely from the shock-bubble interaction without the benefit of target deformation. The addition of a deforming target lowers the flow field pressure, facilitates cavitation and cavitation collapse, as well as reduces the impulse of the initial shock loading.

A cylinder expansion test for high explosives and numerical simulations of the underwater shock waves were carried out by Itoh, Liu, Nadamitsu [70]. It was shown that the behavior of the underwater shock waves at the vicinity of the explosives differs greatly from that far from the explosives. In particular, the strength of the underwater shock wave nearby the explosive rapidly decreases due to the effect of the expansion of the gas products.

Menon and Lal have studied experimentally the dynamics of explosive bubbles formed during underwater detonations of exploding fuel (hydrogen and/or carbon monoxide–oxygen mixture) in a laboratory water tank [71]. Sub-scale explosions are instrumented to provide detailed histories of bubble shape and pressure. Using geometric and dynamic scaling analyses they have shown that these sub-scale bubbles are reasonable approximations of bubbles formed during deep sea underwater explosions. The explosion bubble undergoes pulsation and loses energy in each oscillation cycle. The observed energy loss, which cannot be fully explained by acoustic losses, is shown here to be partly due to the excitation of Rayleigh–Taylor instability excited near the bubble minimum at the interface between the gaseous bubble and the surrounding water.

The time evolution of high-temperature vapor and plasma bubbles generated by underwater electrical discharges was described by Cook et al. [72]. The oscillations of these bubbles generate acoustic signatures similar to the signatures generated by air guns, underwater explosions, and combustible sources. A set of model equations is developed that allows the time evolution of the bubble generated by a spark discharge to be calculated numerically from a given power input. The acoustic signatures produced by the model were compared to previously recorded experimental data, and the model was found to agree over wide ranges of energy and ambient pressure on several characteristic values of the acoustic signatures. The bubble period and the minimum rarefaction pressure were found to depend on depth, while the peak pressures in the expansion and collapse pulses and the acoustic energy in the

2.4 HE-Nuclear-Tests, Explosive Acoustics and Earthquakes 77

expansion pulse were not found to depend on depth over the parameter ranges investigated.

The advantages and performance of an impulse sonar that uses the short, energetic pulses emerging from underwater explosions as its sources were studied in [73]. This transient analysis was carried out for various simple targets of interest, in the time-domain, in the frequency domain and in the combined time-frequency domain. The last processing approach exhibits the time evolution of the target resonances in a more informative way and it is seen to offer additional advantages for target identification purposes.

Propagation histories of underwater shock waves generated by underwater explosion of high explosive in the range close to explosive have been obtained by processing photographs [74]. In order to obtain pressure distributions of these shock waves, Kira, Fujita, and Itoh applied the non-linear curve fitting technique to these histories. Underwater explosions have been simulated by an arbitrary Lagrangian–Eulerian (ALE) method and calculated results agree well with experimental results in both propagation histories and pressure distributions.

Smoothed particle hydrodynamics (SPH) is a gridless Lagrangian technique that is appealing as a possible alternative to numerical techniques currently used to analyze high-deformation impulsive loading events. In [75], Swegle and Attaway have subjected the SPH algorithm to detailed testing and analysis to determine the feasibility of using PRONTO/SPH for the analysis of various types of underwater explosion problems involving fluid-structure and shock-structure interactions. Of particular interest are the effects of bubble formation and collapse and the permanent deformation of thin-walled structures due to these loadings. Coupling SPH into the finite element code PRONTO represents a new approach to the problem. Results show that the method is well-suited for the transmission of loads from underwater explosions to nearby structures, but the calculation of late time effects due to acceleration of gravity and bubble buoyancy will require additional development, and possibly coupling with implicit or incompressible methods.

The results of numerical studies of the problems of underwater explosion such as the transient pressures in both explosive and surrounding water were presented in [76] by Molyneaux and Li. Comparisons with experimental data for both cast (Trinitrotoluene) TNT and (Pentaerythritol Tetranitrate) PETN explosives confirmed that the suggested numerical model is capable of providing good predictions for both the magnitude and the form of the pressure transient and it can be applied to a variety of underwater explosion scenarios, e.g., arbitrary charge shape and orientation; multiple charges; and reflections of shock waves at boundary surfaces.

The investigation of the feasibility of using approximate acoustic methods to describe the transient pressure field produced by the rapid expansion of a high pressure vapor bubble generated by an energetic steam explosion have been carried out in [77] by Frost, Lee, and Thibault. The approximate methods were validated by comparing with exact finite difference calculations. The calculations illustrated the importance of compressibility effects in reducing the peak overpressure between the leading shock and subsequent bubble collapses. The acoustic method was applied to estimate the pressure transients following a hypothetical large-scale corium–water explosion. The calculations provided a more realistic estimate of the pressure transients than methods that scale the blast overpressures and impulses in terms of the total thermal energy available. To obtain a more accurate estimate of the pressure generated, the finite pressurization time, transient phase change effects and bubble instabilities must be considered.

A semiconductor sensor for the study of pressure fields in underwater explosions was described in [78] by Sigeikin et al. The frequency of the sensor's natural mechanical oscillations $f = 800$ kHz, the error of shock wave-pressure measurements is less than or equal to $\pm 10\%$.

References

1. R. Cole: *Underwater Explosions* (Dover, New York, 1965)
2. V.P. Korobeinikov, B.D. Khrisoforov: *Underwater Explosion*, Itogi Nauki Tekhn. Gidromekh. **9**, pp. 54–119 (1976)
3. B.V. Zamyshlyaev, Yu.S. Yakovlev: *Dynamic Loads at Underwater Explosion* (Sudostroenie, Leningrad, 1967)
4. V.K. Kedrinskii: *Kirkwood–Beth Approximation for Cylindrical Symmetry of Underwater Explosion*, Fiz. Goren. Vzryva **8**, 1, pp. 115–123 (1972)
5. V.K. Kedrinskii: On parameters of weak shock waves at great distance from the charge. In: *Dynamics of Continuous Medium*, vol. 10, pp. 212–216 (Institute of Hydrodynamics, SB, the USSR AS, Novosibirsk, 1972)
6. V.K. Kedrinskii: *About Some Models of One Dimensional Pulsation of Cylindrical Cavity in Incompressible Liquid*, Fiz. Goren. Vzryva **12**, 5, pp.768–773 (1976)
7. V.K. Kedrinskii, V.T. Kuzavov: *Dynamics of Cylindrical Cavity in Compressible Liquid*, Zh. Prikl. Mekh. i Tekhn. Fiz. **18**, 4, pp. 102–106 (1977)
8. B.D. Khristoforov, E.A. Shirokova: *Parameters of Shock Wave at Underwater Explosion of Cord Charge*, Zh. Prikl. Mekh. i Tekhn. Fiz. **3**, 5, pp. 147–149 (1962)
9. L.D. Landau: *About Shock Waves for Far Distances From Charge*, Applied Math. Mech. **9**, 4 (1945)
10. S.A. Khristianovich: *Shock Wave in Water Far From Explosion Point*, Appl. Math. Mech. **20**, 5, pp. 599–605 (1956)
11. W.S. Filler: *Propagation of Shock Waves in a Hydrodynamic Conical Shock Tube*, Phys. Fluids **7**, 5, pp. 664–667 (1964)
12. I.I. Glass, L.E. Heuckroth: *Hydrodynamic Shock Tube*, Phys. Fluids **6**, 4, pp. 543–549 (1963)
13. M.H. Rice, J.M. Walsh: *Equation of State Water to 250 kbar*, J. Chem. Phys. **26**, 4, pp. 814–830 (1957)

14. J.G. Kirkwood, J.M. Richardson: *The Pressure Wave Produced by an Underwater Explosion III*, Office of Sc. Research and Development Report No. 813 (1942)
15. V.K. Kedrinskii: *Features of Spherical Gas Bubble Dynamics in a Liquid*, Zh. Prikl. Mekh. i Tekhn. Fiz. **8**, 3, pp. 120–125 (1967)
16. A.S. Besov, V.K. Kedrinskii, E.I. Palchikov: *Study of Initial Stage of Cavitation Using Diffraction-Optiocal Mathod*, Pis'ma Zh. Exp. Teor. Fiz. **10**, 4, pp. 240–244 (1984)
17. M.I. Vorotnikova, V.K. Kedrinskii and R.I. Soloukhin: *Shock Tube for Study of One Dimensional Wave in a Liquid*, Fiz. Goren. Vzryva **1**, 1, pp. 5–15 (1965)
18. F. Kheiman: *Shock Wave Velocity and Pressure Under a Collision of Liquid and Solid at High Velocity*, Theor. Fundamentals of Eng. Calc. **90**, 3 (1968)
19. Berngardt A.R.: Dynamics of cavitation zone under pulse loading of liquid. PhD Thesis (Lavrentyev Inst. of Hydrodynamics, SB RAS, Novosibirsk, 1994)
20. V.K. Kedrinskii, R.I. Soloukhin, S.V. Stebnovskii: *Semiconductor of Pressure Gauge for Measurement of Strong Shock Waves in a Liquid ($> 10^2$ MPa)*, Zh. Prikl. Mekh. i Tekhn. Fiz. **10**, 4, pp. 92–94 (1969)
21. V. Paul, D.M. Warshauer: The role of pressure in studying semiconductors. In: W. Paul, D.M. Warschauer (eds.) *Solids Under High Pressure* (Mir, Moscow, 1966) pp. 205–283
22. V.K. Kedrinskii, N.K. Serdyuk, R.I. Soloukhin, S.V. Stebnovskii: *Study of Fast Reaction in a Solution Behind the Front of Shock Waves*, Dokl. Akad. Nauk SSSR **187**, 1, pp. 130–133 (1969)
23. N.M. Kuznetsov: *State Equation and Thermocapacity of a Water for Wide Spectrum of Thermodynamic Parameters*, Zh. Prikl. Mekh. i Tekhn. Fiz. **2**, 1 (1961)
24. H. Hoffmann, E. Yeager, J. Stuehr: *Apparatus with Laser Formation of Temperature Jump for Relaxation Study of Reactions in Solutions*, Pribory Dlya Nauch. Issled. 5 (1968)
25. H. Hoffmann, E. Yeager: *Relaxation Study of Chemical Reactions by Pressure Jump Method*, Pribory Dlya Nauch. Issled. 8, pp. 9–10 (1968)
26. A. Jost: Berich. Phys. Chem. **70**, (1966)
27. P. Hurwitz, K. Kustin: Inorg. Chem. **3**, 6, pp. 823 (1964)
28. S.W. Benson: *The Foundations of Chemical Kinetics*, (McGraw-Hill, New York, 1960)
29. V.K. Kedrinskii: Underwater explosive sound sources. In: M. Crocker (ed.) Encyclopedia of Acoustics, vol. 1 (John Wiley & Sons, New York, 1997) pp. 539–547
30. F. Baum at al.: *Explosion Physics* (Nauka, Moscow, 1975)
31. N. Roy, D. Frolov: Generation of sound by spark discharges in Water. In: Proc. 3rd Intern. Congress on Acoustics (Elsevier, Stuttgart, Germany, 1961) pp. 321–325
32. R. Laws, L. Hatton, G. Parkes: Energy interaction in marine airgun arrays. In: E. A. E. G., paper No. 1756 (18 Oct. 1989) pp. 1–37
33. A. Laake, G. Meier: Sound generation by the watershock. In: Proc. ICA-12 (Toronto, Canada, 1986) pp. 1–3
34. S. Mitchell, N. Bedford, M. Weinstein: *Determination of Source Depth From the Spectra of Small Explosives Observed at Long Ranges*, J. Acoust. Soc. Am. **60**, 4, pp. 825–828 (1976)

35. M. Blaik, E. Christian: *Near-Surface Measurements of Deep Explosives*, J. Acoust. Soc. Am. **38**, 1, pp. 50–62 (1965)
36. D. Weston: *Underwater Explosions as Acoustic Sources*, Proc. Phys. Society **76**, 2, 488, pp. 233–249 (1960)
37. A. Arons: *Underwater Explosion Shock Wave Parameters at Large Distances From the Charge*, J. Acoust. Soc. Am. **26**, 3, pp. 343–346 (1954)
38. E. Christian: *Source Levels for Deep Underwater Explosions*, J. Acoust. Soc. Am. **42**, 4, pp. 905–907 (1967)
39. A. Kibblewhite, R. Denham: *Measurements of Acoustic Energy From Underwater Explosions*, J. Acoust. Soc. Am. **48**, 1, pp. 346–351 (1970)
40. E. Tree, R. Lugg, Y. Brummitt: *Why Waterguns?*, Geophys. Prospecting. **34**, pp. 302–329 (1986)
41. J. Barger, W. Hamblen: *The Airgun Impulsive Underwater Trancducer*, J. Acoust. Soc. Am. **68**, 4, pp. 1038–1045 (1980)
42. M. Badashkand: *Comparison of Acoustic Effect of Some Explosive Sound Sources in Water*, Dokl. Akad. Nauk SSSR **194**, 6, pp. 1309–1312 (1970)
43. R. Bailey, P. Garces: *On the Theory of Airgun Bubble Interaction*, Geophys. **53**, 2, pp. 192–200 (1988)
44. R. Laws, G. Parkers, L. Hattou: *Energy Interaction. The Long-Wave Interaction of Seismic Waves*, Geophys. Prospecting **36**, pp. 333–348 (1989)
45. H.A. Wright, J.P. Tobey: *Acoustic Generator of the Spark Discharge Type*, J. Acoust. Soc. Am. **45**, 1 (1969)
46. N.D. Smith, W.L. Roever: *Liquid Seismic Explosive and Method of Using*, J. Acoust. Soc. Am. **44**, 4 (1968)
47. Method and device for echo ranging. US Patent No. 3514748, 26 May 1970
48. W.P.J. Filler: Directional explosive echo ranging device. US Patent 3521725, 18 May 1962; publ. 28 July 1970
49. R.M. Johnson, C.A. Axelson: Deep depth line charge. US Patent 3276366, 4 Oct. 1966
50. V.M. Lyuboshits: *Wave Field of Directive Explosion*, Izv. Akad. Nauk SSSR, Mekh. Zhidk. Gaza **3**, 1 (1968)
51. L.G. Kilmer: *Underwater Gas Explosion Seismic Wave Generator*, J. Acoust. Soc. Am. **45**, 2 (1969)
52. A.A. Maksakov, N.A. Roy: *About Underwater Explosion of Gas with High Initial Volume Density of Energy*, Acoust. J. **25**, 2 (1979)
53. R.S. Brand: *Shock Wave Generated by Cavity Collaps*, J. Fluid Mech. **2**, 1 (1965)
54. R.J. Urick: *Implosion as Souce of Underwater Sound*, J. Acoust. Soc. Am. **35**, 12 (1963)
55. R.G. Turner, J.A. Scrimger: *On the Depth Variation in the Energy Spectra of Underwater Explosive Charge*, J. Acoust. Soc. Am. **48**, 3 (1970)
56. B.E. Parkis, R.D. Worley: *Measurement of Spectrum Levels for Shallow Explosive Source*, J. Acoust. Soc. Am. **49**, 1 (1971)
57. B.M. Buck: *Relative Measurements of Pulse Component Source Energies of the USN Explosive Sound Signal MK61 Detonated at 60 ft*, J. Acoust. Soc. Am. **55**, 1 (1974)
58. M. Sieffert: Method of obtaining of directed explosive wave using underwater charges. Patent 2541582 (Germany, 1977)
59. G. Noddin: *Sonic Pulse Generator*, J. Acoust. Soc. Am. **36**, 4 (1964)

60. E. Lavrentiev, O. Kuzyan: *Explosions in the Sea* (Sudostroenie, Leningrad, 1977)
61. V.K. Kedrinskii: On oscillation of a toroidal gas bubble in a Liquid. In: Continuum Mechanics, vol. 16 (Novosibirsk 1974) pp. 35–43
62. V.K. Kedrinskii: *About One Dimensional Pulsation of Toroidal Gas Cavity in Compressible Liquid*, Zh. Prikl. Mekh. i Tekhn. Fiz. **18**, 3, pp. 62–67(1977)
63. V.K. Kedrinskii: *Features of Shock Wave Structure at Underwater Explosions of Spiral Charges*, Zh. Prikl. Mekh. i Tekhn. Fiz. **21**, 5, pp. 51–59 (1980)
64. D. Reymond, O. Hyvernaud, J. Talandier, E.A. Okal: T-Wave detection of two underwater explosions off Hawaii on 13 April 2000. Bull. Seismol. Soc. Am. **93**, 2, 804–816 (2003)
65. M. Eneva, J.L. Stevens, B.D. Khristoforov, J. Murphy, V.V. Adushkin: Analysis of Russian hydroacoustic data for CTBT monitoring. Pure Appl. Geophys. 2. **158**, 3, 605–626 (2001)
66. Y. Gitterman, V. Pinsky, A. Shapira: Spectral classification methods in monitoring small local events by the Israel seismic network. J. Seismol. **2**, 3, 237–256 (1998)
67. Y. Gitterman, Z. Ben-Avraham, A. Ginzburg: Spectral analysis of underwater explosions in the Dead Sea. Geophys. J. Int. **134**, 2, 460–472 (1998)
68. Y. Gitterman, A. Shapira: Spectral characteristics of seismic events off the coast of the Levant. Geophys. J. Int. **116**, 2, 485–497 (1994)
69. A.B. Wardlaw, J.A. Luton: Fluid-structure interaction mechanisms for close-in explosions. Shock Vib. **7**, 5, 265–275 (2000)
70. S. Itoh, Z. Liu, Y. Nadamitsu: An investigation on the properties of underwater shock waves generated in underwater explosions of high explosives. J. Press. Vessel Technol.-Trans. ASME **119**, 4, 498–502 (1997)
71. S. Menon, M. Lal: On the dynamics and instability of bubbles formed during underwater explosions. Exp. Thermal Fluid Sci. **16**, 4, 305–321 (1998)
72. J.A. Cook, A.M. Gleeson, R.M. Roberts, R.L. Rogers: A spark-generated bubble model with semi-empirical mass transport. J. Acoustical Soc. Am. 68, **101**, 4, 1908–1920 (1997)
73. G.C. Gaunaurd, H.C. Strifors: Time-frequency processing of underwater echoes generated by explosive sources. Ultrasonics **33**, 2, 147–153 (1995)
74. A. Kira, M. Fujita, S. Itoh: Underwater explosion of spherical explosives. J. Mater. Process. Technol. **85**, 1–3, 64–68 (1999)
75. J.W. Swegle, S. Attaway: On the feasibility of using smoothed particle Hydrodynamics for underwater explosion calculations. Computational Mechanics **17**, 3, 151–168 (1995)
76. T.C.K. Molyneaux, L.Y. Li, Firth N: Numerical-Simulation of Underwater Explosions. Comput. Fluids **23**, 7, 903–911 (1994)
77. D.L. Frost, J.H.S. Lee, P. Thibault: Numerical Computation of Underwater Explosions due to Fuel Coolant Interactions. Nuclear Eng. Des. **146**, 1–3, 165–179 (1994)
78. I.M. Sigeikin, S.G. Zhilenis, V.S. Sotnikov, A.A. Gritsyus: Algaas-film-based pressure-pulse sensor. Instrum. Exp. Tech. **34**, 3, 720–723, part 2 May–June 1991

3 Explosion of Cylindrical and Circular Charges

3.1 Dynamics of a Cylindrical Cavity in a Compressible Liquid

The equation of one-dimensional oscillation of a cylindrical cavity [1,2],

$$R\left(1 - \frac{\dot{R}}{c}\right)\ddot{R} + \frac{3}{4}\left(1 - \frac{\dot{R}}{3c}\right)\dot{R}^2 = \frac{1}{2}\left(1 + \frac{\dot{R}}{c}\right)H + \frac{R}{c}\left(1 - \frac{\dot{R}}{c}\right)\frac{dH}{dt}, \quad (3.1)$$

was derived assuming a possibility of the asymptotic approximation of the invariant along the characteristics $c + u$ by the function $G = r^{1/2}\Omega$,

$$\frac{\partial}{\partial t}\left[r^{1/2}\left(\omega + \frac{u^2}{2}\right)\right] + (c + u)\frac{\partial}{\partial r}\left[r^{1/2}\left(\omega + \frac{u^2}{2}\right)\right] = 0,$$

and using the equations of continuity and conservation of momentum to replace partial derivatives for total derivatives, as well as the transitions $r \to R$ and $u \to \dot{R}$. Here, $\Omega = \omega + u^2/2$ is the kinetic enthalpy, $\omega = \int \frac{dp}{\rho}$ is the enthalpy, r is the coordinate, u is the speed of liquid particles, and c is the local speed of sound.

The enthalpy H on the cavity wall from the side of the liquid was determined from the Tait equation [3],

$$H = \frac{nB}{(n-1)\rho_0}\left[\left(1 + \frac{p(R) - p_\infty}{B}\right)^{(n-1)/n} - 1\right],$$

where $B = 305$ MPa and $n = 7.15$ are constants, $p(R)$ is the pressure in detonation products, and the local speed of sound is defined by the formula

$$c = c_0\left(1 + \frac{p(R) - p_\infty}{B}\right)^{(n-1)/2n}$$

(c_0 is the speed of sound in undisturbed liquid).

Let us analyze some features of the asymptotic approximation accepted to define the invariance of the function G for the case $\nu = 1$. This approximation allows for a certain arbitrariness in the selection of the numerical coefficient in front of the squared velocity in the oscillation equation. Let us introduce the coefficient β (instead of 3/4) in front of the inertial term of the basic equation (1), rewriting it in the dimensionless form so that the cavity dynamics is studied on the scale of the initial radius of the charge R_{ch} depending on the time $\tau = tc_0/R_{\text{ch}}$:

$$y\left(1 - \frac{\dot{y}}{\bar{c}}\right)\ddot{y} + \beta\dot{y}^2\left(1 - \frac{\dot{y}}{3\bar{c}}\right) = \frac{\bar{H}}{2}\left(1 + \frac{\dot{y}}{\bar{c}}\right) + \frac{y}{\bar{c}}\dot{\bar{H}}\left(1 - \frac{\dot{y}}{\bar{c}}\right). \quad (3.2)$$

Here the dot denotes the initial derivative with respect to τ, $\bar{c} = c/c_0$, $y = R/R_{\text{ch}}$, and $\bar{H} = H/c_0^2$. The equation was studied numerically for the range of the coefficient $\beta = 0.75$–1.25 in order to select the closest solution to the experimental data on the oscillation of a cavity with detonation products at underwater explosion of cylindrical charges.

The results were obtained in laboratory experiments with non-standard cylindrical charges as detonation cord (DC) made of hexogen wrapped in a copper casing. The diameter of HE in the DC was 0.65 and 1.65 mm [4]. The charge density $\rho_{\text{ch}} = 1.55$ g/cm^3 and the detonation rate $D = 7.7$ km/s were determined experimentally, and the charge length to diameter ratio was not less than 10^3.

To calculate the dynamics of an explosive cavity for these charge types, one can restrict oneself to the case of "instantaneous" detonation, where the initial parameters of the problem (density and speed of sound in detonation products and on the cavity boundary from the side of the liquid, the pressure at the contact discontinuity and the velocity of the contact discontinuity) were determined from the condition of discontinuity of arbitrary decay. Calculations were performed for two types of isentropes: $\gamma = 3$ and the variable γ, whose range as a function of the density of detonation products was taken from [5].

We approximated the data of Kuznetsov and Shvedov [5] for $\rho_{\text{ch}} = 1.6$ g/cm^3 that were the closest to the experimental data:

$$\begin{aligned}
p &\sim \rho^{-2.78} & (0.625 \leq \rho^{-1} \leq 1.66)\,; \\
p &\sim \rho^{-2.14} & (1.66 \leq \rho^{-1} \leq 2.51)\,; \\
p &\sim \rho^{-1.73} & (2.51 \leq \rho^{-1} \leq 5.0)\,; \\
p &\sim \rho^{-1.36} & (5.0 \leq \rho^{-1} \leq 20.0)\,; \\
p &\sim \rho^{-1.26} & (20.0 < \rho^{-1})\,.
\end{aligned}$$

It should be noted that in this case, $\rho^{-1} = 0.625$ cm^3/g corresponds to $p = 1.295 \cdot 10^4$ MPa.

Calculation results for the cavity expansion are shown in Fig. 3.1, where $y = R/R_{\text{ch}}$, while the curves (1), (1′), (1″) ($\gamma = 3$) and (2), (2′), (2″) (variable γ) corresponded to $\beta = 0.75$ (data without primes); 1.0 (one prime); 1.25 (double prime).

The dashed line shows the slope of the experimental curve $y(\tau)$ and cross marks the experimental value of the position of $R_{\max}/R_{\text{ch}}$. The dependence (2′) ($\beta = 1$) is shown by dots to mark out the numerical result

3.1 Dynamics of a Cylindrical Cavity in a Compressible Liquid   85

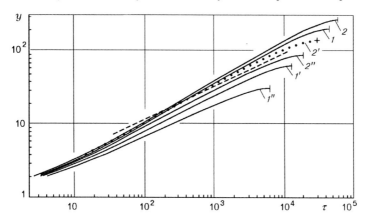

Fig. 3.1. Expansion of a cylindrical cavity with detonation products (calculation): curves 1 and 2 correspond to the constant and variable values of the adiabatic index, respectively; dashed and dotted curves 1, 2 correspond to different values of β

that is the closest to the experimental data: one can see the complete coincidence of the data in time and the expansion of the explosive cavity and quite satisfactory coincidence in the slope. From the right each dependence is bound by a vertical line, which shows the moment when the cavity reached its maximum size.

As one would expect, the closest to the real parameters were those calculated for the case of variable γ. One can see from the diagrams that as the coefficient β increased, the slope of the curves $y(\tau)$ decreased: for curves (2) it was 0.55 at $\beta = 0.75$, then it decreased to 0.5 at $\beta = 1.0$ and to 0.49 at $\beta = 1.25$. The slope of the experimental curves was 0.45.

Figure 3.2 shows the calculation results from Fig. 3.1 for the dependence of $(2')$. The diagram of the dependence is expanded to include three oscillations of the cylindrical cavity. For comparison, experimental data for charges of different diameter are plotted here: crosses for $d = 1.65$ mm and dots for $d = 0.65$ mm. The agreement of the calculation results with the experimental data in the field $y \geq 10$ was satisfactory. Within $30 \leq \tau \leq 10^4$, the dynamics of the expanding explosive cavity could be described by the simple dependence

$$R_{\exp}/R_{ch} \simeq 1.5\tau^{0.45} \ .$$

Table 3.1 summarizes basic numerical and experimental characteristics of oscillations of the cavity containing detonation products of a cylindrical HE charge. Extreme values for the internal energy of detonation products E allow one to write the expressions for oscillation periods of a cylindrical cavity in a general form in terms of the initial energy of HE per unit length $Q_1 = Q\rho_{ch}\pi R_{ch}^2$

$$\tau_{*,i} \simeq 1.635 \frac{\sqrt{\rho_0 \alpha_i Q_1}}{p_\infty} \ ,$$

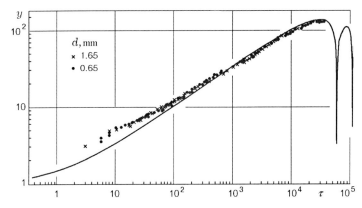

Fig. 3.2. Oscillation of a cylindrical cavity with detonation products; *dots* and *crosses* show experimental data

where $\alpha_i = 0.218, 0.14, 0.11$ was the share of HE energy spent for radial motion of the liquid flow during the first, the second, and the third oscillations, respectively.

Comparison of experimental and numerical results for underwater explosions of cord charges, including the energy distribution between the detonation products, the shock wave, and the oscillations proves the validity of the proposed equation of dynamics of a cylindrical cavity (3.2). The equation allows one to make conclusions about the expediency of utilization of the coefficient $\beta = 1$ instead of $3/4$ in front of the inertial term.

Table 3.1. Basic numerical and experimental characteristics of oscillations in a cavity containing products of a cylindrical HE charge

Characteristics	Experiment [6]	Calculation
$R_{\max}/R_{\text{ch}}$	$\simeq 135$	141
$\tau_{\max}$	$\simeq 3 \cdot 10^4$	$3 \cdot 10^4$
E_1	$\simeq 0.22 \cdot Q_1$	$0.218 \cdot Q_1$
E_2	—	$0.14 Q_1$
E_3	—	$0.11 Q_1$
$\tau_{*,1}$	$\simeq 6 \cdot 10^4$	$6 \cdot 10^4$
$\tau_{*,2}$	—	$4.95 \cdot 10^4$
$\tau_{*,3}$	—	$4.5 \cdot 10^4$

Note: Indices 1, 2, and 3 denote the oscillation period τ_* number; Q_1 is the HE explosion heat per charge unit length; E is the energy remaining in the detonation products after expansion of the cavity; $\tau_{\max}$ is the time of the first expansion of the explosive cavity.

3.2 Cylindrical Cavity: Approximated Models for Incompressible Liquid

As opposed to the case of spherical symmetry, the dynamics of cylindrical cavity in an unlimited incompressible liquid is not described by an exact equation due to the logarithmic peculiarity at infinity. Indeed, the equation of motion of such a cavity can be transformed to the form

$$\frac{\mathrm{d}}{\mathrm{d}R}\left(R^2 \dot{R}^2 \ln \frac{r}{R}\right) = 2R \frac{p_\mathrm{g} - p_\mathrm{r}}{\rho_0},$$

which yields the first integral

$$R^2 \dot{R}^2 \ln \frac{r}{R} - R_0^2 \dot{R_0}^2 \ln \frac{r_0}{R_0} = \frac{1}{\rho_0} \int_{R_0}^{R} 2R p_\mathrm{g}\, \mathrm{d}R - \frac{p}{\rho_0}(R^2 - R_0^2)$$

and expresses the law of energy conservation which for $r \to \infty$ ($p_\mathrm{r} = p_\infty$) makes no sense. Here p_r is the pressure in a liquid at the point with coordinate r and p_g is pressure in gas.

Nevertheless, it is desirable to have at least an approximate model, since a series of practical problems on underwater and underground explosions of distributed charges, in which the model of incompressible liquid is often used, requires simple estimates of the dynamics of cylindrical cavity with detonation products. One can consider three fairly apparent approaches to this kind of approximated estimates [6]. One of them deals with a formal ultimate transition ($c_0 \to \infty$) in Eqs. (3.1–3.2), which can result in approximated equations of oscillation of a cylindrical cavity in an incompressible liquid [1]. Below we will show that this approach enables the estimates of oscillation period of the cavity with detonation products and comparison with experimental data for DC charges of different diameter.

Two other approaches are concerned with the restriction of the space in which the cavity is considered, a free surface of liquid is introduced. Kuznetsov [7] noted that real experimental statements always involve the free surface, which makes it possible to obtain an exact equation for incompressible liquid, which involves the explosion depth as a parameter. Below we consider different approximate models and compare them with experimental data [6].

3.2.1 Generalized Equation (for $c \to \infty$)

Following the above approach to the derivation of the oscillation equation of a one-dimensional cavity in a compressible liquid, one can easily get its acoustic analog:

$$R\left(1 - 2\frac{\dot{R}}{c_0}\right)\ddot{R} + \frac{3}{4}\nu\left(1 - \frac{4\dot{R}}{3c_0}\right)\dot{R}^2 = \frac{\nu}{2}H + \frac{R}{c_0}\left(1 - \frac{\dot{R}}{c_0} + \frac{\dot{R}^2}{c_0^2}\right)\dot{H}.$$

3 Explosion of Cylindrical and Circular Charges

It should be noted that the asymptotic approximation was used in the cylindrical version only when deriving the basic equation for the invariant $r^{\nu/2}\Omega$,

$$\frac{\partial r^{\nu/2}\Omega}{\partial r} + c_0^{-1}\frac{\partial r^{\nu/2}\Omega}{\partial t} = 0,$$

while in transition to total derivatives the change was made on the basis of exact conservation laws. The model of incompressible liquid follows from the ultimate transition ($c_0 \to \infty$)

$$2(R\ddot{R} + \dot{R}^2) - \dot{R}^2/2 = H \tag{3.3}$$

or

$$R\ddot{R} + 3\dot{R}^2/4 = H/2 . \tag{3.3a}$$

Here $H = (p_g - p_\infty)/\rho$.

Considering the fact that following the calculation results in coincidence with the experimental data for two basic parameters ($y_{\max}$, $\tau_{\max}$) occurs when $\beta = 1$, it is reasonable to study the ultimate transition in (3.2), which results in the following oscillation equation:

$$R\ddot{R} + \dot{R}^2 = H/2 . \tag{3.3b}$$

The change of the internal energy of detonation products (gas) per unit length is defined by the obvious expression

$$\Delta U_1 = \frac{p(0)S_0}{\gamma - 1} - \frac{pS}{\gamma - 1}$$

or, with regard to the adiabate $pS^\gamma = \mathrm{const}$,

$$\Delta U_1 = \frac{p(0)\pi R_0^2}{\gamma - 1}\left[1 - \left(\frac{R_0}{R}\right)^{2\gamma-2}\right].$$

The change of the potential energy of the liquid per unit length is

$$\Delta W_1 = \pi p_\infty (R_0^2 - R^2) .$$

The first integral of Eq. (3.3b) at zero initial velocity becomes

$$\frac{p(0)R_0^2}{\gamma - 1}\left[1 - \left(\frac{R_0}{R}\right)^{2\gamma-2}\right] = 2\rho R^2 \dot{R}^2 + p_\infty(R^2 - R_0^2) .$$

Experience in studying the dynamics of spherical cavities suggests that the first integral of the oscillation equation is the law of energy conservation. Multiplying both parts of the latter equality by π, we see that the left-hand part of the equality presents a decrease in the internal energy of detonation products, while the second term from the right is the increment of the potential energy of the liquid. Thus, there are strong grounds to conclude that the remaining term of the equality (multiplied by π)

$$2\pi \rho R^2 \dot{R}^2 = E_l$$

is the specific kinetic energy of the liquid surrounding an oscillating cylindrical cavity, if $\dot R(0) = 0$.

One can define the velocity potential corresponding to the above definition of the kinetic energy of a liquid. Let us use the traditional approach and consider the potential as $\varphi \simeq r^{-1/4}\Phi(t)$. Let us find the value of $\Phi(t)$ from the kinematic condition $\dot R = -\varphi_r\big|_{r=R}$, which will finally define the potential and its gradient as

$$\varphi \simeq \frac{4R^{5/4}\dot R}{r^{1/4}} \quad \text{and} \quad \frac{\partial \varphi}{\partial r} \simeq -\frac{R^{5/4}\dot R}{r^{5/4}}.$$

For the case of cylindrical symmetry, the kinetic energy of a mass unit $(\mathrm{d}m = 2\pi\rho l r\mathrm{d}r)$ should be considered as energy per unit length l, reasoning from the obvious definition $\mathrm{d}E_\mathrm{l} = (v^2/2)\mathrm{d}m_\mathrm{l}$. The final result

$$E_\mathrm{l} = \pi\rho \int_R^\infty \frac{\dot R^2 R^{5/2}}{r^{5/2}} r\mathrm{d}r = 2\pi\rho R^2 \dot R^2,$$

as we see, corresponds to the above expression. However, substitution of the "guessed" potential in the Cauchy–Lagrange integral does not lead to the oscillation equation of the type (3.3b).

3.2.2 Cylindrical Cavity under Free Surface

Let us consider a cylindrical cavity in a liquid, the cavity diameter being R_0 and its center being placed at a depth h from the free surface. It is supposed that $h \gg R_0$, the free surface is horizontal, and its velocity potential is equal to zero. In this statement, it is convenient to use the bipolar system of coordinates with variables γ and β, where $\gamma = \mathrm{const}$ and $\beta = \mathrm{const}$ are the orthogonal circumferences. The coordinates of the Cartesian (x, y) and flat bipolar systems are linked by the relationships

$$x = \frac{a\,\mathrm{sh}\,\gamma}{\mathrm{ch}\,\gamma + \cos\beta} \quad \text{and} \quad y = \frac{a\cdot\sin\beta}{\mathrm{ch}\,\gamma + \cos\beta},$$

where $a = \sqrt{h^2 - R^2} \simeq h$, while $\gamma = 0$ corresponds to the free surface. The Laplace equation in the adopted system has the form

$$\varphi_{\gamma\gamma} + \varphi_{\beta\beta} = 0.$$

In view of the above-mentioned conditions, one can select the coordinate surface γ_0 as the surface of the cylindrical cavity. The solution of the Laplace equation has the simple form $\varphi = \varphi_0\gamma/\gamma_0$. The value of φ_0 is sought from the kinematic condition $\zeta_t + \nabla\varphi\nabla\zeta = 0$, where $\zeta = 0$ $(\gamma - \gamma_0 = 0)$ is the equation of the cavity boundary. The Lame coefficients in the adopted coordinate system are equal to $L_{\gamma,\beta} = a(\mathrm{ch}\,\gamma + \cos\beta)^{-1}$. Then $\varphi_0 =$

3 Explosion of Cylindrical and Circular Charges

$a^2 (d\gamma_0/dt)\gamma_0 (\operatorname{ch}\gamma_0 + \cos\beta)^{-2}$ and, in view of $\operatorname{ch}\gamma_0 = h/R \gg 1$, the final expression for the potential becomes

$$\varphi \simeq a^2 \frac{d\gamma_0}{dt} \gamma \operatorname{ch}^{-2}\gamma_0 .$$

Hence

$$\nabla\varphi = L_\gamma^{-1} \frac{\partial\varphi}{\partial\gamma} = a \frac{d\gamma_0}{dt} \operatorname{ch}^{-2}\gamma_0 (\operatorname{ch}\gamma + \cos\beta)$$

and

$$\frac{\partial\varphi}{\partial t} \simeq a^2 \frac{d^2\gamma_0}{dt^2} \frac{\gamma}{\operatorname{ch}^2\gamma_0} - 2a^2 \left(\frac{d\gamma_0}{dt}\right)^2 \frac{\gamma}{\operatorname{ch}^2\gamma_0} \operatorname{th}\gamma_0 .$$

In the last expression one can use the condition $h \gg R$, then $\operatorname{th}\gamma_0 \simeq 1$. Substituting expressions for $\nabla\varphi$ and $\partial\varphi/\partial t$ in the Cauchy–Lagrange integral and meeting the condition $\gamma = \gamma_0$, we obtain the following equations of oscillation of a cylindrical cavity, given the availability of the free surface:

- In the bipolar system,

$$-a^2 \frac{d^2\gamma_0}{dt^2} \frac{\gamma_0}{\operatorname{ch}^2\gamma_0} + a^2 \left(\frac{d\gamma_0}{dt}\right)^2 \frac{\gamma_0 \operatorname{th}\gamma_0 - 1/4}{\operatorname{ch}^2\gamma_0} = H$$

- On the physical plane,

$$[R\ddot{R} + (\dot{R})^2] \ln\left(\frac{2h}{R}\right) - \frac{1}{2}(\dot{R})^2 = H \qquad (3.4)$$

Here we used the relationship $\gamma_0 = \ln(2h/R)$.

The latter equation was first derived by Kuznetsov [7] using the method of conformal mappings. The form of this equation points to the essential dependence of the dynamics of cylindrical cavity on the charge depth.

3.2.3 The Model of Liquid Cylindrical Layer

The model follows from the natural assumption that under real conditions, the cavity oscillation involves only a finite mass of the liquid surrounding it. In this model, the external radius of the cylindrical layer r_0 is not defined. It can be easily shown that the form of the equation of oscillation of a cylindrical cavity in this statement is as follows:

$$(R\ddot{R} + (\dot{R})^2) \ln\left(\frac{r_0^2 - R_0^2 + R^2}{R^2}\right) + \left(\frac{R^2}{r_0^2 - R_0^2 + R^2} - 1\right)\dot{R}^2 = H .$$

The equation is considerably simplified for $r_0^2 \gg R^2$:

$$(R\ddot{R} + (\dot{R})^2) \ln\left(\frac{r_0}{R}\right) - \frac{1}{2}\dot{R}^2 = H , \qquad (3.5)$$

where the subscript 0 stands for the initial values of the relevant parameters.

3.2 Approximated Models, Incompressible Liquid

Thus, to check the reliability of Eqs. (3.3)–(3.5) representing the three approaches, one should compare them with experimental data. Usually two basic parameters are compared: the maximum expansion (compression) of the cavity under the effect of explosive load and the oscillation period of the cavity with detonation products. However, the former depends significantly on the amount of energy released under explosion as a shock wave, which makes pointless the comparison of approximate models of incompressible liquid with respect to this parameter. As to the oscillation period, the estimates for spherical (e.g., [3]) and cylindrical symmetries [1,2] prove that it is fairly well defined by the doubled collapse time of an empty cavity in incompressible liquid, the cavity initial radius R_0 being consistent with the maximum radius of the first expansion of the cavity with detonation products. The radius can be determined experimentally (for example, by means of optical registration).

Upon substitution of variables $R = yR_0$ and $t = \tau\sqrt{\rho/p_\infty}\,R_0$ the equations for the model of an empty cylindrical cavity in an incompressible ideal liquid will be rewritten as:

$$2\cdot(yy_{\tau\tau}+y_\tau^2)-\frac{1}{2}y_\tau^2=-1, \qquad (a)$$

$$2\cdot(yy_{\tau\tau}+y_\tau^2)=-1, \qquad (b)$$

$$(yy_{\tau\tau}+y_\tau^2)\cdot\ln\left(\frac{2h}{yR_0}\right)-\frac{1}{2}y_\tau^2=-1, \qquad (c)$$

$$(yy_{\tau\tau}+y_\tau^2)\cdot\ln\left(\frac{r_0}{yR_0}\right)-\frac{1}{2}y_\tau^2=-1. \qquad (d)$$

3.2.4 Oscillation Period of a Cylindrical Cavity

The equation (a) yields the first integral. We multiply both parts by $y^{1/2}$, yielding

$$\frac{d}{dy}(y^{3/2}\dot y^2)=-y^{1/2},$$

whence upon easy transformations we obtain the collapse time

$$\tau_a=-\int_1^0\frac{dy}{\sqrt{2[y^{-3/2}-1]/3}}=0.817F\left(\frac{1}{2},\frac{7}{6},\frac{13}{6},1\right)\simeq 1.49.$$

Here $F(\alpha,\beta,\gamma,z)$ is the hypergeometric Gauss function. In the dimensional form, the oscillation period of a cylindrical cavity with detonation products (through its maximum radius and hydrostatic pressure p_∞) will become

$$T_a\simeq 2.98 R_{\max}\sqrt{\frac{\rho}{p_\infty}}.$$

The most interesting results are obtained when considering the model (b), whose first integral is presented above. If the cavity is empty, it practically

corresponds to the state that achieves a maximum size upon explosion. After the replacement of variables at zero initial velocity, we obtain the simple equation
$$2y^2 y_\tau^2 = 1 - y^2 ,$$
which allows for the exact solution
$$\tau = \sqrt{2(1 - y^2)} .$$

Hence the dimensionless collapse time $\tau_{\max}$ and the first oscillation period of the cavity $\tau_{*,1}$ are
$$\tau_{\max} = \sqrt{2} \quad \text{and} \quad \tau_{*,1} = 2\sqrt{2} \simeq 2.83 ,$$
which are fairly close to the result of the model a.

Taking into account the known character of cavity collapse with abrupt decrease in the radius only in the vicinity of its minimum and the fact that it is in this region that the kinetic energy of liquid becomes a noticeable part of the initial potential energy of the system, one can consider an approximated version of the oscillation equations (c) and (d), neglecting the term $y_\tau^2/2$ and assuming that $y = 1$ under the logarithm sign. It is obvious that under these conditions the models (c and d) become analogous to the models of ultimate transition and their solutions are of the same form
$$\tau_{c,d} \simeq \sqrt{\ln C(1 - y^2)} ,$$
where $C = 2h/R_0$ (for a cavity in a half-space) or r_0/R_0 (for cylindrical layer). Naturally, the period of the first oscillation for these models is written as $\tau_{*,(c,d)} \simeq 2\sqrt{\ln C}$.

Hence follows the evident conclusion that the results of the estimates coincide, if in the models (c) and (d) the value $\ln C \simeq 2$, which is not realistic for the model (c) in which the explosion depth can be adjusted arbitrarily. For the model (d) this means that only a limited portion of the liquid is involved in the oscillation, the portion being defined by the condition according to which the layer radius and the cavity radius should differ by approximately one order of magnitude.

The oscillation period of a cylindrical cavity with detonation products, as a time interval between the moment of arrival of the shock wave to the pressure gauge and the moment of recording of the maximum pressure of the fist oscillation, was studied experimentally using a hexogen DC ($d = 0.65$, 1.65, and 3.0 mm) exploded at depths $h = 0.21$–3.5 m from the free surface. The charge diameter and the explosion depth were selected reasoning from the possibility of clear determination of the dependence of the oscillation period on the explosion depth assumed in [7]. In experiments with 1.5 m-long charges, the gauges were placed at the midline of the charge at a distance of 0.5 m from it. The accuracy of determination of the period was at least 1 µs.

Experimental data on the oscillation period $\tau_{*,\exp}$ (dimensional) and calculation results obtained using the above models are summarized in Table 3.2.

One can see that the depth did not have the influence that could have been expected according to [7] and that application of the model (c) was inexpedient. On the contrary, the models (a,b) provided sufficiently reliable estimates of the parameter under study. It should be noted that the value calculated from experimental data using the formula for $\tau_{*,d}$ varied (for fairly wide ranges of explosion depth and charge radius) relative to the average value of 2.05 and differed in r_0 by a factor of 1.5–2 from the above estimate of the real liquid volume involved in cavity oscillations.

The experiment showed that the first maximum expansion of the cavity with detonation products $(R_{\max}/R_{\text{ch}})$ was a constant independent of the charge radius R_{ch} and was within 130–140 for the studied HE types, whose detonation velocity was 7.7 km/s and HE density was 1.55 g/cm^3.

The maximum relative dispersion of experimental data shown in Table 3.2 was noted for the thinnest charge due to the infinitesimal amplitude of its first oscillation and the difficulties in determining the exact position of its maximum on the pressure oscillogram. This was fairly reasonable, since for example, the initial potential energy of the liquid, which was partially spent for the first oscillation during the collapse of the cavity with detonation products, was approximately 20 times lower for the charge with $d = 0.65$ mm than for that with $d = 3$ mm.

Table 3.2. Experimental and calculation results of cylindrical cavity oscillations

d, mm	h, m	$\tau_{*,\exp}$, μs	$\tau_{*,a}(\tau_{*,b})$, μs	$\tau_{*,c}$, μs	$\ln(r_0/R_0)$
0.65	0.21	13.2–14.2	13 (12.34)	13.3	2.26–2.62
	2.10	13.0–13.8	13 (12.34)	18.7	2.20–2.47
1.65	0.54	31–32	33.1 (31.43)	33.4	1.94–2.1
	2.10	29–30	33.1 (31.43)	42.3	1.7–1.8
	0.54	58	60.1 (57.1)	52.1	2.05
3.00	2.10	57–57.5	60.1 (57.1)	70.3	1.98–2.01
	3.50	54–55	60.1 (57.1)	76	1.78–1.84

3.3 Circular Charges

Some problems of explosion hydrodynamics are concerned with the field structure of the shock waves produced by underwater explosions of charges of complicated geometry or with changing flow topology, as for example, in the case of transformation of an initially spherical cavity with detonation products into a torus in the gravity field. The topological effects take place, in particular at large-scale underwater explosions, when the maximum cavity

94   3 Explosion of Cylindrical and Circular Charges

radius reaches tens and thousands of meters, and the cavity is at a sufficient depth. When the cavity emerges and its lower part collapses due to the high pressure gradient along the vertical cross section, an upward cumulative flow is formed. Calculations performed by Pritchett [8] showed that at a later point the cumulative jet "perforates" the upper part of the cavity surface, shaping it as a ring. The dynamics of this cavity near the free surface is concerned with other jet effects [9] related to sultans that will be considered separately (see Chap. 7 for definition and detailed discussions).

It is appropriate to study the features of the flow appearing in the above geometry. In experimental simulation of the flows, two types of sources were used: a circular DC charge initiated from one end and a circular conductor simulating instantaneous explosion [10]. The wave structure of the first type was studied experimentally using DC (0.65, 1.65, and 3 mm in diameter), with circle radii varying from 3 to 30 cm and the distances to the charge were in the range 0.5–5 m. Typical patterns of the formation of the wave field in the near zone of a circular charge are shown in Figs. 3.3–3.4.

Figure 3.3 illustrates the shock wave front formation in a liquid (calculations in the acoustic approximation) in the plane of a circular charge for two successive positions of the detonation front at a detonation rate of 7.5 km/s.

Experimental studies confirmed the validity of such constructions and gave grounds for applying the acoustic model in further estimates of the structure of the wave field in a liquid produced by the explosion of a system of circular charges. Figure 3.4a presents frame-by-frame scanning (figures show the frame numbers) with a time interval of 4 µs, while frames 7 and 12 interval's of 20 µs). The records in Figs. 3.4b and 3.4d showed continuous scanning of the had time development of the wave process in the inner region of the circle (the circle diameter was 5.4 cm) and the coil turn. The plane of the turn was parallel to the observation window, which was closed by opaque

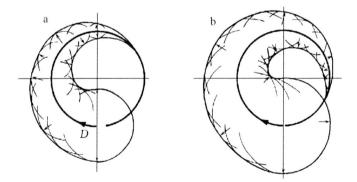

Fig. 3.3a,b. Dynamics of the formation of a shock wave front in a liquid under the explosion of a circular charge for two (**a**, **b**) subsequent positions of the detonation front (calculation)

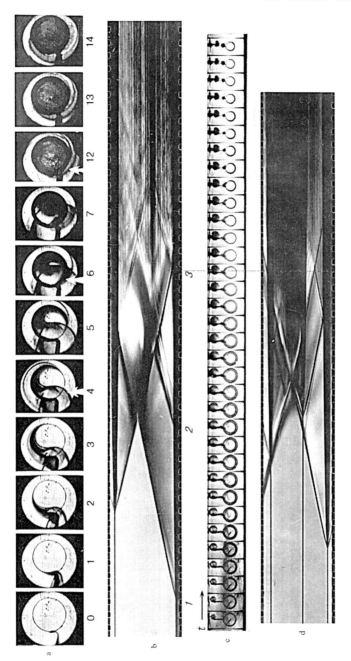

Fig. 3.4a–d. Shock wave formation in liquid a during the explosion of a circular charge (**a**), wave interaction inside the circle (**b**), dynamics of toroidal cavity with detonation products (**c**), wave interaction inside the coil wind (**d**)

paper with a slit oriented along the circle diameter. The charge was placed so that the focusing point was on the slit, since otherwise the phase velocity of converging waves and the apparent increase in the front velocity would be recorded. Indeed, the experiments proved that for the HE types and parameters used there occurred "pure" acoustics. The slit "cut" two diametrically opposite sections of the DC circle: the distant section from the initiation point can be seen on the top left of the photo as a thin dark line, a closer one merged with the boundary of the frame. For the case of a coil, the slit "cut" three elements (Fig. 3.4d): initial, intermediary, and final sections of the coil.

One can see that three waves were radiated in the space, the order of which depended on the position of the registration point with respect to the section of detonation initiation. The first shock wave always arrived from the section of the turn that was the closest to the gauge, then came waves 2 or 3. In this case, wave 2 was excited behind the wave front 1 by the final section of the circle, when the detonation wave front completed the full turn. This moment, if continuous rotation of the detonation front over spiral charge turns were considered, corresponded to the beginning of next turn. Wave 3 arose from the wave focusing in the inner part of the circle.

Axial asymmetry of the charge (Fig. 3.4d) led to essential changes in the wave interaction pattern. The photo shows that the shock wave produced by detonation of the final element of the coil ("thread" in the photo center) interacted with the explosive cavity: an attenuated shock wave 2 was recorded. This interaction yielded a rarefaction wave that attenuated the shock wave 3, the result of the interaction of waves from the initial and intermediary elements of the coil. These effects should be taken into account when using the flat spiral charges considered in Chap. 2.

Since the detonation velocity of a standard HE is approximately five times as high as the speed of sound in a liquid, one can expect that liquid flow for circular charges should be close to axisymmetrical. Indeed, the filming of the oscillation of a circular explosive cavity (Fig. 3.4c) showed that the cavity retained the toroidal shape, at least during the first oscillation. Here the circle diameter was 30.5 cm, the HE charge diameter was $d_0 = 0.65$ mm, the detonation velocity was 7.7 km/s, and the frequency was 1500 frames per second. According to these records, the maximum cross-section of the cavity with detonation products at the moment of stoppage was $\approx 120 R_{\mathrm{ch}}$. One can see that there was strong cavitation inside the circle.

As was mentioned above, the detonation rate of a standard DC is significantly higher than the speed of sound in a liquid, which gives grounds for applying the model of instantaneous explosion to preliminary estimates. For this purpose, one can use the explosion of a circular conductor.

The experiments were run at a high-voltage set-up with a capacitor bank capable of storing the energy of up to several kJ and releasing a required amount on the circle whose diameter was about 5 cm (0.15-mm thick nichrome wire). Exploding the wire in a liquid made it possible to reveal one characteristic feature of the wave field formation. A typical pattern registered by a high-speed photorecorder is shown in Fig. 3.5.

One can clearly see that the shock wave front had a toroidal surface, near the circle axis the wave was focused and reflected, which was recorded by the pressure gauge as a compression wave. During further development of the

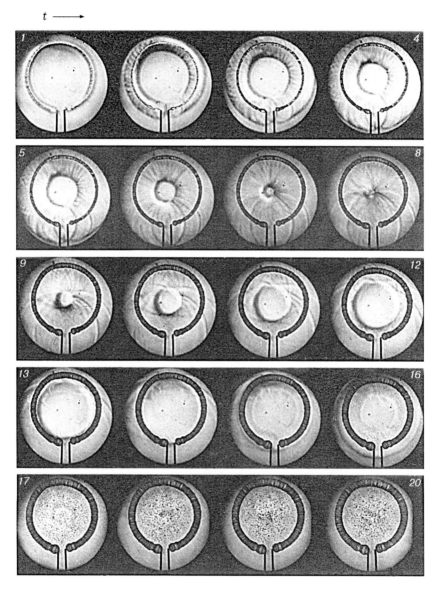

Fig. 3.5. Dynamics of toroidal cavity and shock wave focusing under the explosion of a circular conductor in a liquid

process due to the interaction of the reflected wave with the explosive cavity, a converging rarefaction wave propagated inside the circle, with intense bubble cavitation developing behind the wave front. This points to the fact that at the moment of interaction between the shock wave reflected from the axis and the cavity pressure inside, the latter was noticeably lower than the hydrostatic value.

Pressure measurements confirmed the above dependence of the registered radiation structure on the position of the registration point with respect to the closest to the charge zone. As the registration point moved away from the charge, a transition occurred from distinct stratification of the wave into the three mentioned types, followed by gradual degeneration into one wave. The character of change of the wave field parameters was different as well. Figure 3.6 shows the changes of maximum amplitudes of the first wave $p_{1,\max}$ depending on the relative distance along the axis for the charge with $R_0 = 0.0325$ cm and different charge radii $a = 30$, 20, and 10 cm (curves 1–3 respectively).

All three dependences had identical order character $p_{1,\max} \simeq A(\bar{r})^{-1.3}$, where the coefficient A depends on the charge mass, i.e., defined by the values of the charge radius R_{ch} and the circle radius a for fixed charge density ρ_* and HE type. However, it should be noted that using generalized data for all three values of HE circle radii beyond the specified range can result in significant discrepancies. Nevertheless, the simple approximate estimate of the amplitude of the first wave on the circle axis presented,

$$p_{1,\max} \simeq 1.410^7 \left(\frac{R_0}{a}\right)^{1.5} \left(\frac{a}{r}\right)^{1.3},$$

can be of practical significance. It is noteworthy that the amplitude of the second wave increased with the decrease of the circle radius a at fixed r/a.

Experimental data for the distribution p_1 in the circle plane are shown in Fig. 3.7 ((1) $a = 30$ cm, (2) $a = 10$ cm). It appears that the pressure in the shock wave front of a charge with greater radius a decreased with

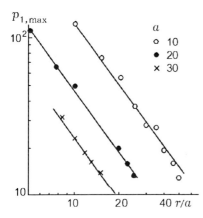

Fig. 3.6. Change of the first shock wave amplitude $p_{1,\max}$ along the circular change axis in liquids for charge radii $a = 30(1)$, $20(2)$, and $10(3)$ cm

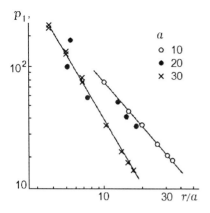

Fig. 3.7. Pressure distribution for the first wave p_1 in the plane of the circular charge

distance faster (power ~ -1.65) than for small circles (~ -1.15). The data for $a = 20$ cm were intermediate. The obtained result made it possible to assume that the character of pressure distribution in the turn plane at distances $\bar{r} = r/a \gg 1$ was similar to the case of condensed charges whose weight was equal to the weight of HE in the circle.

Experimental checks confirmed this assumption: satisfactory compliance between the data for equivalent concentrated charges was observed in the circle plane from distances $\bar{r} \approx 10$ and along the generatrix (*) of the cylindrical surface of the radius a from distances $\bar{r} \approx 30$–40 (Fig. 3.8).

Measurements of maximum amplitudes of the second waves (Fig. 3.9) showed that they could reach high values and sometimes exceeded the amplitude of the first wave, but they decreased with distance faster in the circle plane than along the axis. This effect might have resulted from the wave focusing in the region of the axis and the attenuation due to the interaction of the reflected waves with the explosive cavity in the circle plane.

These experimental data make it possible to study the dynamics of a toroidal cavity with axisymmetry, assuming that the cavity retains its form

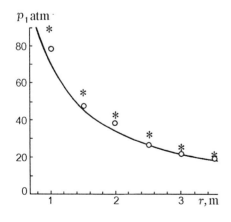

Fig. 3.8. Asymptotic behavior of the wave amplitude produced by a circular charge in a liquid at large distances from the charge (experiment)

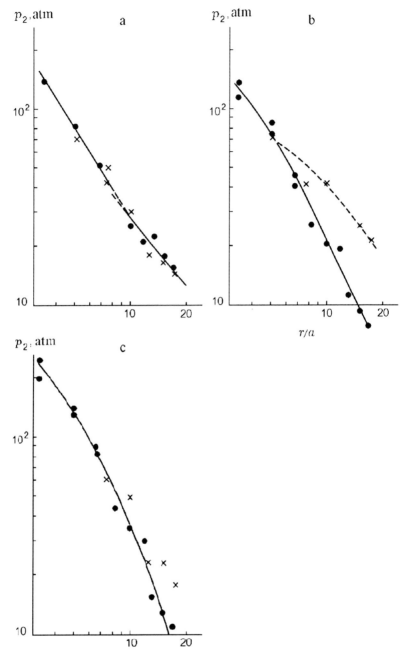

Fig. 3.9a–c. Pressure distribution for secondary shock waves for the explosion of a circular charge in a liquid at different distances and positions of pressure gauges (experiment): (**a**) on the axis, (**b**) in the circle plane, (**c**) at the generatrix

during oscillations. They also make it possible to apply the appropriate approximation in mathematical modeling of the process. Thus, for example, one can consider that the cavity section remains circular. The results obtained can be applied to analyze the structure of the wave field of the charges with complicated spatial structure. Experiments with DC showed that the model of instantaneous detonation is quite applicable for studying the cavity dynamics because it simplifies the calculations of the initial parameters of the problem.

3.4 Dynamics of a Toroidal Cavity, Numerical Models

3.4.1 Ideal Incompressible Liquid

Let us consider an ideal incompressible imponderable liquid containing a toroidal gas cavity, whose radius is a and the cross-section is a circle of radius R_0. We assume that pressure inside the cavity changes following the adiabatic law $p = p(0)(\bar{V})^{-\gamma_*}$, where $\bar{V} = V/V_0$ is the relative volume of the torus. We need to know the equation of motion of the boundary of the torus. For this purpose, we consider the orthogonal toroidal system of coordinates,

$$x = \frac{c\,\text{sh}\,\alpha \cos\gamma}{\text{ch}\,\alpha - \cos\beta}, \quad y = \frac{c\,\text{sh}\,\alpha \sin\gamma}{\text{ch}\,\alpha - \cos\beta}, \quad \text{and} \quad z = \frac{c\sin\beta}{\text{ch}\,\alpha - \cos\beta},$$

where c is the scaling multiplier (the radius of the "basis" circle). Coordinate surfaces in such a system will be toruses with $\alpha = \text{const}$ formed by rotation of a circle of radius $R = c/\text{sh}\,\alpha$ about the z-axis, with the center placed at a distance $a = c\,\text{cth}\,\alpha$ from the axis; and the spheres $\beta = \text{const}$ formed by rotation of a circle of radius $R_1 = c/\sin\beta$, with the center on the z-axis located at a distance $a_1 = c\,\text{ctg}\,\beta$ from the coordinate origin about the z-axis and the plane $\gamma = \text{const}$.

Let us consider the boundary value problem [11]. In the region $\Omega(t)$ bound by the closed smooth toroidal surface $\sigma(t)$ we find the potential $\varphi(\alpha, \beta, \gamma, t)$ such that at $t \geq 0$ in $\Omega(t)$:

$$\Delta\varphi = 0, \quad \varphi \to 0 \quad \text{as} \quad \alpha \to 0 \quad \text{and} \quad \beta \to 0;$$
$$\zeta(\alpha, \beta, \gamma, t) = 0 \quad \text{is the equation of the torus boundary};$$
$$\zeta_t + \nabla\varphi\nabla\zeta = 0 \quad \text{is the kinematic condition } (*);$$

in $\sigma(t)$:

$$-\varphi_t + 1/2(\nabla\varphi)^2 + p/\rho = p_\infty/\rho, \quad \text{where} \quad p = p(0)(V_0/V)^{\gamma_*}.$$

For $t = 0$, we assume $p = p(0)$, $\alpha = \alpha_*$, $\dot{\alpha} = 0$, where $\sigma(0)$ is the torus surface.

Let us seek the solution of the Laplace equation as

$$\varphi = \psi\sqrt{2\operatorname{ch}\alpha - 2\cos\beta}\,. \tag{3.6}$$

Then for ψ it becomes

$$\psi_{\alpha\alpha} + \psi_{\beta\beta} + \operatorname{sh}_\alpha^{-2}\psi_{\gamma\gamma} + \operatorname{cth}\alpha\,\psi_\alpha + \frac{1}{4}\psi = 0\,. \tag{3.7}$$

Hereafter the subscripts α, β, and γ denote the appropriate partial derivatives. The solution of this equation for the axisymmetrical case (ψ is independent of γ) is known and is expressed in terms of spherical functions. Thus, one can show that

$$\varphi = \sqrt{2\operatorname{ch}\alpha - 2\cos\beta}\,[A\,P_{\nu-1/2}(\operatorname{ch}\alpha) + BQ_{\nu-1/2}(\operatorname{ch}\alpha)] \\ \times (C\sin\nu\beta + D\cos\nu\beta)\,. \tag{3.8}$$

We are interested in solving the boundary value problem for the region $\Omega(t)$ that is external with respect to $\sigma(t)$. In this case, α and β change within the ranges $0 \le \alpha \le \alpha_0$ and $-\pi \le \beta \le \pi$ ($\alpha = 0$, $\beta = \pi$ is the coordinate center and $\alpha = 0$, $\beta = 0$ is infinity). Since the problem is symmetrical with respect to the torus surface, the solution (3.9) is significantly simplified:

$$\varphi = \sqrt{2\operatorname{ch}\alpha - 2\cos\beta}\sum_{n=0}^{\infty} A_n P_{n-1/2}(\operatorname{ch}\alpha)\cos n\beta\,. \tag{3.9}$$

Let us consider the case $n = 0$. Then (3.10) is written as

$$\varphi = A\sqrt{2\operatorname{ch}\alpha - 2\cos\beta}\,\frac{2K(k)}{\pi\operatorname{ch}(\alpha/2)}, \tag{3.10}$$

where $K(k)$ is the total elliptical integral of the first kind and $k = \operatorname{th}(\alpha/2)$ is its modulus. The coefficient A is determined from the kinematic condition, see (*), which in view of the assumption on the toroidal shape of the cavity and, consequently, the condition that the equation of the boundary is $\alpha - \alpha_0 = 0$, accepts the simple form

$$\dot\alpha_0 H_{\alpha_0}^2 = \varphi_\alpha\big|_{\alpha=\alpha_0}, \tag{3.11}$$

where $H_{\alpha,\beta} = c(\operatorname{ch}\alpha - \cos\beta)^{-1}$ are Lamé coefficients and α_0 is the coordinate of the cavity boundary. Assuming that $a \gg R$ ($\operatorname{ch}(\alpha_0/2) \gg 1$) for A, we obtain

$$A \simeq \frac{\pi c^2 \dot\alpha_0 \operatorname{sh}(\alpha_0/2)}{\sqrt{2}\operatorname{ch}\alpha_0^{5/2}}\,. \tag{3.12}$$

It should be noted that for $\operatorname{ch}(\alpha_0/2) \gg 1$, $\operatorname{th}(\alpha_0/2) \to 1$ and $E(k_0) \to 1$ (the value of total elliptic integral on the cavity boundary $k_0 = \operatorname{th}(\alpha_0/2)$), while $K(k_0)$ can be approximated to a sufficient degree of accuracy by the expression $K \simeq \ln(4\,\mathrm{ch}(\alpha_0/2))$ used to derive (3.13). Velocity components of liquid particles and φ_t are defined by the expressions

$$\varphi_\alpha = \frac{c^2 \dot\alpha_0 \,\mathrm{sh}(\alpha_0/2)}{\mathrm{ch}^{5/2}\alpha_0} \cdot \frac{E(\mathrm{ch}\,\alpha - \cos\beta) - K(1 - \cos\beta)}{\mathrm{sh}(\alpha/2)(\mathrm{ch}\,\alpha - \cos\beta)^{1/2}},$$

$$\varphi_\beta = \frac{c^2 \dot\alpha_0 \,\mathrm{sh}(\alpha_0/2)}{\mathrm{ch}^{5/2}\alpha_0} \cdot \frac{K\,\mathrm{th}(\alpha/2)\sin\beta}{\mathrm{sh}(\alpha/2)(\mathrm{ch}\,\alpha - \cos\beta)^{1/2}}, \qquad (3.13)$$

and

$$\varphi_t \simeq \frac{2c^2 \,\mathrm{sh}(\alpha_0/2)}{\mathrm{ch}^{5/2}\alpha_0}(\ddot\alpha_0 - 2\dot\alpha_0^2)\frac{K(k)}{\mathrm{ch}(\alpha/2)}(\mathrm{ch}\,\alpha - \cos\beta)^{1/2}.$$

In the last expression we ignored $(E - K)/E\,\mathrm{sh}\,\alpha_0$ as compared to 2.

Substituting (3.14) in the Cauchy–Lagrange integral (see problem statement), we obtain the relationship

$$p - p_\infty = -A_1\ddot\alpha_0 + (B_1 - C_1)\dot\alpha_0^2 \qquad (3.14)$$

for pressure at any point of the liquid, where

$$A_1 = \frac{2\rho c^2 \,\mathrm{sh}(\alpha_0/2)}{\mathrm{ch}^{5/2}\alpha_0}\frac{K(k)}{\mathrm{ch}(\alpha/2)}(\mathrm{ch}\,\alpha - \cos\beta)^{1/2}, \quad B_1 = 2A_1,$$

and

$$C_1 = \frac{c^2\,\mathrm{sh}^2(\alpha_0/2)}{2\,\mathrm{ch}^5\alpha_0}\frac{\mathrm{ch}\,\alpha - \cos\beta}{\mathrm{sh}^2(\alpha/2)}\{(K\,\mathrm{th}(\alpha/2)\sin\beta)^2 +$$
$$+ [E(k)(\mathrm{ch}\,\alpha - \cos\beta) - K(k)(1 - \cos\beta)]^2\}.$$

Thus, assuming that $\sigma(t)$ is the coordinate surface, on the basis of the Cauchy–Lagrange equation one can obtain the equation of cavity oscillation in toroidal systems,

$$\rho\frac{c^2 \ln(8\,\mathrm{ch}\,\alpha_0)}{\mathrm{ch}^2\alpha_0}(-\ddot\alpha_0 + 2\dot\alpha_0^2) - \frac{\rho c^2 \dot\alpha_0^2}{2\,\mathrm{ch}\,\alpha_0^2} = p(0)\left(\frac{\mathrm{ch}\,\alpha_0}{\mathrm{ch}\,\alpha_{00}}\right)^{2\gamma_*} - p_\infty, \qquad (3.15)$$

and polar systems,

$$(R\ddot R + \dot R^2)\ln\frac{8a}{R} - \frac{1}{2}\dot R^2 = \frac{p(0)(R_0/R)^{2\gamma_*} - p_\infty}{\rho}.$$

It is obvious that the last equation has a common form with different variants of the equations for cylindrical symmetry. Naturally, one should not expect asymptotic transition (at $a \to \infty$) of the toroidal cavity into a cylindrical one, since an exact solution of this problem, as shown above, is nonexistent.

The requirement would be analogous to the transition from a half-space to an unbound liquid in the Kuznetsov statement [5], which, as is known, does not lead to desirable result.

Let us write down the law of energy conservation

$$T + E = \text{const},$$

where $T = -\rho/2 \int_\sigma \varphi(\partial\varphi/\partial n)\, d\sigma$ is the kinetic energy of liquid, $E = U_1 + U_2$, U_1 is the work against the forces of external pressure, and U_2 is the internal energy of gas in the cavity. Expressions for T and E are as follows:

$$T = \frac{4\pi^2 \rho c^5 K(k_0) \dot{\alpha}_0^2}{\text{ch}^5 \alpha_0}, \quad E = \frac{2\pi^2 \rho c^3}{(\gamma_* - 1)\,\text{sh}^2 \alpha_{00}} + \frac{2\pi^2 p_\infty c^3}{\text{sh}^2 \alpha_{00}},$$

where $2\pi^2 c^3 / \text{sh}^2 \alpha_{00}$ is the initial volume of the torus. Then the law of energy conservation becomes

$$\rho \frac{c^2 \ln(8\,\text{ch}\,\alpha_0)}{\text{ch}^2 \alpha_0} \dot{\alpha}_0^2 + p_\infty \left(1 - \frac{\text{sh}^2 \alpha_0}{\text{sh}^2 \alpha_{00}}\right)$$

$$= \frac{p(0)}{\gamma_* - 1}\left[\frac{\text{sh}^2 \alpha_0}{\text{sh}^2 \alpha_{00}} - \left(\frac{\text{sh}\,\alpha_0}{\text{sh}\,\alpha_{00}}\right)^{2\gamma_*}\right].$$

One can easily show that the law corresponds exactly to the first integral of (3.16).

To determine the accuracy of Eq. (3.11) for φ with respect to the expansion (3.10), one can consider the Dirichlet problem aimed at determining the pressure in the region Ω, which results in the following relationship:

$$p = \frac{2p(0)}{\pi}(2\,\text{ch}\,\alpha - 2\cos\beta)^{1/2}\left[\frac{Q_{-1/2}(\text{ch}\,\alpha_0)}{2P_{-1/2}(\text{ch}\,\alpha_0)} P_{-1/2}(\text{ch}\,\alpha) + \right.$$

$$\left. + \sum_{n=1}^\infty \frac{Q_{n-1/2}(\text{ch}\,\alpha_0)}{P_{n-1/2}(\text{ch}\,\alpha_0)} P_{n-1/2}(\text{ch}\,\alpha)\cos n\beta\right].$$

Here p and $p(0)$ are excessive pressures with respect to the pressure at infinity. The first addend in brackets corresponds to the value of p defined by the dependence (3.15) at $\dot{\alpha}_0 = 0$. Let us estimate the following term of the expansion ($n = 1$). We have

$$Q_{1/2}(\text{ch}\,\alpha_0) \simeq -e^{-3\alpha_0/2}, \quad P_{1/2}(\text{ch}\,\alpha_0) \simeq \frac{2}{\pi}e^{\alpha_0/2},$$

and

$$P_{1/2}(\text{ch}\,\alpha) \simeq \frac{2}{\pi}e^{\alpha/2} E(\sqrt{1 - e^{-2\alpha}}).$$

3.4 Dynamics of a Toroidal Cavity, Numerical Models

Now for relative pressure $\bar{p} = (p - p_\infty)/(p(0) - p_\infty)$ with regard to two expansion terms one can get the following expression:

$$\bar{p} \simeq \frac{(2\operatorname{ch}\alpha - 2\cos\beta)^{1/2}}{2}\left[\frac{K(k)}{K(k_0)\operatorname{ch}(\alpha/2)} - \frac{4}{\pi}e^{-2\alpha_0+\alpha/2}E(\sqrt{1-e^{-2\alpha}})\cos\beta\right].$$

Since α varies within $0 \le \alpha \le \alpha_0$, the second addend in brackets will be maximum at $\alpha = \alpha_0$, namely $4e^{-3\alpha_0/2}/\pi$. But under the above conditions $a \gg R$ ($\operatorname{ch}(\alpha_0/2) \gg 1$) and one can consider that $\exp(\alpha_0/2) \simeq 2\operatorname{ch}(\alpha_0/2)$, thus the expression for pressure is simplified as

$$\bar{p} \simeq \frac{(2\operatorname{ch}\alpha - 2\cos\beta)^{1/2}}{2}\left[\operatorname{ch}^{-1}(\alpha_0/2) - \frac{1}{2\pi}\operatorname{ch}^{-3}(\alpha_0/2)\right].$$

It is clear that under the adopted conditions the second term inside the brackets can be ignored, which implies that the expression (3.11) can be considered sufficiently exact.

Let us cite some estimates for pressure at $t = 0$:

– Beyond the torus in the torus plane ($\beta = 0$),

$$\bar{p} \simeq \frac{\pi}{\ln(8a/R_0)}\frac{a}{r} \quad \text{at} \quad r \gg a$$

– At the torus center ($\beta = \pi$, $\alpha = 0$),

$$\bar{p} \simeq \frac{\pi}{\ln(8a/R_0)}$$

– On the symmetry axis ($\alpha = 0$, $0 \le \beta \le \pi$),

$$\bar{p} \simeq \frac{\pi a}{\sqrt{z^2 + a^2}\ln(8a/R_0)}$$

One can see that at greater distances, the character of pressure change in the first and the last examples is the same and is determined analogously to the case of spherical symmetry, within an accuracy of a scaling multiplier. This result correlates well with the conclusion on the feasibility of estimating the shock wave amplitude at a great distance from the ring based on the explosion of spherical HE charge of equivalent weight.

3.4.2 Compressible Liquid

Let us consider the problem of the oscillation of a toroidal cavity produced by an explosion of a circular HE charge within the framework of acoustic approximation, assuming that $a \gg R$ [12]. In this case, according to the experimental data, the cross section of a toroidal cavity practically retains the shape of a regular circle throughout the first oscillation period at $a \approx 10^3 R_{\text{ch}}$, and at $a \approx 10^2 R_{\text{ch}}$ throughout the first half-period (R_{ch} is the charge radius). The above problem of torus oscillation in incompressible liquid does not allow one to estimate such important parameters as maximum radius of the cavity with detonation products and energy distribution between them and the shock wave.

Solution of this axisymmetrical problem involves many difficulties, including the complexity of solution of the wave equation. Therefore, by first restricting oneself to the known frames, one should find the way of constructing the equation of one-dimensional oscillation to simplify the problem. Since an expression for the velocity potential can be found under certain assumptions for a series of spatial potential problems for ideal incompressible liquid, it can be used to proceed to acoustic models. The validity of this method will be estimated below by the example of the equation of one-dimensional oscillation of bubbles.

1. Let the velocity potential be $\varphi = \Phi(t)/f(r)$ for the case of incompressible liquid, the form of the function $f(r)$ being dependent on the symmetry type. Then its acoustic variant can be presented as $\varphi = \Phi(t - r/c_0)/f(r)$. We consider a potential liquid flow $u = -\nabla \varphi$, where u is the mass velocity of liquid, thus

$$u = \frac{\Phi f_r}{f^2} + \frac{\Phi'}{c_0 f}, \qquad (3.16)$$

where the subscript denotes the appropriate partial derivative and the prime marks a derivative with respect to $\zeta = t - r/c_0$. Taking into account the form of φ, the Cauchy–Lagrange potential can be expressed as

$$\Phi' = f(\omega + u^2/2). \qquad (3.17)$$

Here $\omega = \int \mathrm{d}p/\rho$, where p and ρ are the pressure and density of the liquid, respectively. Using the two equations and substituting $\Omega = \omega + u^2/2$, one can find the expression for Φ:

$$\Phi = f^2 \frac{u - \Omega/c_0}{f_r},$$

whose derivative with respect to t has the form

$$\Phi' = f^2 \frac{u_t - \Omega_t/c_0}{f_r}. \qquad (3.18)$$

Here $\Omega_t = \omega_t + u u_t$ and it is taken into account that $\Phi_t = \Phi'$ and the function f depends only on r. On the basis of the continuity equations (acoustic variant) and conservation of momentum,

$$u_r + \nu \frac{u}{r} = -\left(\frac{\mathrm{d}\omega}{\mathrm{d}t}\right) c_0^{-2} \quad \text{and} \quad \frac{\partial \omega}{\partial r} = -\frac{\mathrm{d}u}{\mathrm{d}t},$$

3.4 Dynamics of a Toroidal Cavity, Numerical Models

($\nu = 0, 1, 2$ for flat, cylindrical, and spherical cases, respectively), we find the expressions for partial derivatives u_t and w_t in the expression for Φ':

$$u_t = \frac{du}{dt} + \frac{\nu u^2}{r} + u\frac{dw}{dt}c_0^{-2}; \qquad w_t = \frac{dw}{dt} + u\frac{du}{dt}. \qquad (3.19)$$

Substituting (3.19) into (3.18) and equating the resulting expression to Eq. (3.17), we eventually get

$$f\left(1 - \frac{2u}{c_0}\right)\frac{du}{dt} + \frac{\nu f u^2}{r}\left(1 - \frac{u}{c_0} - \frac{r f_r}{2\nu f}\right) = w f_r + \frac{f}{c_0}\frac{dw}{dt}\left(1 - \frac{u}{c_0} - \frac{u^2}{c_0^2}\right). \qquad (3.20)$$

In this equation, which is valid for any point in the liquid, one can proceed to the cavity wall assuming $r = R$, $u = dR/dt$, and $H = (p(R) - p_\infty)/\rho_0$, where $p(R)$ is the pressure in detonation products, p_∞ is the pressure in liquid at infinity, and ρ_0 is the density of undisturbed liquid. Thus, we have

- For flat cavity ($\nu = 0$, $f = 1$, $f_r = 0$),

$$\left(1 - 2\frac{\dot R}{c_0}\right)R\ddot R = \left(1 - \frac{\dot R}{c_0} + \frac{\dot R^2}{c_0^2}\right)\frac{R\dot H}{c_0}$$

- For spherical cavity ($\nu = 2$, $f = r$, $f_r = 1$),

$$\left(1 - 2\frac{\dot R}{c_0}\right)R\ddot R + \frac{3}{2}\dot R^2\left(1 - 4\frac{\dot R}{3c_0}\right) = H + \left(1 - \frac{\dot R}{c_0} + \frac{\dot R^2}{c_0^2}\right)\frac{R\dot H}{c_0}$$

- For cylindrical cavity ($\nu = 1$, assume $f = \sqrt{r}$, then $f_r = r^{-1/2}/2$),

$$\left(1 - 2\frac{\dot R}{c_0}\right)R\ddot R + \frac{3}{4}\dot R^2\left(1 - 4\frac{\dot R}{3c_0}\right) = \frac{H}{2} + \left(1 - \frac{\dot R}{c_0} + \frac{\dot R^2}{c_0^2}\right)\frac{R\dot H}{c_0}$$

All three equations are completely equivalent to the equations obtained on the basis of the exact solution of the problem on the invariance of the function G along the characteristics c_0:

$$\left(\frac{\partial}{\partial t} + c_0\frac{\partial}{\partial r}\right)G = 0 .$$

Here $G = r^{\nu/2}\Omega$.

The proposed method is fairly simple; however, one more problem is still to be solved: in two-dimensional problems the equation of continuity does not permit substitution of partial derivatives of velocity components for total derivatives. Analysis of the method has shown that it enables the following simplification. Let us assume that in the continuity equation we can neglect the term $(\dot{\omega}/c_0^2)$, i.e., we can consider the relation between the velocity components as primarily defined by the frames of the approximation for an ideal incompressible liquid. If the solution of the Laplace equation is found for the velocity potential and given the appropriate statement of the problem, the boundary conditions permit separation of variables, each velocity component will be expressed in terms of the total derivative of the cavity radius with respect to t.

It can be shown that the above assumption slightly changes Eq. (3.20): the terms u/c_0 and $(u/c_0)^2$ disappear from the right-hand part in brackets at $\dot{\omega}$. But from the acoustics viewpoint they can be neglected as well, since main losses for radiation under cavity oscillation in incompressible liquid are defined by the term $R\dot{\omega}/c_0$. The results will be used to find the equation of the cavity dynamics under explosion of a circular charge in the acoustic approximation.

2. Let the liquid contain a toroidal cavity formed as a result of "instantaneous" detonation of a circular charge and the linear dimensions of the cavity satisfy the condition $a \gg R$. Then within the framework of the circular source approximation one can write the following expression for the velocity potential:

$$\varphi = -\frac{a}{2\pi} \int_0^\pi \frac{\Phi(t - f/c_0)}{f} \, d\alpha, \qquad (3.21)$$

where $\Phi = dS/dt$; $S = \pi R^2$ is the cross-sectional area of the torus cavity; $f = \sqrt{z^2 + r^2 + a^2 - 2ar\cos\alpha}$ in the cylindrical coordinate system (r, α, z), where α is from an arbitrary direction which in the case of a real DC charge should be selected with regard to the point of detonation initiation. Following item 1, one could put an explicit form of the equation for φ for the case of incompressible liquid. However, in this case it is unclear how the argument of the function Φ under the transition to the acoustic model should be written. Therefore, we keep the expression for the velocity potential as (3.21), but agree that one can take the function Φ' from under the integral sign assuming that it possesses the properties of the invariant G. This assumption is insignificant for constructing the oscillation equation of a cavity and can influence only the estimate of the fine structure of the shock wave in the near-charge zone.

By analogy with the above-stated one can write

$$\Phi' = \frac{2\pi}{a}\left(\omega + \frac{V^2}{2}\right) \bigg/ \int_0^\pi \frac{d\alpha}{f} \quad \text{and} \quad \frac{a}{2\pi} \int_0^\pi \frac{\Phi \cdot (f_r + f_z)}{f^2} d\alpha$$

$$= u + v - \left(\left(\omega + \frac{V^2}{2}\right)\bigg/c_0 \int_0^\pi \frac{d\alpha}{f}\right) \int_0^\pi \frac{(f_r + f_z)}{f} d\alpha, \qquad (3.22)$$

3.4 Dynamics of a Toroidal Cavity, Numerical Models 109

where u, v, and V are the components and the full speed of a liquid particle. Let us take the partial derivative with respect to t from the second equation in (3.22). Then, taking Φ' from under the integral sign, we obtain

$$\frac{a\Phi'}{2\pi}\int_0^\pi \frac{f_r+f_z}{f^2}\,d\alpha = u_t + v_t - \left((\omega_t + VV_t)\Big/c_0\int_0^\pi \frac{d\alpha}{f}\right) \times \int_0^\pi \frac{f_r+f_z}{f}\,d\alpha\,. \tag{3.23}$$

One can show that $\omega_t = \dot{\omega} + V\dot{V}$. The partial derivatives u_t and v_t are sought from the solution for incompressible liquid. In this case, using for the velocity potential the expression $\varphi = -(a/2\pi)\Phi(t)\int_0^\pi d\alpha/f$, we eventually get

$$u_t = \dot{u} - \frac{a}{2\pi}\Phi\left\{u\int_0^\pi\left(\frac{f_{rr}}{f^2} - \frac{2f_r^2}{f^3}\right)d\alpha + v\int_0^\pi\left(\frac{f_{rz}}{f^2} - \frac{2f_rf_z}{f^3}\right)d\alpha\right\}$$

and

$$v_t = \dot{v} - \frac{a}{2\pi}\Phi\left\{u\int_0^\pi\left(\frac{f_{rz}}{f^2} - \frac{2f_rf_z}{f^3}\right)d\alpha + v\int_0^\pi\left(\frac{f_{zz}}{f^2} - \frac{2f_z^2}{f^3}\right)d\alpha\right\}.$$

Substitute ω_t, u_t, v_t in (3.23) and express Φ' from the first equation of the system (3.22). We obtain the general equation

$$(1 - 2F_1u)\dot{u} + (1 - 2F_1v)\dot{v} - \frac{\pi}{a}\frac{F_0}{F}(u^2 + v^2)+$$

$$+\frac{a}{2\pi}\Phi F_1 F_2 u^2 + +\frac{a}{2\pi}\Phi F_1 F_3 v^2 + \frac{a}{\pi}\Phi F_1 F_4 uv - \frac{a}{2\pi}\Phi(F_2+F_4)u-$$

$$-\frac{a}{2\pi}\Phi(F_3+F_4)v = \frac{2\pi}{a}\frac{F_0}{F}\omega + F_1\dot{\omega}\,, \tag{3.24}$$

where

$$F_0 = \frac{a}{2\pi}\int_0^\pi \frac{f_r+f_z}{f^2}\,d\alpha\,; \quad F = \int_0^\pi \frac{d\alpha}{f}\,; \quad \Phi = 2\pi R\dot{R}\,;$$

$$F_1 = \frac{1}{c_0 F}\int_0^\pi \frac{f_r+f_z}{f}\,d\alpha\,; \quad F_2 = \int_0^\pi \frac{f_{rr}-2f_r^2/f}{f^2}\,d\alpha\,;$$

$$F_3 = \int_0^\pi \frac{f_{zz}-2f_z^2/f}{f^2}\,d\alpha\,; \quad \text{and} \quad F_4 = \int_0^\pi \frac{f_{rz}-2f_zf_r/f}{f^2}\,d\alpha\,.$$

110   3 Explosion of Cylindrical and Circular Charges

If we introduce the notation $\xi_1 = z^2 + (r+a)^2$, $\xi_2 = z^2 + (r-a)^2$, $\xi_3 = z^2 + a^2 - r^2$, $\xi_4 = z^2 + a^2 + r^2$, $\xi_5 = 2r^2 - a^2 - (z-r)^2$, and $\xi_6 = 3(z^2 + a^2) - r^2$, the functions F will finally be written as follows:

$$F_0 = \frac{a}{2\pi r \sqrt{\xi_1}} \left[\frac{\xi_5}{\xi_2} E(k) + K(k) \right], \quad F = \frac{2K(k)}{\sqrt{\xi_1}},$$

$$F_1 = \frac{\pi \sqrt{\xi_1}}{4rc_0 K(k)} \left[1 + \frac{\xi_5}{\sqrt{\xi_1 \xi_2}} \right], \quad k = \sqrt{\frac{4ar}{\xi_1}},$$

$$F_2 = -\frac{3K(k)}{2r^2 \sqrt{\xi_1}} + \frac{\xi_6}{r^2} \frac{E(k)}{\xi_2 \sqrt{\xi_1}} - \frac{\xi_3^2 \sqrt{\xi_1}}{2r^2 \xi_2^2 \xi_1^2} \{4\xi_4 E(k) - \xi_2 K(k)\},$$

$$F_3 = \frac{2E(k)}{\xi_1 \sqrt{\xi_1}} - \frac{2z^2 \sqrt{\xi_1}}{\xi_2^2 \xi_1^2} \{4\xi_4 E(k) - \xi_2 K(k)\},$$

and

$$F_4 = -\frac{3zE(k)}{r\xi_2 \sqrt{\xi_1}} + \frac{z\xi_3 \sqrt{\xi_1}}{r\xi_2^2 \xi_1^2} \{4\xi_4 E(k) - \xi_2 K(k)\},$$

where $K(k)$ and $E(k)$ are the total elliptic integrals of the first and the second kind, respectively, and k is their modulus. Proceeding to the torus wall, after a series of transformations in expressions for the coefficients F from (3.24) we obtain the equation of oscillation of toroidal cavity in compressible liquid:

$$\left\{ \left[1 - 2\pi \frac{\dot{R}}{c_0 \ln(8a/R)} \right] R\ddot{R} + \left[1 - \pi \frac{\dot{R}}{c_0 \ln(8a/R)} \right] \dot{R}^2 \right\} \ln(8a/R) - \frac{\dot{R}}{2} =$$

$$= \omega + \frac{\pi R}{c_0} \dot{\omega}. \quad (3.25)$$

3.4.3 Comparison with Experiments

Experimental studies compared below with the calculation data from (3.25) were run with copper-casing charges. The high-speed filming showed that the final detonation velocity does not affect the cavity shape. The initial parameters of the problem were determined from the condition of discontinuity of arbitrary decay and instantaneous detonation at constant volume. For isentropic index we used the data of [5]: initial charge density of 1.55 g/cm, detonation velocity of 7.7 km/s, and charge diameter of 0.65 mm. Numerical and experimental results are compared in Table 3.3:

The following conclusuions were drawn from analysis of the data presented in Table 3.3:

1) For $a \geq 3 \cdot 10^2 R_{ch}$, the calculation results are in satisfactory agreement with experimental data.

Table 3.3. Comparison of calculated and experimental results of oscillation of a toroidal cavity

a/R_{ch}	Calculation			Experiment		
	y_{max}	y_{min}	τ_*, s/cm	y_{max}	τ_*, s/cm	$E, \%$
$1.54 \cdot 10^2$	125.83	4.02	0.21	103	0.21	11.8
$3.08 \cdot 10^2$	127.8	3.82	0.24	119.5	0.237	15.9
$4.60 \cdot 10^2$	128.8	3.72	0.256	123.6	0.256	17.0
$1.54 \cdot 10^3$	131.5	3.48	0.3	—	—	—
$3.08 \cdot 10^3$	132.8	3.37	0.323	—	—	—
$3.08 \cdot 10^4$	136.4	3.09	0.394	—	—	—
∞	140.9	3.15	0.2	135	0.2	22

Note: $y_{max} = R_{max}/R_{ch}$, $y_{min} = R_{min}/R_{ch}$, $\tau_* = t_{max}/R_{ch}$, t_{max} is the time of expansion of the cavity with detonation products to the first maximum, E is the portion of energy that remains in the detonation products by the moment of the first expansion of the cavity, $a/R_{ch} = \infty$ corresponds to cylindrical charge, R_{max} is the first maximum radius of the cavity with explosion products, R_{min} is the cavity radius at the moment of the first compression, R_{ch} is the charge radius.

2) With increasing circle charge radius the oscillation parameters of the explosive cavity R_{max} and R_{min}, which characterize the energy balance, asymptotically approach the appropriate values of explosion parameters for cylindrical symmetry.

3) Circular geometry of the charge significantly affects the oscillation period of the cavity with detonation products: according to the calculation results, even when the circular charge radius is $a = 10$ m (second line from the bottom), the oscillation period of the torus is twice as long as for the case of cylindrical symmetry. Experimental results confirm the tendency to the increase of the oscillation period with increasing the charge radius.

4) Following the experimental data, as the circular charge radius decreases (to the value when the circular section of toroidal cavity is distorted during of expansion), a portion of energy falling at the shock wave generated by this charge increases and at $a \approx 150 \cdot R_{ch}$ amounts to almost 90 %. As the circle radius increases, the energy balance approaches the results for cylindrical symmetry.

The above results verify the efficiency of the method proposed above and the resulting oscillation equation of toroidal cavity in compressible liquid (3.25).

3.4.4 Basic Characteristics for Toroidal Charges

Analysis of the numerical results (3.25) for a wide range of circle radii a made it possible to obtain an approximate analytic expression for maximum radius of toroidal cavity with detonation products,

$$y_{\max} \simeq 141\left[1 - \frac{2}{3}\ln^{-1}\left(\frac{8a}{R_{\text{ch}}}\right)\right],$$

which is in satisfactory agreement with experimental data on real detonation cord (DC) and complies with the above conclusions. Both experiments and calculations showed that the oscillation period of a cylindrical cavity is not an asymptotic for the oscillation period of the torus as the circular charge radius increases. The period can be estimated using the traditional approach (the model of incompressible liquid): ignoring the term $\dot{R}^2/2$ and substituting R for $R_{\max}$ under the logarithm sign, we consider that these approximations should be justified by the experiment. Thus, Eq. (3.25) is reduced to the form

$$\rho_0(R\ddot{R} + \dot{R}^2)\ln\left(\frac{8a}{R_{\max}}\right) \simeq -p_\infty$$

and can be integrated, yielding

$$R^2 \simeq R_{\max}^2 - \frac{p_\infty}{\rho_0 \ln(8a/R_{\max})}t^2.$$

Assuming that $R = 0$ in this expression, we obtain for the collapse time of an empty toroidal cavity in an incompressible liquid

$$\tau \simeq R_{\max}\sqrt{\frac{\rho_0 \cdot \ln(8a/R_{\max})}{p_\infty}}.$$

The collapse time determined from this expression, to within the accuracy of a certain constant, coincides with the experimental data and numerical results for the initial equation. With allowance for this fact, the final expression for the first oscillation period of the explosive cavity of the circular charge becomes

$$T_{\text{cal}} \simeq 2R_{\max}\sqrt{\frac{\rho_0 \cdot \ln(8a/R_{\max})}{p_\infty}} + 0.04 R_{\text{ch}},$$

where R is measured in centimeters, T, in seconds, and the coefficient of the second summand is dimensional.

3.5 Comparative Estimates for Spherical Charges

The model of an ideal incompressible liquid is the most frequent and simple approximation for many problems of explosion hydrodynamics. It suggests that the liquid density $\rho = $ const and enables investigation of certain

3.5 Comparative Estimates for Spherical Charges

important characteristics of the explosion process within the framework of conservation laws. The equation of continuity in this case allows for the solution in the form $vr^\nu = f(t)$, which at fixed t is invariant and holds for every point of the liquid. The model makes it possible to obtain the exact equation only for a spherical cavity:

$$R\frac{d^2R}{dt^2} + \frac{3}{2}\left(\frac{dR}{dt}\right)^2 = \rho^{-1}[p(R) - p_\infty],$$

which is known as the Rayleigh equation [13]. At $p_\infty = $ const its first integral can be found easily, given the equation of state of detonation products (or gas) inside the cavity. When multiplied by $2\pi\rho$, the equation gives the law of energy conservation: the change of the internal energy of detonation products,

$$\Delta E = \frac{2p(0)R_0^3}{3\rho(\gamma-1)}\left[1 - \left(\frac{R_0}{R}\right)^{3(\gamma-1)}\right],$$

is equal to the increment of kinetic energy $T_k = 2\pi\rho R^3 \dot{R}^2$ (if $\dot{R}(0) = 0$) and potential energy $U = (4/3)\pi p_\infty(R^3 - R_0^3)$ of the liquid. Detonation products are considered as an ideal gas with variable (in general case) adiabatic index γ. The most convenient form is writing the equation in dimensionless variables for $y = R/R_0$ and $\tau = tR_0^{-1}\sqrt{p_\infty/\rho}$. The integral makes it possible to determine two major parameters: the oscillation period τ_* and the maximum radius of the cavity, with detonation products $y_{\max} t$ (corresponds to the condition $y_\tau = 0$) whose estimate is simplified provided that $y_{\max} \gg 1$:

$$y_{\max}^3 \approx \frac{p(0)}{p_\infty(\gamma-1)} + \frac{3}{2}y_\tau^2(0),$$

and defines the above mentioned energy balance but relates it to the initial potential energy $p_\infty V_0$.

The parameters $p(0)$ and $y_\tau(0)$ define the initial conditions of the problem on the contact discontinuity and are found from the formulas for shock transition in a liquid and for transition in the unloading wave in detonation products (Chap. 1). It should be noted that their definition naturally does not account for the model of incompressible liquid, and the results for $p(0)$ and $y_\tau(0)$ do not depend on the symmetry type.

The charge detonation is assumed instantaneous, the volume V_* being constant and the density of detonation products being equal to the initial density of the HE charge. In this case, pressure in detonation products can be determined from the known relationship $p_* = (\gamma_* - 1)\rho_* Q_\alpha$, which follows from the condition of equality of specific internal energy of detonation products to the portion of the explosion heat Q_* per HE mass unit, which accounts for detonation products. The speed of sound c_* in the products of instantaneous detonation is determined from the detonation rate D_*:

$$c_*^2 = \frac{\gamma_* D_*^2}{2(\gamma_* + 1)}.$$

Let us consider a specific case for a hexogen change with $\rho_* = 1.65$ g/cm^3, $Q_* = 1.32$ kcal/g, and $D_* = 8.35$ km/s. Following the data of Kuznetsov and Shvedov [5], the expansion isentrope of detonation products of this HE is characterized by a practically invariable γ_*, which, as the density of the products decreases during their expansion from 2.2 to 0.8 g/cm^3, varies in the narrow range of 2.95–2.75. A change is noticeable within the pressure range 0.6–0.05, where γ_* decreases from 2.5 to its minimum value 1.26.

Assume $\gamma_* = 2.85$, then $p_* = 69$ kbar and $c_* = 5.08$ km/s. The solution of the problem on discontinuity of an arbitrary decay makes it possible to find $p(0)$ and $\dot{R}(0)$ (or $y_\tau(0)$). Thus, if we assume $\gamma = \gamma_* = $ const, the result for $y_{\max}$ can be obtained immediately by equating the appropriate portion of the explosion energy $\alpha \cdot Q_* \rho_* V_*$ (where $\alpha = 0, 41$) to the potential energy of the liquid $p_\infty V_{\max}$ at the moment of maximum expansion of the cavity:

$$y_{\max} = \left(\frac{\alpha Q_* \rho_*}{p_\infty}\right)^{1/3}.$$

When the cavity radius reaches maximum ($y = y_{\max}$, the time of cavity expansion $\tau = \tau_+$), the pressure in detonation products is several orders of magnitude lower than the hydrostatic pressure p_∞ and the cavity can be considered empty. Under the effect of this pressure difference, the cavity collapses with initial zero velocity over the time $\tau = \tau_-$, while the Rayleigh equation takes the form of Besant's equation:

$$y^3 y_\tau^2 = \frac{2}{3}(1 - y^3).$$

It should be noted that in this equation the cavity radius is normalized by $R_{\max}$. The time τ_- is determined from the appropriate integral of this equation under the assumption that the cavity collapses to zero

$$\tau_- = \sqrt{3/2} \int_0^1 y^{3/2} (1-y^3)\, dy = \frac{1}{\sqrt{6}} B(5/6.1/2) \simeq 0.915.$$

Here B is the beta function. Experimental studies proved that oscillation of explosive cavity is symmetrical: the time of its expansion, τ_+, of R to $R_{\max}$ equals to a high degree of accuracy to the time of its subsequent collapse τ_-. Thus, the period of the first oscillation of the cavity τ_* with detonation products can be defined as a sum of τ_+ and τ_-:

$$\tau_* = 1.83 \quad \text{or} \quad T_1 = 1.83 R_{\max} \sqrt{\frac{\rho}{p_\infty}}.$$

It should be noted that the interval between the shock wave front and the maximum of the first oscillation (or oscillations, if the cavity form remains stable) equals to the period T_1. If one manages to register the pressure profile $p(t)$ experimentally, and consequently T_1 as well, $R_{\max}$ automatically becomes known, which allows one, provided that the coefficient of energy transfer into radiation is known, to estimate the explosion energy. Likewise, if the explosion heat is known, it is possible to estimate the energy adsorbed by the shock wave.

If the density ρ, hydrostatic pressure p_∞, and explosion energy $Q_1 = Q_* \rho_* V_*$, where $\rho_* V_*$ is the HE charge mass, are accepted as governing parameters of the problem on underwater explosion in an ideal incompressible liquid, they can be combined with the time dimensionality. It is sufficient to express $R_{\max}$ in Eq. (3.7) in terms of the expression Q_1/p_∞. The result is the well-known Willis formula [3],

$$T_i = 1.14 \alpha_i^{1/3} \frac{Q_1^{1/3} \rho^{1/2}}{p_\infty^{5/6}},$$

in which the coefficient α_i reflects the degree of HE energy transition into liquid under detonation and subsequent oscillations. The subscript i denotes the number of oscillations.

3.6 Oscillation Parameters

In this section we distinguish the characteristic parameters (maximum radius and oscillation period) of the dynamics of explosive cavities containing detonation products [14]:

- For spherical cavity,

$$R_{\max}^i \simeq \left(\frac{3}{4\pi} \frac{\alpha_i E_0 W}{p_0} \right)^{1/3}, \quad \alpha_i = 0.41;\ 0.14;\ 0.076\ ;$$

$$T_i \simeq 1.14 \rho^{1/2} (\alpha_i E_0 W)^{1/3} p_0^{-5/6}$$

- Empirical expression for the current radius,

$$R/R_0 \simeq 665 (t/R_*)^{0.4}, \quad 10^{-4} \leq t \leq T_1/4$$

- For cylindrical cavity,

$$R_{\max}^i \simeq \left(\frac{\beta_i E_0 W_1}{\pi p_0} \right)^{1/2}, \quad \beta_i = 0.218;\ 0.14;\ 0.11\ ;$$

$$T_i \simeq 1.635 \rho_0^{1/2} (\beta_i E_0 W_1)^{1/2} p_0^{-1}$$

- For toroidal cavity,

$$R_{\max} \simeq \left(\frac{0,218 E_0 W_1}{\pi p_0} \right)^{1/2} \left[1 - \frac{2}{3 \ln(8a/R_{\text{ch}})} \right],$$

$$T_1 \simeq 2 R_{\max} \sqrt{\frac{\rho_0 \ln(8a/R_{\max})}{p_0}} + 0,04 R_{\text{ch}}$$

Here E_0 is the HE explosion heat; W and W_1 are the charge mass; $i = 1$, 2, and 3 are the numbers of oscillations; and α and β are HE energy portions remaining in the detonation products.

3.7 Explosions of Spatial Charges in Air

Spiral charges as hydroacoustic sources were considered in Chap. 2. It was shown that the wave packets radiated by such charges are characterized by long duration, weak attenuation, and given frequency of the sequence of shock waves in the packet. The wave packet can keep its profile and duration owing to the rarefaction phases between shock waves, which prevent the waves from overlapping. One can expect that the structure of the wave field produced by explosion of a spiral charge in air will differ.

Below we consider the results of experimental study of these problems for the near zone of spatial charges shaped as separate coils and annular spirals exploded in air.

3.7.1 Experimental Arrangement

In test experiments, linear charges made of standard detonating cord (DC) consisting of RDX 3 mm in diameter and two to three meters long were suspended horizontally. The density and specific weight of HE were 1.6 g/cm^3 and 11–12 g/m, respectively, the specific heat of explosion was $Q = 1400$ kcal/kg, and the detonation velocity was 7.5 km/s. Pressure transducers (piezoelectric cells of radius $r_0 = 1$ mm) were placed in the central plane, normal to the linear charge axis. Shock waves were recorded at distances $r = 15, 20, 50$, and 100 cm from the charge axis along its radius. A spherical cast charge of 50/50 TNT/RDX (radius of 12 mm, weight equivalent of a 1-m long DC charge) was initiated from the center by an ÉD-637 detonator. The pressure was measured at distances $r = 10, 15, 30, \ldots, 150$ cm from the spherical charge center. The shock-wave velocity in air was measured from the time it takes a wave front to pass a base of 15 mm.

In test experiments, the pressure gauges were calibrated and the correlation between the shock-wave parameters in air was determined for linear and spherical charges of equivalent weight.

The studies were primarily aimed at analyzing the features and structure of the fields generated by explosions of plane annular charges (diameter $D_0 = 300, 450$, and 600 mm), bulk spiral coils (pitch $h = 150$ mm, $D_0 = 300$ mm, and length $L = 1$ m), plane spiral coils ($h = 150$ mm and $L = 1$ m), and spiral charges ($h = 150$ mm and $L = 10$ m). The charges were fixed on a rigid cylindrical frame made of thin steel wire. Detonation was excited from the furthermost charge end. Shock waves were registered along the charge axis within the range of relative distances $r^* = r/R_0 = 1$–10 or $r^* = r/h = 1$–10 counted from the central plane of symmetry. For bulk coils, the center of symmetry was at the axis at a distance of 75 mm from the butt ends. The signals of the pressure transducers were consecutively supplied to a two-beam digital oscillograph and to a computer. The oscillograph was triggered with a delayed pulse generator switched into a high-voltage current circuit generating pulses of 5 or 10 kV to initiate detonation.

3.7.2 Test Measurement Results

3.7.2.1 Linear and Spherical Charges

The characteristic pressure profiles of the shock waves induced by blasting linear (a) and spherical (b) charges are shown in Fig. 3.10 for $r \approx 20$ cm and $r \approx 30$ cm, respectively. The measurements showed that the characteristic shock wave length τ varied with increasing distance to the charge: for $r = 150$, 200, 500, and 1000 mm, τ increased to 150–250, 190–280, 350–420, and 500–580 µs, respectively.

The pressure in the front of the transient shock wave, $p_{1,\text{calc}}$, calculated from the mean shock wave velocity, D, and the experimental data for $p_{1,\exp}$ agreed with the well-known experimental data within the entire measurement region. The pressure could be approximated by the relation

$$p_1 \simeq bx^{-a}, \qquad (3.26)$$

where $a = 1.37$ and $b = 5.65 \cdot 10^{-2}$, if p_1 was measured in megapascals and x was measured in meters. One could introduce the characteristic size for a cylindrical charge $\lambda = (q_\text{l}/p_0)^{0.5}$ (where $q_\text{l} = Qm/L$ is the explosion heat per unit length) and compare the data for cylindrical and spherical geometry with the results obtained in [15, 16].

The greatest dispersion in p_1 around the mean value was $\pm(25\text{–}30)\%$ and was observed near the charge. For $r = 1$ m, the dispersion in p_1 decreased to

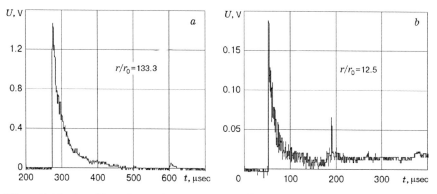

Fig. 3.10a,b. Experimental pressure profiles obtained by blasting linear ($r \approx 20$ cm) (**a**) and spherical ($r \approx 30$ cm) (**b**) charges

±(5–7)% and was almost equal to the measurement error (slightly less than 5%). Obviously, during expansion of explosion products, the inhomogeneous character of failure of the DC inert shell had a significant effect on the formation and parameters of the shock wave. Attenuation of the shock wave by the shell was mentioned in [15, 17], though the instability and dispersion of shock-wave parameters from one experiment to another were not reported.

For spherical charges, the experiments showed that the dispersion in p_1 in the shock wave front at distances $r = 0.1$ and 0.15 m was of the same order of magnitude as for linear charges. The instability of the boundary of expanding products that influenced the early stage of shock wave generation (in the near zone) was noticed even in the experiments without the shell. At the same time, for $r > 0.3$ m the dispersion in p_1 did not exceed the measurement error. Within the entire range of measurements for spherical charges, the relative pressure p_1/p_0 in the transient shock wave as a function of r/r_0 could be described by the same Eq. (3.26) only for $a = 2.04$ and $b = 3.56 \cdot 10^3$.

3.7.2.2 Circular and Spiral Charges

The pressure in transient shock waves p_1 during the explosion of spatial charges was determined using the curves calibrated in the incident wave during the test measurements.

Rings and coils

At the first stage of experiments, charges of equal mass but different shapes were compared. The characteristic pressure oscillograms for different relative distances r^* are plotted in Fig. 3.11 for 1-m long DC charges. The pressure profiles were identical for all three charge types. A two-wave structure was observed in the near zone ($r^* = 1$). Note that the amplitude of the second wave could be higher than that of the first wave by a factor of 1.5 to 2 for a ring (Fig. 3.11a) and a bulk spiral (Fig. 3.11b), and by a factor of six to seven for a plane spiral (Fig. 3.11c). The first wave was generated by explosion of the charge end closest to the transducer, while the second wave was generated by focusing a curved shock wave converging to the inner region of the charge. At a greater distance from the charge, the bow shock wave had a smoothly decreasing triangular profile (oscillograms on the right of Fig. 3.11a–c). The wave duration τ for the considered charge geometry was almost the same for identical r^*, and their value changed almost linearly from 60–125 to 650–700 μs with r^* increasing from one to ten.

The experimental data on the decay of the first and the second waves and the dependences $p(r^*)$ showed that for these charge types, the pressure in the shock wave front was almost invariant beginning from $r^* \geq 3$ irrespective of the shape of the spatial charge. The wave amplitude in this region is described by Eq. (3.26) with $a = 2.31$ and $b = 48.6$.

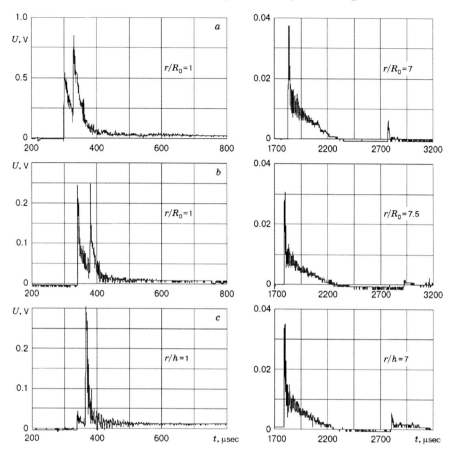

Fig. 3.11a–c. Typical pressure oscillograms along the axes after the explosion of one-meter long circular DC charge ($R_0 = 150$ mm) (**a**), a coil of a bulk spiral ($R_0 = 150$ mm and $h = 150$ mm) (**b**), and a plane spiral ($h = 150$ mm) (**c**).

Comparison of experimental data for spatial and spherical charges of an equivalent HE mass showed that spatial charges yielded a pressure rise by a factor of 2 to 3 as compared with spherical charges concentrated at equal absolute distances from its geometric center. Note that, among the above types of spatial charges, the best directivity and focusing along the axis (in the zone nearest to the charge) was exerted by a bulk coil in which the pressure reached 6.5 MPa on the axis for $r^* = 1$ in the second wave front.

Spiral charges

As was shown in Chap. 2 [29, 63], the explosion of a spiral charge in water generated a shock wave packet whose duration depended on the DC length and detonation velocity. Similar effects in air were complicated by strong

120   3 Explosion of Cylindrical and Circular Charges

nonlinear dependence of the shock wave velocity on distances to the charge. One could expect that both the duration of the wave packet and the structure of the wave field in air would differ essentially from those in water.

Characteristic examples are presented in Figs. 3.12 and 3.13 [18, 19]. Figure 3.12 shows the structure of the wave field resulting from explosion of a multiturn spatial spiral as the sequence of shock waves with attenuating amplitudes. Figure 3.13 demonstrates the case of explosion of a plane Archimedes' spiral. In both cases, the charges were made of a 10-m long DC and a pitch of 15 cm.

The longest signals as the shock wave series recorded by the gauge (the maximal number of secondary shock waves behind the front of the first shock

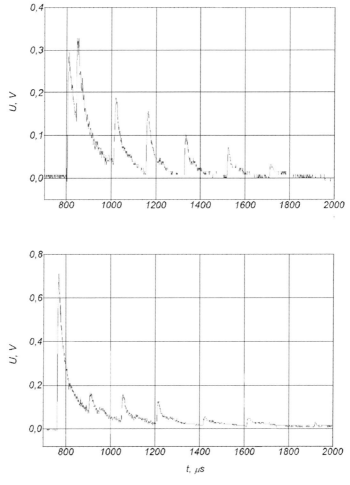

Fig. 3.12. Shock wave packet radiated under explosion of a spatial spiral charge (DC, $L = 10$ m, $R_0 = h = 150$ mm, $\Delta p_1 = 0.6$ MPa, and $r = 50$ cm)

wave was 7–8) were observed in the central cross-section plane of the charge (Fig. 3.12). Here the first pulse was caused by explosion of a spiral coil nearest to the gauge measuring.

Amplitudes of the shock wave sequence in the packet exponentially decreased as the detonation wave moved away from the central section plane where the gauge was placed. One could see that sometimes bifurcation of the first pulse of the shock wave set was registered (Fig. 3.12, upper oscillogram). Thus, the amplitudes of both pulses evidently decreased. This effect was obviously related to the formation of the wave field generated by explosions of the previous turn and the turn in the central section. For example, let the detonation front move in the charge from right to left. In this case, if the gauge was slightly shifted to the right with respect to central section, the signal from the previous turn could be recorded earlier than that of the wave from central turn. Thus, the latter signal propagated in the field of pressure and density gradients behind the first pulse front. The inverse situation occured, when the first shock wave was overtaken by the wave from the central turn. The two waves merged and the amplitude of the first resulting pulse increased essentially (Fig. 3.12, lower oscillogram).

For the Archimedes' spiral, the pressure oscillograms (Fig. 3.13) recorded along the axis are presented for different distances and scanning times (2, 4, and 8 ms, respectively) for the case when the detonation was initiated from the external turn. They show a complex structure characterized by an

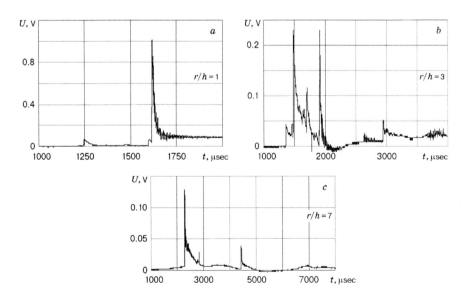

Fig. 3.13a–c. Typical pressure oscillograms recorded under explosion of a plane Archimedes' spiral ($h = 150$ mm and $L = 10$ m) for scanning times of 2 (**a**), 4 (**b**), and 8 ms (**c**)

obvious inhomogeneity of the pressure field. The oscillogram in Fig. 3.13a corresponds to a distance of 15 cm from the spiral plane, which was equal to the spiral pitch h. For $t \approx 1600$ μs, a strong wave was registered, and its amplitude was higher by an order of magnitude than that of the other waves in the packet. As the distance to the charge increased and the amplitude decreased, we observed the fine structure of wave packet (Fig. 3.13b).

The experiment showed that sometimes the wave packet generated by blasting of a spiral charge had longer duration than the time necessary for the detonation wave to cover the DC length in the charge (Fig. 3.13c). Obviously this "extension" of the signal was associated with the focusing times.

Shock wave velocity in air depends considerably on the distance to the charge. This fact complicates the analysis of experimental data on the distribution of pressure amplitudes in wave packets generated by blasting charges in the form of a plane coil and Archimedes' spiral. The data are compared in Fig. 3.14: (1) pressure $\Delta p_{3,4}$ in the third or fourth waves (the highest amplitude being in the case of Archimedes' spiral); (2) pressure Δp_1 in the bow shock wave of the same charge; and (3) pressure Δp_1 in the case of a plane coil. The solid curves represent experimental data (dots). For comparatively short distances $r^* = 1$ (Fig. 3.13a), the greatest amplitude in the wave packet with a total (resolvable) duration of $\tau \geq 750$ μs was reached in the third or fourth waves (curve 1, Fig. 3.14). The amplitude of the first shock wave (curve 2) was lower by almost an order of magnitude. In the intermediate region $2 < r^* < 5$, several waves with close pressure amplitudes ($\tau \approx 650$–750 μs) were observed, since the effect of outer spiral turns was more explicit as the distance from the charge plane changed (see Fig. 3.13b). For $r^* > 5$, as a result of merging of a series of shock waves (curves 1 and 2 overlap in Fig. 3.14), a bow shock wave was formed ($\tau \approx 700$–750 μs), which was followed by the second, weaker shock wave approximately 2.1 ms later (see Fig. 3.13c).

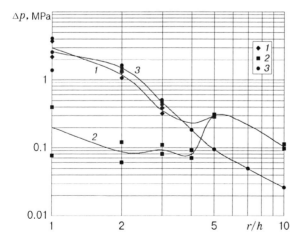

Fig. 3.14. Shock wave decay in a plane coil and an Archimedes' spiral

The experimental dependences $p(r/h)$ in the shock wave for a coil and a plane spiral (curves 3 and 1 in Fig. 3.14, respectively) showed that the spiral length in the zone closest to the charge had practically no effect on the maximum pressures in the shock wave, which were rather close for both charge types. An abrupt change was noticed for $r^* > 4$: the wave amplitudes in the packet induced by explosion of the spiral were significantly higher than the wave amplitudes in the packet generated by explosion of a single coil. Naturally, the wave duration and its total impulse increased. Comparison of the pressure oscillograms in Fig. 3.11c and Fig. 3.13 allows one to conclude that weaker shock waves with gradually increasing pressure in the shock wave front were formed by the large-radius outer coils with detonation propagating to the center along them, and the maximum pressure amplitude in the last shock wave was reached due to the focusing of a converging shock wave generated by the explosion of the central part of a plane spiral (approximately 1 m long).

This effect was more pronounced in experiments with three annular charges ($R_0 = 150$, 225, and 300 mm). As the radius of an individual ring (or a given coil of Archimedes' spiral) increased, the pressure in the shock wave in the near zone at a fixed distance decreased.

Summing up, we may say that experimental studies proved that the considerable dispersion in pressure and shock wave velocities registered in the near zone in the case of spatial, spherical, and cylindrical charges during explosion in air is due to the instability of the boundary of expanding detonation products. It was found experimentally that the decay of a shock wave induced by blasting spatial charges is less intense and the amplitudes in the shock wave front are two to three times higher than those in spherical charges of equal mass.

It should also be noted that the duration of the shock wave generated by blasting spiral charges in air is longer than that in concentrated charges of equal mass. As the DC length increases, so do the duration (approaching 3 ms for a DC length of about 10 m) and impulse of the wave packet in a plane spiral without significant increase in pressure in the shock waves in the near zone of the spiral plane (at distances from one to several pitches). The greatest jump in the shock wave is observed in the case of focusing of a converging shock wave in the central region of the charge.

References

1. V.K. Kedrinskii: On oscillation of cylindrical gas cavity in dimensionless liquid. In: Continuum Dynamics, vol. 8 (Institute of Hydrodynamics, SB of the USSR AS, Novosibirsk, 1971) pp. 163–168
2. V.K. Kedrinskii: *Kirkwood–Bethe Approximation for Cylindrical Symmetry of Underwater Explosion*, Fiz. Goren. Vzryva **8**, 1, pp. 115–123 (1972)
3. R. Cole: *Underwater Explosions* (Dover, New York, 1965)

4. V.K. Kedrinskii, V.T. Kuzavov: *Dynamics of the Cylindrical Cavity in Compressible Liquid* Zh. Prikl. Mekh. i Tekhn. Fiz. **38**, 4, pp. 102–106 (1977)
5. N.M. Kuznetsov, K.K. Shvedov: *Isentropic Expansion of Detonation Products of Hexagen*, Fiz. Goren. Vzryva **3**, 2 (1967)
6. V.K. Kedrinskii: *About Some Models of One Dimensional Pulsation of Cylindrical Cavity in Incompressible Liquid*, Fiz. Goren. Vzryva **12**, 5, pp. 768–773 (1976)
7. V.M. Kuznetsov: On expansion of a cylindrical cavity in an ideal incompressible liquid. In: Continuum Dynamics, vol. 14 (Institute of Hydrodynamics, SB of the USSR AS, Novosibirsk, 1973)
8. J.W. Pritchett: Incompressible calculations of underwater explosion phenomena. In: Proc. 2nd Internat. Conf. on Numerical Methods in Fluid Dynamics (Springer, Berlin, Heidelberg, New York, 1971) pp. 422–428
9. V.K. Kedrinskii, V.T. Kuzavov: *Underwater Explosion of Coil Charge Near Free Surface*, Zh. Prikl. Mekh. i Tekhn. Fiz. **24**, 4, pp. 124–130 (1983)
10. V.K. Kedrinskii: Dynamics of a cylindrical cavity in a boundless compressible liquid. Zh. Prikl. Mekh. i Tekhn. Fiz. **21**, 5 (1980)
11. V.K. Kedrinskii: On oscillation of a toroidal gas bubble in a liquid. In: Continuum Dynamics, vol. 16 (Institute of Hydrodynamics, SB of the USSR AS, Novosibirsk, 1974) pp. 35–43
12. V.K. Kedrinskii: *About One Dimensional Pulsation of Toroidal Gas Cavity in Compressible Liquid*, Zh. Prikl. Mekh. i Tekhn. Fiz. **18**, 3, pp. 62–67 (1977)
13. J.W. Rayleigh: *On the Pressure Developed in a Liquid During the Collapse of a Spherical Cavity*, Phil. Mag. **34**, pp.94–98 (1917)
14. V.K. Kedrinskii: *Hydrodynamics of Explosion (Review)*, Zh. Prikl. Mekh. i Tekhn. Fiz. **28**, 4, pp.23–48 (1987)
15. M.A. Tsykulin: *Air Shock Wave in Explosion of a Cylindrical Charge of Large Length*, Zh. Prikl. Mekh. i Tekh. Fiz., **1**, 3, pp. 188–193 (1960)
16. V.V. Adushkin: *Formation of a Shock Wave and Spreading of Explosion Products in Air*, Zh. Prikl. Mekh. i Tekh. Fiz., **4**, 5, pp. 107–1114 (1963)
17. A.A. Vasil'ev and S.A. Zhdan: *Shock-Wave Parameters on Explosion of a Cylindrical Charge in Air*, Zh. Fiz. Gorenia i Vzryva, **17**, 6, pp. 99–105 (1981)
18. A.V. Pinaev, V.T. Kuzavov and V.K. Kedrinskii: *Shock-Wave Structure in the Near Zone Upon Explosion of Spatial Charges in Air*, Zh. Prikl. Mekh. i Tekh. Fiz., **41**, 5, pp. 81–90 (1980)
19. Alexandr Pinaev, Vasily Kuzavov and Valery Kedrinskii: *Structure of Near Wave Field at Explosion of Spatial Charges in Air*, Proc. of 23d Intern. Symp. on Shock Waves, July 22–27, 2001, ed. by F.Lu, Fort Worth, Texas, USA. pp. 158–164 (2001)

4 Single Bubble, Cumulative Effects and Chemical Reactions

Numerous experimental and theoretical studies of the dynamics of a cavity in an unbounded liquid are concerned with a number of problems of underwater explosion, hydroacoustics, bubble cavitation, and shock initiation of detonation in liquid high explosives. The research results are used to analyze the structure of shock and rarefaction waves in two-phase media, play an important role in the development of the appropriate mathematical models describing spherical (cylindrical) cumulation in a viscous liquid, and are directly related both to the dynamics of shock-wave destruction of nephroliths (lithotripsy) and to cavitation erosion. The dynamics of a spherical cavity with a reactive gas mixture is of particular importance not only for the development of detonation of liquid HE, but also for the formation of conditions for large-scale explosions of tanks with easily ignitable gases. One of the difficult problems in this field is the formation of a reactive mixture in a bubble if the carrier phase of the bubbly media is fuel, while the gaseous phase contains an oxidizer.

4.1 Passive Gas Phase

4.1.1 Short Shock Waves

As a rule, the experiments with bubbles involve pulsed shock or compression waves, which are characterized by an explicit, relatively steep front with a pressure jump and an exponent-type profile. The duration of the positive phase of such waves can be up to tens of microseconds and is usually on the same order as the collapse time of bubbles. The model of bubble–shock wave interaction is based on the assumption, following which the time of the wave front propagation can be neglected against the characteristic time of its oscillation and the bubble collapse can be considered spherically symmetrical, while the incident wave is simulated by pressure at infinity, whose time variation is assigned in one-to-one correspondence to its profile [1, 2].

Analytical dependences for the collapse time t_{col} and the relative minimum bubble radius R_{min}/R_0 are represented by

$$t_{col} = 0.915 \cdot R_0 \sqrt{\rho/p_\infty}$$

$$\left(\frac{R_0}{R_{\min}}\right)^{3\gamma-3} = 1 + (\gamma-1)\frac{p_\infty}{p_0},$$

which follow from the Rayleigh equation for $p = \text{const}$, and are not valid for the case of variable pressure at infinity. Here γ is the gas adiabatic index in the bubble, R_0 and $R_{\min}$ are the initial and minimum bubble radii, p_0 is the initial gas pressure in the bubble, and p_∞ is the amplitude of the incident stationary shock wave.

Soloukhin [1] analyzed numerically the Rayleigh equation for exponential pressure change corresponding to the classical profile of a shock wave produced by underwater explosion. This equation was reduced to the convenient dimensionless form

$$y\frac{d^2 y}{d\tau^2} + \frac{3}{2}\left(\frac{dy}{d\tau}\right)^2 = \mu[y^{-3\gamma} - A\exp(-\tau)],$$

where $y = R/R_0$, $\tau = t/\theta$, $\mu = (\theta/R_0\sqrt{\rho/p_0})^2$, and $A = p_\infty/p_0$. One can easily notice a new dimensionless parameter μ that defines the relation of the time constant of the pressure differential θ to the characteristic compression time $t_* = R_0\sqrt{\rho/p_0}$ of an empty bubble of radius R_0 compressed by constant pressure p_0. The calculation was performed for a wide range of parameters $A = 10\text{--}10^3$, $\gamma = 1.33\text{--}1.67$, and $\mu = 10^{-2}\text{--}10^3$.

Based on these calculations, an approximate analytical dependence of the minimum bubble radius on the amplitude of the incident shock wave and the parameter μ was established in [2]:

$$\left(\frac{R_0}{R_{\min}}\right)^{3(\gamma-1)} \simeq 1 + (\gamma - 1)\frac{\mu A^2}{1 + \mu A}.$$

One can see that for $\mu \to \infty$ the minimum radius will be determined only by the amplitude A and the adiabatic index, which follows from the Rayleigh equation for stepwise waves at zero initial velocity ($y_\tau(0) = 0$).

It is obvious that the estimate of the bubble collapse time under the effect of a short shock wave should contain the same combinations of the parameter μ, so that for $\mu \to \infty$ ($\theta \to \infty$) the value t_{col} should achieve the appropriate asymptotics. Let us use the combination μA instead of p_∞, then

$$t_{\text{col}} \simeq 0.915 \cdot R_0 \cdot \sqrt{\frac{\rho}{p_0} \cdot \frac{1 + \mu A}{\mu A^2}}.$$

Dividing both parts of the expression by θ, we obtain the simple analytical dependence for $\tau_{\text{anl}} = t_{\text{col}}/\theta$:

$$\tau_{\text{anl}} \simeq 0.915 \cdot \frac{\sqrt{1 + \mu A}}{\mu A}.$$

Its data and those for the minimum radius are compared in Table 1 with the calculation results ($\tau_{\text{num}}, \bar{R}_{\text{num}}$).

Table 4.1. Compariso of numerical and analytical results of bubble-shock wave interaction

A	μ	γ	$\theta,$	τ_{num}	τ_{anl}	$\bar{R}_{num}$	$\bar{R}_{anl}$
10	10	1.4	$3.16\cdot 10^{-3}$	0.105	0.092	0.262	0.262
10	1	1.4	$1.0\cdot 10^{-3}$	0.35	0.303	0.307	0.28
10	0.1	1.4	$3.16\cdot 10^{-4}$	1.27	1.294	0.404	0.4
10	0.01	1.4	$1.0\cdot 10^{-4}$	5.11	9.59	0.738	0.772
100	10	1.4	$3.16\cdot 10^{-3}$	0.029 4	0.028 9	0.046	0.046
100	1	1.4	$1.0\cdot 10^{-3}$	0.092	0.095	0.048	0.045 7
100	0.1	1.4	$3.16\cdot 10^{-4}$	0.308	0.303	0.053	0.049
100	0.01	1.4	$1.0\cdot 10^{-4}$	1.096	1.294	0.074	0.079 3
100	10	1.67	$3.16\cdot 10^{-3}$	0.029 5	0.028 9	0.123	0.121
100	1	1.67	$1.0\cdot 10^{-3}$	0.094 4	0.092	0.126	0.122
100	0.1	1.67	$3.16\cdot 10^{-4}$	0.309	0.303	0.136	0.127
100	0.01	1.67	$1.0\cdot 10^{-4}$	1.1	1.294	0.167	0.17
1 000	1	1.33	$1.\cdot 10^{-3}$	0.029	0.028 9	0.003 04	0.003 02
1 000	0.1	1.33	$3.16\cdot 10^{-4}$	0.093	0.092	0.003 17	0.003 02
1 000	0.01	1.33	$1.0\cdot 10^{-4}$	0.304	0.303	0.003 58	0.003 33

One can see that the analytical dependences are quite suitable for the approximate estimates of the behavior of a gas bubble affected by pressure strongly changing with time. The most substantial are the discrepancies of the collapse time τ_{anl} for the least considered values of A and μ. It should be noted that the values $\mu = 0.01$ and $\theta = 10^{-4}$ correspond to large bubbles of radius $R_0 \simeq 1$ cm.

4.1.2 Formation of a Cumulative Jet in a Bubble (Experiment)

Experimental studies show that the interaction of a flat shock wave with a single gas bubble in a liquid is often far from the above ideal one-dimensional scheme. The process of streamlining of a bubble by a shock wave front, no matter how short it is with respect to the collapse time, is accompanied by interaction of the front with the surface. Detailed analysis of the interaction proved that it is of principal importance for the real process of the bubble deformation [3,4].

The wave pattern was recorded in the frame-by-frame regime by means of an optical device IAB-451, a high-speed photorecorder and a continuous-scan photorecorder through a horizontal slit on a film rotating with linear velocity of up to 100 m/s. A special frame-by-frame lighting scheme (Fig. 4.1) with a flash frequency of several tens of kHz was used in the experiments. The lighting scheme was based on a double discharge and included an accumulative $C_1 = 1$ mF and discharge $C_2 = 0.1$ mF capacitors, as well as a hydrogen spark discharger [3].

After breakdown of the gap, being a high-voltage switch, the capacitor C_1 discharges through the resistance $R_1 = 2 600$ Ohm on the capacitor C_2,

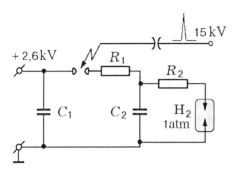

Fig. 4.1. Schematic view of the frequency hydrogen flash tube

at which the voltage increases to the breakdown voltage for the discharger placed in a special transparent casing. The casing was sealed, vacuumed, and filled with hydrogen ($p = 1$ atm). The discharger was made of nichrome wire ($d \simeq 1$ mm), and the distance between the electrodes was about 2 cm. Hydrogen provided short deionization time and prevented arcing. The inductance of the condenser and conducting wires was on the order of 1–2 µH, which provided the discharge duration (and consequently, the exposition time) that was no longer than 1 µs for the resistance $R_2 \simeq 1$–2 Ohm. The frequency of flashes reached 40 kHz at the beginning of the discharge process and reduced noticeably as the capacitor C_1 discharged. The accuracy of determining the rotation speed of the photorecorder film was about 1%.

Oscillations of a single spherical bubble caused by a shock wave with amplitudes within 1–3 MPa and durations of the positive pressure phase of the order of 100 µs) are shown in Figs. 4.2 and 4.3 [3].

Figure 4.2 shows the schlieren frame scanning of the bubble collapse under the effect of a shock wave propagating from left to right. The process was filmed using a pulse hydrogen lamp. One can readily see an explicit asymmetry of the bubble and the angular profile of its end wall in the last frame. The reason for this deformation of the bubble becomes apparent in Fig. 4.3 (the frame lapse time was 40 µs): a cumulative jet developing inside the bubble in the direction of the propagating shock wave impinged on the opposite wall of the bubble (frames 9–14). The interaction of a jet with a wall resulted in the formation of a local concavity at the wall [3]. Analysis of the data on the jet dynamics showed that its velocity reached 30 m/s when the back wall velocity was 10–15 m/s. The cumulative jet and the shape asymmetry were retained for the stage of bubble expansion as well.

As is evident from analysis of the experimental data and the numerical estimates [3, 4], the cumulative effect developed due to the peculiar distribution of initial velocities of the bubble wall arising during the propagation of the plane shock front through it. The front interacted with the bubble wall as a free surface and imparted different initial velocities to different points of the wall, the velocities being distributed over the surface as $(2\Delta p/\rho_0 c_0) \cos \alpha$, where α was the angle between the radius at the bubble surface in the con-

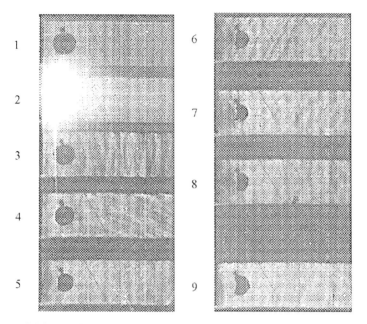

Fig. 4.2. Schlieren frame scanning of the shape dynamics of the bubble during interaction with the shock wave

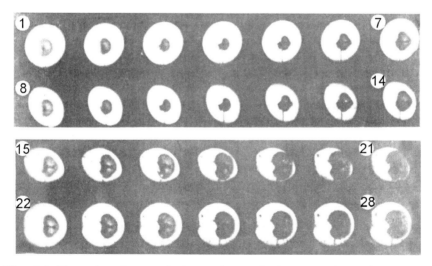

Fig. 4.3. Formation of a cumulative jet during shock-wave collapse of the bubble

tact point with the shock wave front and the direction of motion of the front [3,4]. Therefore, in spite of the uniform distribution of pressure in the liquid surrounding the bubble, when the wave front streamlined it (the bubble size was much smaller than the shock wave length), the spherical symmetry of the process was distorted. This in turn caused the collapse, with formation of a cumulative jet having the same direction as the wave: the classical cumulative jet occurred as the plane wave achieved the cumulative hollow.

The cumulative effect found in [3,4] led to bubble destruction, increased the dissipative losses, and distorted the structure of the wave reradiated by the bubbly medium. These effects would vanish if the front was steep and the pressure amplitude in the front decreased.

4.1.3 Real State of a Gas

Estimating the real state of a gas is an important aspect of bubble dynamics studies in a pressure field. It is important both for understanding the shock-wave transformation by a bubbly medium on the whole and for constructing the appropriate mathematical models. Therefore, the statement of direct experiments aimed at determining the state of gas in the bubble in dynamics is of doubtless interest. Below we describe the experiment arranged by Kedrinskii and Soloukhin in 1960 [3,5].

A bubble of very thin rubber filled with stoichiometric gas mixture $2H_2 + O_2$ or $2C_2H_2 + 5O_2$ was placed on the propagation trajectory of a shock wave in water. The bubble oscillations were recorded through a slit by a continuous-scanning streak camera with pulsed illumination and the appropriate synchronization. It was expected that during the adiabatic collapse of the bubble behind the shock front, the temperature of the mixture inside the bubble would rise to the ignition point, which would result in explosion of the bubble. The assumption was confirmed. The ignition was recorded both by spontaneous luminescence of the mixture in the cavity at the ignition moment (pulsed illumination was not used to prevent shading of the explosion flash) and by abrupt change of the cavity radius in time (Fig. 4.4) that evidenced the explosive character of expansion, which as a rule led the beginning of expansion in an inert gas. The dark band at the left center was the initial bubble diameter, the light vertical band was the moment of explosion of the wire

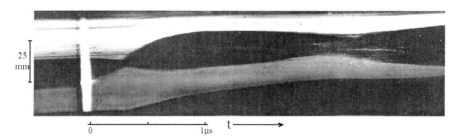

Fig. 4.4. Bubble oscillation: explosive character of expansion after ignition of the gaseous detonation inside the bubble

generating the shock wave in the liquid. The photo recorded several oscillation periods. The explosion was marked by a sharp deviation of the back bubble wall (see the top of the photo). It should be noted that due to the formation of a cumulative jet, the front wall was always smoothed in the scan.

The temperature as a function of the compression of the bubble was calculated using the adiabatic compression law $T_d = T_0(V_0/V_d)^{\gamma-1}$. A comparison of experimental ignition temperatures of the above mixtures showed that they were within the known values corresponding to adiabatic heating [6]. The use of stoichiometric hydrogen-oxygen mixture as a filler may be of interest for studying the dynamics of the collapsing cavity. Since the detonation product of this mixture was water vapor, which was cooled down and condensed during the expansion of the cavity, the cavity was practically empty at the collapse stage, and its boundary could acquire rather high velocities.

4.1.4 Viscosity and the Effect of Unbound Cumulation

Theoretical investigation of the dynamics of cavitation bubbles using the Navier–Stokes equations was carried out in [7–9]. It was shown that analysis of these equations can result in a paradoxical conclusion that the equation of motion of an ideal liquid describes the spherically symmetrical flow of a viscous liquid [10]. Indeed, in the right-hand part of the Navier–Stokes equation the term conditioned by the effect of viscosity formally disappears,

$$\frac{d\mathbf{v}}{dt} = -\frac{1}{\rho}\text{grad}\, p + \nu \left[\frac{4}{3}\text{grad}\,(\text{div}\,\mathbf{v}) - \text{rot}\,(\text{rot}\,\mathbf{v})\right]$$

and

$$\frac{d\rho}{dt} + \rho(\text{div}\,\mathbf{v}) = 0\,,$$

if a potential spherically symmetrical motion of incompressible liquid ($\mathbf{v} = \text{grad}\,\varphi$, $\Delta\varphi = 0$) is considered: $\text{rot}\,\mathbf{v} = 0$, and $\text{div}\,\mathbf{v} = 0$.

Considering the boundary conditions eliminates this paradox. The equality to zero of the resultant of viscous forces in the liquid volume does not mean disappearance of tangent stresses at the interface. For viscous liquids, the normal stress acting on the plate perpendicular to the radius r is defined by the relationship

$$p_{rr} = -p + 2\mu\frac{\partial v}{\partial r}\,,$$

where $\mu = \rho\nu$ is the dynamic coefficient of viscosity. The velocity derivative with respect to radius can be expressed in terms of the parameters of the bubble wall

$$\frac{\partial v}{\partial r} = -2\frac{\dot{R}R^2}{r^3}\,.$$

The result can be substituted in the expression for p_{rr}, then in the Cauchy–Lagrange integral, and proceeding to the cavity wall, the following equation of the wall dynamics in viscous incompressible liquid is obtained:

$$R\ddot{R} + \frac{3}{2}\dot{R}^2 = \frac{p(R) - p_\infty}{\rho_0} - \frac{4\nu\dot{R}}{R}.$$

For an empty cavity $p(R) = 0$, we introduce the dimensionless radius (y), the time (τ), and the Reynolds number (Re),

$$y = \frac{R}{R_0}, \quad \tau = \frac{t}{R_0}\sqrt{\frac{p_\infty}{\rho_0}}, \quad \text{Re} = \frac{R_0}{\nu}\sqrt{\frac{p_\infty}{\rho_0}},$$

and obtain the Poritsky–Shu equation (see [7–10]):

$$yy_{\tau\tau} + \frac{3}{2}y_\tau^2 = -1 - \frac{4}{\text{Re}}\frac{y_\tau}{y}.$$

One can see that the solution of the equation depends on the parameter $4/\text{Re}$ defined by Shu as a reduced coefficient of kinematic viscosity ν'. The point is that the radial velocity of the collapsing bubble is negative, thus the right-hand part of the equation can rapidly become positive and begin attenuating inertial effects. Then a moment comes, when the radial acceleration becomes positive, the absolute value of the collapse rate decreases, and thus the viscosity can stop the process, preventing closing of the cavity in a point: the initial potential energy of the liquid $(p_0 V_0)$ is totally converted into heat.

The equation can be easily transformed by the change of variables and integrated at zero initial velocity. We obtain

$$y^3 y_\tau^2 = -\frac{2}{3}(y^3 - 1) - 2\nu' \int_1^y yy_\tau \, dy.$$

Hence at $\nu' \int_1^y yy_\tau \, dy \to 1/3$ the radial velocity of the cavity wall tends to zero for any arbitrary small nonzero value of y.

From the data of [7], the critical value of the reduced viscosity is estimated as $\nu' = 0.46$, which makes it possible to find the critical bubble radius, for which this effect can occur:

$$R_0^* = \frac{4\nu}{0.46\sqrt{p_\infty/\rho_0}}.$$

For water, this value is equal to $R_0^* \simeq 0.87$ μm. It should be noted that according to experimental data, it is close to the radii of microbubbles of free

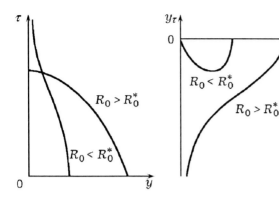

Fig. 4.5. Physical and phase (y_τ, y) planes of bubble collapse in a viscous liquid

gas that are practically always available even in thoroughly purified distilled water. The collapse on the phase plane (y_τ, y) is very clear from the perspective of this result (see Fig. 4.5): for bubbles with radii smaller than the critical value, there is an asymptotic closing and the velocity in the vicinity of the focus point tends to zero, while for $R_0 \geq R_0^*$ has unrestricted growth.

Zababakhin [9] proved that if cumulation of an empty cavity with a certain initial velocity $\dot{R}(0)$ is considered in the absence of an external pressure ($p_\infty = 0$), the equation under study is simplified and in the dimensional form transforms into

$$R\ddot{R} + \frac{3}{2}\dot{R}^2 = -\frac{4\nu \dot{R}}{R}.$$

The first integral of this equation,

$$R^{3/2}\dot{R} - R_0^{3/2}\dot{R}(0) = -8\nu(R^{1/2} - R_0^{1/2}),$$

allows an analytical estimate of the condition of unbound cumulation (asymptotic closing of the cavity in an infinitely long time). We introduce the Reynolds number

$$\text{Re}_* = \frac{R_0|\dot{R}(0)|}{\nu},$$

considering that the collapse rate is negative and assuming $\dot{R} = 0$, we obtain the condition of termination of the cumulation process by viscosity at a certain finite radius $R = R_*$:

$$\sqrt{\frac{R_*}{R_0}} = 1 - \frac{\text{Re}_*}{8}.$$

This implies that for $\text{Re}_* < 8$ the cavity stops collapsing before reaching the center, for critical value $\text{Re}_* = 8$ it is filled over an unlimited time, and for greater values it collapses with unbound velocity. If pressure is taken into account, the critical Reynolds number is equal to 8.4, which corresponds to the above result [7].

4.1.5 Spherical Cumulation in a Compressible Liquid

Let us consider the collapse of an empty cavity in a compressible liquid. This statement allows comparison of a "model" dynamics [2] with exact results [11] over a practically unlimited range of collapse velocities of the cavity wall. If the cavity is empty, $H = -p_\infty/\rho_\infty = \text{const}$, $c = c_\infty$, and the equation for spherical cavity,

$$R\left(1 - \frac{\dot{R}}{c}\right)\ddot{R} + \frac{3}{2}\left(1 - \frac{\dot{R}}{3c}\right)\dot{R}^2 = \left(1 + \frac{\dot{R}}{c}\right)H + \frac{R}{c}\left(1 - \frac{\dot{R}}{c}\right)\frac{dH}{dt},$$

becomes

$$R\left(1 - \frac{\dot{R}}{c_\infty}\right)\ddot{R} + \frac{3}{2}\left(1 - \frac{\dot{R}}{3c_\infty}\right)\dot{R}^2 = -\frac{p_\infty}{\rho_\infty}\left(1 + \frac{\dot{R}}{c_\infty}\right). \quad (*)$$

In a certain approximation one can obtain the first integral of this equation [2],

$$\left(\frac{R_0}{R}\right)^3 \simeq \left[1 + \frac{3}{2}\left(\frac{\dot{R}}{H}\right)^2\right]\left[1 - \frac{\dot{R}}{3c_\infty}\right]^4, \quad (4.1)$$

which in the vicinity of the collapse point allows self-similar solution of the type of $R = At^\alpha$ with the index $\alpha = 2/3$. The value of the self-similarity index differs noticeably from the "exact" solution obtained by Hunter [11] using the method of characteristics. His numerical studies of the complete system of hydrodynamics equations proved that near the point of collapse of an empty spherical cavity the solution is self-similar to the index $\alpha = 0.555$. However, this does not mean that in a real range of mass velocities the Kirkwood–Bethe model fails, since the considered range corresponds to the "boundary" condition $\dot{R} \gg c_\infty$.

In [2], an attempt was made to analyze possible solutions within the formal fitting of the condition on the propagation velocity of the disturbance, which eventually can be considered as an approximation of Hunter's numerical solution [11]. In a slightly changed form it is as follows:

$$\left(\frac{R_0}{R}\right)^3 \simeq 1 + \frac{3}{2}\left(\frac{\dot{R}}{H}\right)^2\left[1 - \frac{\dot{R}}{c_\infty}\right]^2. \quad (4.2)$$

Within the Kirkwood–Bethe model one can easily obtain the acoustic variant of the oscillation equation that coincides with the result of [12]. For an empty cavity, it yields the exact solution [2]:

$$\left(\frac{R_0}{R}\right)^3 = 1 + \frac{3}{2}\left(\frac{\dot{R}}{H}\right)^2\left[1 - \frac{4}{3}\frac{\dot{R}}{c_\infty}\right]. \quad (4.3)$$

Recall that for incompressible liquid the solution has the form:

$$\left(\frac{R_0}{R}\right)^3 = 1 + \frac{3}{2}\left(\frac{\dot{R}}{H}\right)^2. \quad (4.4)$$

One can see that all four integrals (4.1)–(4.4) have similar structure and differ in the power index of the square bracket, which for the incompressible variant (4.4) is zero, for acoustics (4.3) it is equal to 1, in the case of the approximation (4.2) it is 2, and for the Kirkwood–Bethe model it is 4. In Table 4.2 and Figure 4.6, Hunter's numerical results (2) for relative radius $y = R/R_0$ are compared with the above solutions (4.1–4.4) for the range of velocities $\dot{R}$ exceeding c_∞: (4.1) – (1), (4.2) – (3), (4.3) – (4), (4.4) – (5).

The analytical approximation (4.2), curve (3) provides a fairly complete reflection of the Hunter solution, including asymptotics for $R \to 0$: the self-similarity index in this candidate solution ($\alpha = 0.5714$) is rather close to the

Table 4.2. Comparison of Hunter's numerical solution for the relative radius with model dynamics over a range of collapse velocities of the cavity wall

$\dot{R}/c_\infty$	y_2	$y_{(3)}$	$y_{(1)}$	$y_{(4)}$	$y_{(5)}$
1.46	0.0148	0.0132	0.0141	0.0168	$2.40 \cdot 10^{-2}$
2.05	1.00	$9.12 \cdot 10^{-3}$	$9.56 \cdot 10^{-3}$	1.24	1.92
2.50	$7.84 \cdot 10^{-3}$	7.29	7.48	1.03	1.68
2.93	6.42	6.07	6.08	$8.90 \cdot 10^{-3}$	1.51
3.56	5.00	4.82	4.66	7.40	1.33
4.00	4.27	4.20	3.97	6.62	1.23
4.62	3.56	3.53	3.22	5.80	1.12
5.50	2.85	2.85	2.49	4.90	1.00
6.88	2.14	2.16	1.75	3.94	$8.55 \cdot 10^{-3}$
9.50	1.43	1.44	1.03	2.96	6.90
11.30	1.14	1.15	$7.66 \cdot 10^{-4}$	2.44	6.13
19.50	$5.70 \cdot 10^{-4}$	$5.70 \cdot 10^{-4}$	2.91	1.43	4.26
50.00	1.72	1.66	$4.90 \cdot 10^{-5}$	$5.62 \cdot 10^{-4}$	2.28
80.00	$9.42 \cdot 10^{-5}$	$8.90 \cdot 10^{-5}$	2.10	3.51	1.67
100.00	7.16	6.62	1.34	2.81	1.44

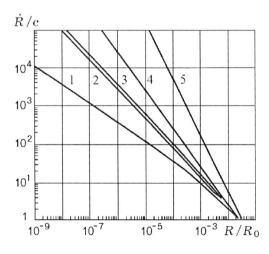

Fig. 4.6. Collapse of an empty bubble at the phase plane in an compressible liquid

exact solution shown above. In [13], a reference is given to Schneider's exact solution (1949), that is the numerical solution of the system of exact differential equations for the problem on the collapse of an empty cavity, which practically coincides with the solution of the equation (*) to the velocities $\dot{R}/c_\infty \simeq 2$. From Eq. (4.2) one can reconstruct the equation of cavity oscillation. To this end, one should differentiate it with respect to R and use the last term in the complete equation of oscillations with dH/dt to reconstruct the missing addend considering the radiation losses. The final result acquires the form

$$R\left(1 - \frac{3\dot{R}}{c} + \frac{2\dot{R}^2}{c^2}\right)\ddot{R} + \frac{3}{2}\left(1 - \frac{\dot{R}}{c}\right)^2 \dot{R}^2 = H + \frac{R}{c}\left(1 - \frac{\dot{R}}{c}\right)\frac{dH}{dt} .$$

and can be used in a wide range of oscillation velocities.

4.1.6 Oscillation Parameters

Let us consider the characteristic parameters of the dynamics of single bubbles affected by shock waves [14].

1. Incompressible liquid (spherical cavity, minimum radius, and collapse time):
   - $p_\infty = \text{const}$,

$$\left(\frac{R_0}{R_{\min}}\right)^{3(\gamma-1)} = 1 + \frac{(\gamma-1)p_\infty}{p_0} , \qquad t_* = 0.915 R_0 \sqrt{\frac{p_0}{p_\infty}} ;$$

   - $p_\infty = \mathbf{p}_{\max} \exp^{-t/\theta}$,

$$\left(\frac{R_0}{R_{\min}}\right)^{3(\gamma-1)} \simeq 1 + (\gamma-1)\frac{\mu P^2}{1 + \mu P} ,$$

$$t_{*,\theta} \simeq 0,915 R_0 \cdot \sqrt{\frac{\rho}{p_0} \frac{1 + \mu A}{\mu A^2}} ,$$

where

$$\mu = \left(\frac{\theta}{R_0 \sqrt{\rho_0/p_0}}\right)^2 .$$

For $P = p_{\max}/p_0 = 10\text{–}10^3$, $\mu \geq 0.1$, and $\gamma = 5/3, 4/3, 1.4$.

2. Compressible liquid (spherical cavity, minimum radius, $p_\infty = \text{const}$), for $p_\infty = 1\text{–}1.8 \cdot 10^3$ MPa we get

$$\frac{R_{\min}}{R_0} \simeq \frac{5}{3} \cdot 10^3 T_* + 0.025 ,$$

where t_* (in s) is specified above.

4.2 Chemical Reactions in Gas Phase

4.2.1 Todes' Kinetics, Initiation of Detonation by a Refracted Wave

The application of stoichiometric hydrogen–oxygen and acetylene–oxygen mixtures as gas in bubbles made it possible to find their capability to detonate at degrees of bubble compression that corresponded to adiabatic heating of the mixtures to the ignition temperatures [3, 5] known from publications on gas detonation.

The dynamics of a spherical cavity with reactive gas mixture is a kind of elementary act in complicated physico-chemical processes that determine the mechanisms of detonation initiation in liquid HE, the nature of large-scale explosions of liquefied-gas containers, and the phenomena of bubbly detonation. The concept of detonation in small bubbles should be somehow justified, so should be the probability of formation of a reactive mixture in the bubble due to heat and mass exchange, if the initial undisturbed bubbly system exists with physically and chemically separated phases: for example, the carrying phase of the bubbly medium is a fuel and the gas phase contains only an oxidizer.

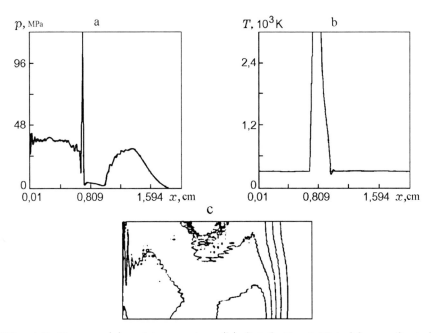

Fig. 4.7. Pressure (**a**) and temperature (**b**) distribution initiated by a refracted wave. (**c**) Density distribution in the vicinity of the bubble: $t = 10.5$ μs, $p_{sh} = 50$ MPa, $R_0 = 2$ mm

138   4 Bubble, Cumulation and Chemical Reactions

The first time the dynamics of a reactive bubble was numerically estimated with regard to the Todes chemical kinetics in [15], the following system of equations presented in the dimensionless form ($\beta = R/R_0$) was used:
The Rayleigh equation,

$$\beta \frac{d^2\beta}{d\tau^2} + \frac{3}{2}\left(\frac{d\beta}{d\tau}\right)^2 = \frac{\bar{T}}{\beta^3} - \bar{p}_\infty - \frac{4}{\beta \mathrm{Re}} \frac{d\beta}{d\tau} ; \qquad (4.5)$$

The temperature equation,

$$\frac{d\bar{T}}{d\tau} = \eta \frac{d\bar{N}}{d\tau} - \frac{3(\gamma-1)}{\beta}\bar{T}\frac{d\beta}{d\tau} ; \qquad (4.6)$$

The kinetics equation,

$$\frac{d\bar{N}}{d\tau} = \zeta \beta^{-3} \sqrt{\bar{T}} e^{-\alpha/\bar{T}} (1-\bar{N})^2 ; \qquad (4.7)$$

where

$$\tau = t\frac{\sqrt{p_0/\rho_0}}{R_0}, \quad \eta = \frac{Q}{c_m T_0}, \quad \alpha = \frac{E_a}{BT_0}, \quad \text{and} \quad \zeta = \frac{az\sqrt{T_0}}{V_0}\frac{R_0}{\sqrt{p_0/\rho_0}} . \qquad (4.8)$$

Here E_a is the activation energy; B is the gas constant; Q is the reaction heat; c_m is the heat capacity; $\bar{N} = n/a$, where a is the original concentration of the component and n is the number density of molecules formed during the reaction; z is the constant related to molecule radii r_1 and r_2 and masses m_1 and m_2 by the relationship $z = N_0(r_1+r_2)^2(8\pi B/N_0)\sqrt{1/m_1 + 1/m_2}$, where N_0 is the Avogadro constant, $\bar{T} = T/T_0$ is the mixture temperature, $\mathrm{Re} = R_0\sqrt{p_0\rho_0}/\mu$, and $\bar{p}_\infty = p_\infty/p_0$ is the external pressure.

The calculations showed that during the interaction of a bubble with a shock wave by adiabatic heating, the active mixture inside it explodes almost instantly at constant volume and a kink occurs on the dynamics curve of the radial velocity of the bubble wall. These results can be considered as a first step in understanding the "elementary" act in the phenomenon of bubble detonation, the formation of a quasi-stationary regime in active bubbly medium. The calculations made it possible to reveal one more unexpected effect: the explosion of the mixture does not stop the process of bubble collapse due to the added mass inertia, thus the pressure and temperature in detonation products continue to increase.

However, one should mention the possibility of the initiation of bubble detonation following a completely different "scenario": calculations for the axisymmetrical problem on the interaction between a strong shock wave and a bubble with acetylene–oxygen mixture made it possible to find that the shock wave refracted into the bubble can be of major importance. It initiates the detonation locally, inside the bubble in the vicinity of its wall from the side of arrival of the shock wave [16] (Fig. 4.7). In this case, the bubble does not have time to acquire radial velocity. It preserves the initial volume, while the detonation wave inside it starts propagating with a delay behind the shock wave front in the liquid: Fig. 4.7 shows the jumps of pressure (a) and temperature (b). Figure 4.7c presents the lower half of the calculation domain with density distribution: a half of the bubble whose boundary coordinates correspond to coordinates in Fig. 4.7a comes forth at the center of the top of the figure. This new unexpected result has not been studied experimentally yet.

4.2.2 Generalized Kinetics of Detonation in Gas Phase

It is believed that for excitation of a self-sustaining (multifront) detonation in a free gas volume, it is required to release a minimum amount of energy in a certain zone and the energy is defined as the critical initiation energy E_*. The process has typical spatial scales, such as the unit cell size a and the formation radius r_{form} determining a region outside of which the detonation propagates in the stationary regime.

One of the best approximate models for describing the detonation initiation in compliance with the experimental data is a multiple point initiation model. It is based on the assumption that transverse wave collisions at the detonation front initiate multifront detonation. The collision energy E_0 is estimated from the cell model proposed by Vasiliev and Nikolaev in [17] and is used for estimating E_*. In the calculations presented below for the mixture $2H_2 + O_2$, it is assumed that the induction period τ_i is completed by instant chemical reaction, since the time of chemical transformation is much less than τ_i and for nonstationary conditions is determined from the integral equality:

$$\int_0^{\tau_i} \frac{dt}{\bar{\tau}} = 1 \quad \text{and} \quad \bar{\tau} = \frac{A}{c_f^{n_1} c_{ox}^{n_2}} \exp\left(\frac{E_a}{RT}\right), \tag{4.9}$$

where c_f and c_{ox} are the fuel and the oxidizer concentrations, respectively, and n_1 and n_2 are the orders of the reaction. The mixture temperature $T(t)$ in a bubble is determined by an adiabate with different values of the index γ before ($\gamma = 1.4$) and after ($\gamma = 1.21$) the chemical reaction.

In a free volume of a specific gas mixture, the critical initiation energy for a spherical detonation and a unit cell size a are determined using the data of [17,18] and are used for calculations of the process parameters in the bubble. For the mixture $2H_2 + O_2$ at $p_0 = 0.1$ MPa and $T_0 = 298$ K ($A = 5.38 \cdot 10^{-5}$ μs·mol/l, $E_a = 17.15$ kcal/mol), $a = 1.6$ mm and $E_* \simeq 6$ J. Clearly, under these conditions the assumption about the multipoint initiation and, consequently, about the detonation regime of the reaction in gas inside the microvolume of 1 mm radius is unacceptable. Moreover, in this case the requirements for the relative characteristic sizes $R \geq r_{form} \gg a$ are not met.

However, with increasing pressure p (which occurs in a collapsing bubble) the conditions for initiating the detonation in the bubble considerably improve with respect to the characteristic scale of the process (a decreases $\sim 1/p$) and the critical initiation energy E_*, which decreases as $1/p^2$ [19]. Parameters of the basic detonation characteristics for the free volume and different initial pressure (corresponding to the pressure p_{ad} achieved during adiabatic compression of the hydrogen–oxygen mixture) are presented in Table 4.3 [17,18]. Hence, based on the data on variation of the relative parameter R/a, one can conclude on the possibility of initiating detonation in the bubble during compression [19].

This point is of principal importance for the bubble system, since the conditions required for detonation with respect to R/a and E_* can be readily realized in the bubbly liquid interacting with shock waves. Thus, the reaction character in bubbles is determined by the number of cells in the bubble microvolume and by the possibility to initiate a detonation process. In this respect the concept of self-initiation of the reaction in a mixture due to the rising temperature under the effect of external factors (e.g., adiabatic compression by an external shock wave) becomes important. Figure 4.8a shows the dynamics of the "bubble–cell" relation (R/a) that depends on the current relative degree of compression of the bubble R_0/R. The dotted line shows the data corresponding to one of the moments, when the increment of the internal energy of gaseous mixture ΔE exceeded the critical initiation energy E_* (Fig. 4.8b, logarithmic scale). The initial conditions were $R_0 = a = 1.6$ mm, $\gamma = 1.4$, and the pressure jump at infinity in the liquid surrounding the bubble was equal to $p_\infty = 10$ MPa.

The increase of R/a and sharp decrease of E_* as the bubble was compressed were evidence that, though the detonation regime was unfeasible at the initial stage, in the course of adiabatic compression the possibility of its occurrence became undoubted [19]. Furthermore, beginning from a certain compression stage, the mixture appeared to be capable of self-initiation due to growing internal energy (ΔE).

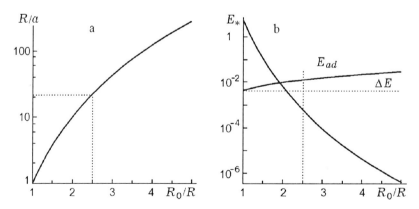

Fig. 4.8a,b. The variation dynamics of the number of cells (**a**) and the values of E_* and E_{ad} in a collapsing bubble under adiabatic compression (**b**)

4.2 Chemical Reactions in Gas Phase   141

Table 4.3. Basic detonation characteristics for the free volume

R_0/R	p_{ad}	T_{ad}	E_{ad}	p_{ch}	T_{ch}	p_*	T_d	E_*	ΔE_{ad}	a	R/a
1	0.1	298.15	$4.3 \cdot 10^{-3}$	0.96	3504.1	18.79	3681.6	5.946	0.0	1.594	1
1.4	0.41	478	$6.9 \cdot 10^{-3}$	2.57	3694.4	12.09	3888.7	0.2001	0.0026	0.375	3
2	1.827	681	$9.9 \cdot 10^{-3}$	8.5	3943.7	8.82	4160.9	0.00651	0.0056	0.0822	9.7
3	9.994	1104	$1.6 \cdot 10^{-2}$	30.9	4278.9	5.68	4522	$9 \cdot 10^{-5}$	0.0117	0.0126	42.2
4	33.37	1555	$2.2 \cdot 10^{-2}$	78.8	4582.1	4.18	4844.4	$4.1 \cdot 10^{-6}$	0.0177	0.00324	123
5	85.03	2028	$2.9 \cdot 10^{-2}$	164.1	4877.9	3.3	5155	$4.4 \cdot 10^{-7}$	0.0247	0.00115	277.2

Notice: p_{ad} – adiabatic pressure, MPa; T_{ad} – adiabatic temperature, K; E_{ad} – internal energy of mixture, J; p_{ch}, T_{ch} – parameters of explosion products at constant volume, according to [18]; $p_* = p_d/p_{ad}$, T_d – parameters for detonation wave.

4.2.3 Dynamics of Bubbles Filled with a Reactive Mixture

The bubble dynamics in the wave field with constant pressure instantly applied at infinity p_∞ is described by the Rayleigh equation, which considering the viscosity in acoustic approximation for dimensionless variables becomes

$$\beta \frac{d^2\beta}{d\tau^2} + \frac{3}{2}\left(\frac{d\beta}{d\tau}\right)^2 = \beta^{-3\gamma} - \bar{p}_\infty - \frac{4}{\beta \mathrm{Re}}\frac{d\beta}{d\tau} - 3\gamma\alpha\beta^{-3\gamma}\frac{d\beta}{d\tau}, \qquad (4.10)$$

where $\alpha = c_0^{-1}\sqrt{p_0/\rho_0}$ (zero index is assigned to initial data), ρ_0 and c_0 are the initial density and speed of sound in the liquid, $\mathrm{Re} = R_0\nu^{-1}\sqrt{p_0/\rho_0}$, and ν is the kinematic viscosity. The last addend in the right-hand part of (4.10) accounts for acoustic losses. The temperature of a mixture in a bubble at every moment is determined from the known equation of state: $pV = mRT/\mu$. In this case, the change in molar mass caused by the reaction from $\mu_0 = 12$ to $\mu_{ch} = 14.71$ is calculated assuming $m = \mathrm{const}$. The induction period τ_* can be calculated in two ways. For example, based on fulfillment of the inequality

$$\bar{E} - 1 \geq \bar{E}_*, \qquad (4.11)$$

where $\bar{E}_*$ and $\bar{E}$, respectively, are the ratios of the initiation energy and the internal energy to the initial internal energy of the gaseous mixture. The magnitude of E_* and the temperature of reaction products are calculated with the interpolation formulas derived by analyzing the calculation data (Table 4.3),

$$E_* = E^0 \beta^\delta, \qquad (4.12)$$

$$T_{ch} = A_0 + A_1\beta^{-1} + A_2\beta^{-2} + A_3\beta^{-3}, \qquad (4.13)$$

where $E^0 = 6.4746$ J, $\delta = 10.2385$, $A_0 = 2925.65947$, $A_1 = 667.16983$, $A_2 = -95.94769$, and $A_3 = 8.1264$.

The induction time τ_i is found from (4.9), while the reaction is initiated at the moment $t_{ch} = t_* + \tau_i$ and proceeds instantly. The second way is easier: since the temperature of the mixture rises instantly under the collapse, one can put $t_* = 0$ in the integral (4.9), which practically does not change the

142   4 Bubble, Cumulation and Chemical Reactions

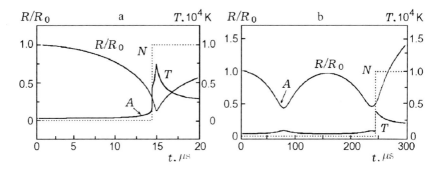

Fig. 4.9a,b. Variation of the mixture temperature T and the relative fraction of reacted molecules N ($t_* = 13.3$ μs) during the compression stage (**a**), a reaction delay to the second oscillation ($t_* = 68.75$ μs) (**b**)

characteristic explosion times. If the reaction is not initiated during the first oscillation, the integral (4.9) continues to be calculated for the next oscillation with regard to the earlier determined part of the induction period.

The calculations thus performed showed that for sufficiently high p_∞ the pressure jump in the reaction products (p) changed the value (not the sign) of the velocity of bubble collapse at the moment of explosion: the inertia of the liquid surrounding the bubble was appreciable and prevented the collapse from stopping. The gas was additionally compressed to 920 MPa at $p_{\text{ch}} = 44$ MPa and the temperature increased up to $7.3 \cdot 10^3$ K at $T_{\text{ch}} = 4.3 \cdot 10^3$ K (Fig. 9a, $p_\infty = 10$ MPa, $R_0 = 1.6$ mm). This effect determined the most probable mechanism of the wave amplification in settling of the stationary regime of bubble detonation.

The analysis of the numerical results showed that the acoustic radiation losses in this case decreased the maximum temperature of the overcompressed reaction products by about 10%.

Figure 4.9b illustrates the temporal evolution of various parameters for weak waves with the amplitude $p_\infty = 0.52$ MPa. The figure shows that the reaction was initiated at the second oscillation. In this case the explosion of the gaseous mixture inside the bubble occurred in the vicinity of its minimum radius or even at the initial expansion stage. Clearly, the maximum temperature of the reaction products was determined only by the energy release in the reaction. This fact could be very important, since the intensity of a weak stepwise shock wave with such an amplitude could be sufficient to initiate the self-sustained regime in a bubbly liquid.

4.3 Mass Exchange and Chemical Reactions

Early studies [20–23] already used the "hot spots" model for detonation initiation in liquid HE and discussed the significance of microdrops. High eplosive microdrops are formed in bubbles due to their shape instability under oscillation behind the shock wave front. This instability results in the formation of microjets and their disintegration into drops.

Alternatively in [24,25] the detonation of bubbly media was determined by heat and mass transfer, the intensity of the processes notably increased with the development of the bubble surface instability resulting in the formation of microdrops [25]. Naturally, the evaporation of microdrops strongly affects the run of a chemical reaction in the gas phase inside the bubble. Moreover, it is impossible to initiate detonation in a system where the fuel and the oxidizer are in different phases [24] without the interphase heat and mass transfer leading to the formation of a reactive mixture inside the bubbles, which can sustain the detonation regime of the reaction run in the bubble.

Evidently, for the model of bubble detonation in such media to be adequate, it should account for the effects of inert and chemically reacting admixtures on the explosive process and reaction dynamics in the process of formation of the mixture. The effects are considered below, depending on the injection time t_{inj}, the initial sizes of microdrops D_0, and the evaporated liquid mass M_L, both in the case of an instant evaporation of microdrops and in the case of a transient evaporation [26].

4.3.1 Instant Evaporation of Microdrops

Let us assume that a gas bubble oscillates in a liquid in the field of instantly applied constant external pressure. The real processes of heat and mass transfer are replaced by instant evaporation of all liquid drops of mass M_L introduced in the bubble at time t_{inj} after the beginning of the compression. The bubble dynamics is described by the Rayleigh equation, assuming that the bubble retains its spherical shape and the gas is ideal:

$$yy'' + \frac{3}{2}(y')^2 = \bar{p} - \bar{p}_\infty \ . \tag{4.14}$$

Here the prime denotes derivatives with respect to the dimensionless time τ (the notation is analogous to earlier calculations), $\bar{p}$ is introduced to account for the acoustic radiation losses of the bubble and the viscous dissipation:

$$\bar{p} = \frac{\rho_\mathrm{g} \Re T}{p_0 \mu} - \frac{4}{y} \frac{y'}{\mathrm{Re}} \tag{4.15}$$

$$+ \frac{1}{c_0} \sqrt{\frac{p_0}{\rho_\mathrm{L}}} \left(1 + \frac{\bar{p} - \bar{p}_\infty}{B_1 + \bar{p}_\infty}\right)^{-1/n} y \frac{\mathrm{d}\bar{p}_\mathrm{g}}{\mathrm{d}\tau} \ . \tag{4.16}$$

Here ρ_g is the gas density, $B_1 = B/p_0$ and n are the Tait equation constants, c_0 is the speed of sound in a liquid, $\Re$ is the universal gas constant, $\mathrm{Re} = (R_0/\nu_\mathrm{d})\sqrt{p_0\rho_0}$ is the Reynolds number, ν_d is the dynamics viscosity of the liquid, and μ is the molar mass of the gas.

It is assumed that the gas–vapor mixture is non-reactive during the induction time and it is instantaneously transformed into a chemical equilibrium afterwards. The induction period (here τ_i) is determined from

$$\int_0^{\tau_i} \frac{dt}{\bar{\tau}} = 1 , \qquad (4.17)$$

where $\bar{\tau} = (A_i/\eta) \exp(E_a/\Re T)$ is the induction period at constant parameters [27], $\eta = \rho\mu^{-1}(\nu_{H_2}\nu_{O_2})^{-1/2}$, ν_{H_2} and ν_{O_2} are the mole fractions of H_2 and O_2, respectively, and A_i and E_a are constants.

Thermodynamic parameters of the mixture are calculated by the high-accuracy kinetics model [28–30] without the adiabate of inert gas that is traditionally used for such problems. The kinetics model involves the *isentrope equation*, which with regard to

$$\frac{dT}{d\rho} = -\frac{U_\mu\mu_\rho - \Re T/\rho\mu}{U_T + U_\mu\mu_T} \quad \text{and} \quad \frac{d\rho}{d\tau} = -3\rho_{g0}\frac{y'}{y^4} , \qquad (4.18)$$

can be rewritten as

$$\frac{dT}{d\tau} = \frac{dT}{d\rho}\frac{d\rho}{d\tau} = \left(-\frac{U_\mu\mu_\rho - \Re T/\rho\mu}{U_T + U_\mu\mu_T}\right)\left(-3\rho_{g0}\frac{y'}{y^4}\right) , \qquad (4.19)$$

and *the chemical equilibrium equation*

$$\frac{\rho_g}{\mu}\frac{\left(1 - \mu/\mu_{\max}\right)^2}{\mu/\mu_{\min} - 1} \exp(E_D/\Re T) = \frac{AT^{3/4}}{4K_+}\left(1 - \exp(-\theta/T)\right)^{3/2} , \qquad (4.20)$$

where *the internal gas energy* is calculated from the formula

$$U = \left[\frac{3}{4}\left(\frac{\mu}{\mu_a}+1\right) + \frac{3}{2}\left(\frac{\mu}{\mu_a}-1\right)\frac{\theta/T}{\exp(\theta/T)-1}\right]\frac{\Re T}{\mu} + E_D\left(\frac{1}{\mu} - \frac{1}{\mu_{\min}}\right). \qquad (4.21)$$

Here T and ρ_{g0} are the gas temperature and initial density; U_μ, U_T, μ_T, and μ_ρ are derivatives with respect to the parameters mentioned in the index; E_D is the mean dissipation energy of the reaction products; μ_a, $\mu_{\min}$, and $\mu_{\max}$ are molar masses in atomic, completely dissociated and fully recombined states, respectively; θ is the effective excitation temperature of oscillating degrees of freedom; and A and K_+ are the rate constants of dissociation and recombination of generalized reaction products. In this case the values of μ_a, $\mu_{\min}$, and $\mu_{\max}$ are determined by the gas composition and remain unchanged up to the moment $t = t_{\text{inj}}$. After injection they undergo a jump, then they stay constant. The magnitude of the jump depends on the chemical composition and the mass of the evaporated liquid.

The above model is applicable to hydrogen–oxygen systems of arbitrary chemical composition (including the case of presence of inert components).

4.3 Mass Exchange and Chemical Reactions

It enables an adequate account for strong changes of the molecular mass, isentropic index, heat capacities, and for thermal effect of the chemical reaction due to recombination and dissociation, as well as the change of the fuel–oxidizer ratio in the gas phase. For example, for a cryogenic system [H_2 (gas)–O_2 (liquid)], given the appropriate initial conditions, the molecular mass of gas can vary by an order of magnitude:

$$\mu_0 = \mu_{O_2}, \quad \mu_a = \mu_{\min} = \frac{1}{2}\frac{M_0 + M_L}{M_0/\mu_{H_2} + M_L/\mu_{O_2}}, \quad (4.22)$$

$$\mu_{\max} = \frac{M_0 + M_L}{M_0/\mu_{H_2}} \quad \text{at} \quad \frac{2M_L}{\mu_{O_2}} \leq \frac{M_0}{\mu_{H_2}},$$

$$\mu_{\max} = \frac{M_0 + M_L}{M_0/2\mu_{H_2} + M_L/\mu_{O_2}} \quad \text{at} \quad \frac{2M_L}{\mu_{O_2}} > \frac{M_0}{\mu_{H_2}}.$$

The equality $2M_L/\mu_{O_2} = M_0/\mu_{H_2}$ corresponds to the evaporation of the proper amount of oxygen at which stoichiometry is established. If the left-hand side is less, the content of oxygen is insufficient; if it is greater, there is an excess of oxygen.

Analogously, for a normal system with a hydrogen–oxygen gas mixture of arbitrary chemical composition in the bubbles [$n_1 H_2 + n_2 O_2$ (gas phase) – H_2O (liquid)] one can get the following relationships:

$$\mu_0 = \frac{n_2 \mu_{O_2} + n_1 \mu_{H_2}}{n_1 + n_2}, \quad \mu_a = \mu_{\min} = \frac{M_0 + M_L}{2M_0/\mu_0 + 3M_L/\mu_{H_2O}};$$

$$\mu_{\max} = \frac{M_0 + M_L}{M_L/\mu_{H_2O} + n_1 M_0/(n_1+n_2)\mu_0} \quad \text{at} \quad 2n_2 \leq n_1;$$

$$\mu_{\max} = \frac{M_0 + M_L}{M_L/\mu_{H_2O} + (n_1/2 + n_2)M_0/(n_1+n_2)\mu_0} \quad \text{at} \quad 2n_2 > n_1.$$

It should be noted that at $\tau = \tau_i$, the gas instantly reaches chemical equilibrium, which is continuously shifted owing to the bubble dynamics.

The instantaneous change of the gas parameters under the jump can be calculated using the chemical equilibrium equation (4.20) and the condition $U_1 = U_2$, where U_1 and U_2 are the internal gas energies before and after the jump. The bubble radius, the gas density, and the parameters μ_a, $\mu_{\min}$, and $\mu_{\max}$ do not change at the instant of the jump. Upon the evaporation of the liquid mass M_L, the thermodynamic parameters and gas composition are abruptly changed, the value of the jump at fixed time and the bubble radius are calculated in three steps:

1. The gas mass M_2 in the bubble and the gas density after the evaporation are calculated using the law of mass conservation, $M_0 + M_L = M_2$ (here M_0 is the initial gas mass in the bubble).

2. The parameters μ_a, μ_{min}, and μ_{max} after the evaporation are calculated from M_2 and the known initial composition of the gas and the liquid using the known algorithm [28].
3. The gas pressure and temperature after the evaporation are determined from the equation of state and the law of energy conservation, $U_1 + U_L = U_2$, where U_1 and U_2 are internal gas energies before and after evaporation, and U_L is the internal energy of the injected liquid. If the evaporation occurs by the end of the induction period, the molecular mass of gas μ is calculated from [28], otherwise μ is calculated from (4.20).

Using the proposed model, we calculated the behavior of the bubble depending on t_{inj} and M_L in cryogenic and normal systems with and without inert diluent. The calculations were performed for the following initial parameters: $T_0 = 87$ K (in cryogenic mixture), $T_0 = 293$ K (in normal mixture), initial gas pressure in the bubble $p_0 = 1.011 \cdot 10^5$ Pa, external pressure of $p_\infty = 100\, p_0$, and $R_0 = 1.6$ mm. Other constants were taken from [30–32]. We verified the validity of the condition concerning the partial pressure of evaporated component that cannot exceed the pressure p_{sat} of the appropriate saturated vapor. The calculation showed that there were domains where the behavior of the solutions had physical sense and enabled analysis of the gas temperature variation depending on M_L for cryogenic and normal mixtures. The temperature behavior appeared to differ essentially. For example, as M_L/M_0 increased, the average gas temperature in the bubble in cryogenic system attained a maximum, while in the normal system it monotonically decreased.

It was shown in [33] that the model describing bubbly detonation based only on pure kinetics of the chemical reactions does not predict formation of a solitary wave. The temperature distribution after the process reaches the stationary regime corresponds to experimental data for the amplitude and velocity, which points to the existence of an unrealistically long "tail" in the detonation wave whose amplitude is close to that of the initiating wave.

It is obvious that a probable mechanism of the formation of a solitary wave is concerned with the intense evaporation of the liquid, which results in the decrease of the temperature and gas pressure in the tail of a bubbly detonation wave. Thus, it follows from the calculations that the evaporation of a small (comparable with the initial gas mass in the bubble) amount of liquid reduces the final gas temperature practically to the initial value. Analysis of the gas temperature dynamics has shown that the values of final (after the oscillations attenuate) temperature T_f can essentially depend on the injection time t_{inj}, if it is close to the moment of maximum compression of the bubble. If not, this dependence is insignificant. For example, the low value of T_f at $t_{inj} \geq 15$ µs in a cryogenic mixture is explained by the fact that the gas temperature decrease in the bubble due to the evaporation results in a considerable increase of the induction period, which causes the "breakdown" of the reaction when combined with the acoustic losses of the bubble.

Figure 4.10 shows the influence of inert diluent (argon) on the time variation of the gas temperature in a cryogenic system. Curve 1 illustrates the process of evaporation of liquid oxygen at $M_{L(1)} = 0.43\,M_0$ (without argon). Introduction of the equal portion of argon in the gas phase for the same ratio M_L/M_0 results in an abrupt increase of the gas temperature (curve 2). The point is that when argon is added, the value of M_0 increases. The mass of evaporated liquid M_L increases as well and the ratio of "hydrogen–oxygen" gas phase approaches the stoichiometric ratio, which results in increased energy release. When the hydrogen/oxygen ratio is analogous to the first case (curve 3), as well as under the evaporation when $M_L = M_{L(1)}$ the presence of argon increases the gas temperature (curve 4). Furthermore, the calculations showed that an increased portion of gaseous argon in the bubble always leads to increased final temperature in cryogenic system and to decreased temperature in the normal system. The pressure and the compression degree decrease in both cases, while the temperature at the first oscillation grows.

Figure 4.11 shows the time variation of the isentropic index γ depending on M_L for normal mixture at $t_{inj} = 14.5$ μs, which implies that the evaporation leads to considerable fluctuations of γ. Allowance for this factor is of principal importance and is one of the advantages of the proposed model.

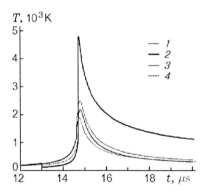

Fig. 4.10. Influence of an inert diluent on the gas temperature (cryogenic mixture)

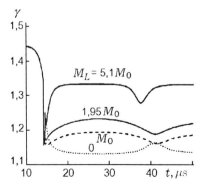

Fig. 4.11. Dynamics of the γ index (normal mixture)

4.3.2 Continuous Evaporation

In real conditions, microdrops do not evaporate instantly. Thus, at the next stage of modeling the gas-phase processes we assumed that at time t_{inj} a system of liquid microdrops of diameter D_0 and the total mass M_{L} were formed in the bubble instantaneously. At each integration step the continuous evaporation of microdrops was simulated by the instant evaporation of the mass Δm, whose value was determined from the current diameter D of the microdrop using the known equation of burning of a liquid drop in gas [34]:

$$\frac{\mathrm{d}D}{\mathrm{d}t} = -k' \frac{\text{Nu}}{4D}, \qquad (4.23)$$

where the evaporation coefficient k' was calculated from the formula

$$k' = \frac{8k_{\text{g}}}{\rho_{\text{L}} c_p} \ln\left(\frac{L + c_p \Delta T}{L}\right). \qquad (4.24)$$

Here k_{g} is the heat conductivity of combustion products, c_p is the thermal capacity at constant pressure, L is the evaporation heat, ρ_{L} is the liquid density, $\text{Nu} = hD/k_{\text{g}}$ is the Nusselt number, and h is the heat transfer coefficient. If the current gas temperature in the bubble falls below the initial temperature, the microdrops are assumed not to evaporate ($\mathrm{d}D/\mathrm{d}t = 0$).

We calculated the jump of the parameters and their successive changes using the same algorithm and under the same assumptions as for the case of instantaneous evaporation of microdrops.

Figure 4.12 shows the dynamics of the mass of microdrops for three initial diameters $D_0 = 1, 5,$ and 15 µm in a normal mixture given that $t_{\text{inj}} = 13$ µs and $M_{\text{L}}/M_0 = 0.45$. To compare the evaporation of a drop with the bubble dynamics, the change of the bubble radius $y(t)$ at $D_0 = 5$ µm is shown by the dashed line. One can see that microdrops whose size was on the order of 1 µm evaporated almost instantly (over a much shorter time period than the bubble oscillation period), while for relatively large microdrops the evaporation dynamics could be of significance. Here, as in the case of instant

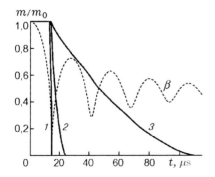

Fig. 4.12. The relative mass of microdrops versus time for D_0: (1), (2), (3)

evaporation of microbubbles, we checked the fulfillment of the condition on saturated vapor pressure. The total volume of microbubbles, which at each moment of time must be much less than the bubble volume, was monitored to avoid recalculation of the pressure in the gas phase.

Figure 4.13 shows the domain of existence of solutions for the normal mixture as the planes (M_L, D_0) for different injection times t_{inj}. It should be noted that for a cryogenic mixture the maximum value of $M_L/M_0 = 12$ in the domain of existence of solutions was achieved at instant evaporation of drops ($D_0 = 0$, vapour mass M_L is instantaneously injected) and $t_{inj} = 14.5$ µs. One can see from the figure that the cross-section of the three-dimensional space by the plane (M_L, t_{inj}) at $D_0 = 0$ was a two-dimensional domain of existence of solutions for the case of instant evaporation of the liquid. An abrupt increase of the maximum possible value of M_L/M_0 for the cross-sections $t_{inj} \geq 16$ µs was caused by the pressure increase of saturated vapors. The increase in pressure was a result of gas temperature increase due to chemical reaction and adiabatic compression of the bubble. The dependence of M_L/M_0 on D_0 for any t_{inj} had a maximum. The initial diameter of microdrops D_0 corresponding to the maximum was 1–1.5 µm at $t_{inj} < ti$, 0 µm at $ti < t_{inj} < t_{max}$, and was about 10 µm at $t_{max} < t_{inj}$, where t_{max} was the moment of maximum compression of the bubble.

For the above problem on the formation of a solitary bubble detonation wave, the data on the gas temperature T_f and the finite values μ and γ versus the injected mass of microdrops M_L/M_0 are of interest. If for the cryogenic mixture we accepted $t_{inj} = 14$ µs and $D_0 = 0.75$ µm, while for the normal mixture $t_{inj} = 14$ we assumed $D_0 = 1$ µm, the calculation gave the following results.

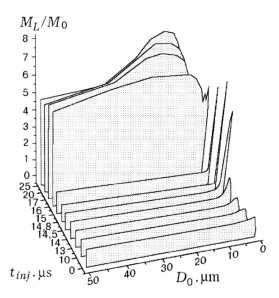

Fig. 4.13. Domain of existence of solutions for the case of continuous evaporation (normal mixture)

The increase of the mass of injected liquid in the normal mixture resulted in a monotone decrease of the final temperature T_f, which in its turn led to a slight monotone increase of the final values of μ and γ. In this case, μ reached its maximum μ_{max} already at $M_L/M_0 = 1$ and remained constant. In the case under consideration, $T_f > T_0$, however, there were certain values of the parameters t_{inj} and D_0 (for example $t_{inj} \simeq 15$ μs and $D_0 \simeq 1$ μm), at which the final gas temperature decreased practically to the initial value and the bubble detonation wave separated from the initiating wave and turned into a solitary wave.

In a cryogenic mixture, the dependences of T_f, μ, and γ on M_L/M_0 are fundamentally different. As M_L/M_0 increases, the value of T_f attains a maximum, which corresponds to the stoichiometric ratio between the fuel and the oxidant. In this case, γ reaches a minimum, while μ increases several fold.

Figure 4.14 shows the dynamics of the gas temperature and bubble radius in the vicinity of the first minimum y_{min} for normal mixture. The calculations were performed for different values of M_L/M_0 at $t_{inj} = 14.5$ μs and $D_0 = 1$ μm. Curves 2–6 present temperature dynamics at M_L/M_0 equal to 0, 0.5, 1, 1.5, and 1.95, respectively. Curve 1 is a set of the curves $y(t)$ (calculated for the specified parameters M_L/M_0) that are barely perceptible in the scales of the figure. The temperature jump at $t = 14.2$ μs was related to the beginning of the chemical reaction, while its subsequent increase was concerned with adiabatic compression of the detonation products due to the inertial properties of the "bubble–liquid" system [19], which in the absence of evaporated liquid brought the temperature of reaction products to 6 200 K (curve 2).

The injection of microdrops in this example occured at the stage of inertial compression of the bubble and clearly decreased the gas temperature T with increasing M_L/M_0. In this case, the temperature maximum was clearly defined at the moment of maximum collapse of the bubble only up to $M_L/M_0 \leq 0.5$ (the region between curves 2 and 3). For other values of M_L/M_0, the temperature decrease resulting from the intense evaporation of the considerable part of the liquid was more substantial than its growth due

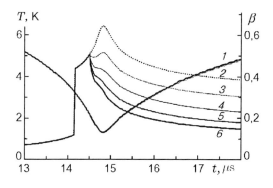

Fig. 4.14. Dynamics of the gas temperature and bubble radius during the first oscillation for different values of M_L/M_0

to the gas compression, while its maximum did not exceed the gas temperature during injection of microdrops at $t = t_{\text{inj}}$.

As D_0 increased, the intensity of evaporation decreased. Therefore, the limiting value of M_{L}/M_0 at which the temperature maximum corresponded to the maximum collapse of the bubble would increase. In the cryogenic mixture, the temperature maximum occured at maximum compression irrespective of M_{L}/M_0.

References

1. R.I. Soloukhin: *About Gas Bubble Pulsation in Incompressible Liquid*, Proceedings of the Scientific Council for Economic Use of Explosion **18**, pp. 21–25 (1961)
2. V.K. Kedrinskii: *Features of the Spherical Gas Bubble Dynamics in Liquid*, Zh. Prikl. Mekh. i Tekhn. Fiz. **8**, 3, pp. 120–125 (1967)
3. V.K. Kedrinskii: Shock-wave compression of a gas cavity in water: MSc Thesis, Lavrentyev Institute of Hydrodynamics, Leningrad Polytechnic Institute (1960)
4. V.K. Kedrinskii, R.I. Soloukhin: *Collapse of a Spherical Gas Cavity in Water by Shock Wave*, Zh. Prikl. Mekh. i Tekhn. Fiz. **2**, 1, pp.27–29 (1961)
5. R.I. Soloukhin: *About Bubble Mechanism of Shock Inflammation in Liquid*, Dokl. Akad. Nauk SSSR **136**, 2 (1960)
6. S.G. Zaitsev, R.I. Soloukhin: *To Question About an Inflammation of Adiabatic Heated Gas Mixture*, Dokl. Akad. Nauk SSSR **112**, 6 (1958)
7. H. Poritsky: The collapse or growth of a spherical bubble or cavity in a viscous fluid. In: Proc. 1st US Nat. Congress on Appl. Mech., (1952) pp. 822
8. S.S. Shu: Note on the collapse of a spherical cavity in a viscous incompressible fluid. In: Proc. 1st US Nat. Congress on Appl. Mech. (1952)
9. E.I. Zababakhin: *Filling of Bubbles in Viscous Liquid*, Appl. Math. Mech. **24**, 6 (1960)
10. A.D. Pernik: *Cavitation Problems* (Leningrad, Sudostroenie, 1966)
11. C. Hunter: *On the Collapse of an Empty Cavity in Water*, J. Fluid Mech. **8**, p. 241 (1960)
12. C. Herring: *Theory of Pulsation of the Gas Bubble Produced by an Underwater Explosion*, OSRD Rep. No. 236 (1941)
13. G. Flinn: Physics of acoustic cavitation in liquids. In: W. Mason (ed.) Physical Acoustics, vol. 1, part B, (1966) pp. 7–138
14. V.K. Kedrinskii: *Hydrodynamics of Explosion*, Zh. Prikl. Mekh. i Tekhn. Fiz. **28**, 4, pp. 23–48 (1987)
15. V.K. Kedrinskii, F.N. Zamarayev: Wave amplification in reactive bubble media (plenary lecture). In: Proc. 17th Int. Symp. on Shock Tubes and Waves, ed. by W. Yong (1989) pp. 51–62
16. V.K. Kedrinskii, Ch. Mader: Accidental detonation in bubbly liquids. In: H. Gronig (ed.) Proc. 16th Int. Symp. on Shock Tubes and Waves, (1987) pp. 371–376
17. A.A. Vasiliev, Yu.A. Nikolaev: *Closed Theoretical Model of a Detonation Shell*, Acta Astraun. **5**, pp. 983–986 (1978)

18. A.A. Vasiliev, A.I. Valishev, V.A. Vasiliev et al: *Parameters of Detonation Waves at Increased Pressure and Temperature*, Chem. Phys. **16**, 9, pp. 113–117 (1997)
19. A.A. Vasiliev, V.K. Kedrinskii, S.P. Taratuta: *Dynamics of Single Bubble with Chemically Reactive Gas*, Fiz. Goren. Vzryva **34**, 2, pp. 121–124 (1998)
20. F.P. Bowden, A.D. Yoffe: *Fast Reaction in Solid* (Butterworths Sci. Publ., London, 1958)
21. K.K. Andreev: *Some Consideration on the Mechanism of Initiation of Detonation in Explosive*, Proc. Roy. Soc. London A **246**, pp. 257–267 (1958)
22. C.H. Johansson: *The Initiation of Liquid Explosives by Shock and the Importance of Liquid Break-Up*, Proc. Roy. Soc. London A **246** (1958)
23. A.V. Pinaev, A.I. Sychev: *Discovery and Study of Self- Sustaining Regimes of Detonation in the Systems of Liquid Fuel-Bubble of Oxidizer*, Dokl. Akad. Nauk SSSR **290**, 3, pp. 611–615 (1986)
24. A.V. Pinaev, A.I. Sychev: *Structure and Properties of Detonation in the Systems "Liquid-Gas Bubbles"*, Fiz. Goren. Vzryva **22**, 3, pp. 109–118 (1986)
25. A.V. Dubovik, V.K. Bobolev: *Shock Sensitivity of Liquid Systems* (Nauka, Moscow, 1978)
26. V.K. Kedrinskii, P.A. Fomin, S.P. Taratuta: *Dynamics of Single Bubble in a Liquid at Chemical Reactions and Interface of Heat-Mass Exchange*, Zh. Prikl. Mekh. i Tekhn. Fiz. **40**, 2, pp. 119–127 (1999)
27. D.R. White: Density induction times in very lean mixture of D_2, H_2, C_2, C_2H_2 and C_2H_4 with O_2, In: XI Int. Symp. on Combustion (Academic Press, Pittsburgh, 1967) pp. 147–154.
28. Yu.A. Nikolaev, P.A. Fomin: *Approximate Kinetic Equation for Heterogenic Systems of Gas-Condensed Phase Type*, Fiz. Goren. Vzryva **19**, 6, pp. 49–58 (1983)
29. Yu.A. Nikolaev, D.V. Zak: *Coordination of Chemical Reaction Models in Gas with Second Law of Thermodynamics*, Fiz. Goren. Vzryva **24**, 4, pp. 87–90 (1988)
30. P.A. Fomin, A.V. Trotsyuk: *Approximated Calculation of Isotropy of Chemically Balanced Gas*, Fiz. Goren. Vzryva **31**, 4, pp. 59–62 (1995)
31. A.V. Trotsyuk, P. A Fomin: *Model of Bubbly Detonation*, Fiz. Goren. Vzryva **28**, 4, pp. 129–136 (1992)
32. I.S. Grigoriev, E.Z. Meilikhova (eds.): *Physical Quantities: Handbook*, (Energoatomizdat, Moscow, 1991)
33. F.N. Zamaraev, V.K. Kedrinskii, Ch. Mader: *Waves in Chemically Active Bubble Medium*, Zh. Prikl. Mekh. i Tekhn. Fiz. **31**, 2, pp. 20–26 (1990)
34. S. Lambarais, L. Kombs: *Detonation and Two-Phase Flow* (Mir, Moscow, 1966)

5 Shock Waves in Bubbly Media

5.1 Nonreactive Media, Wave Structure, Bubbly Cluster Radiation

In the 1950–70s, a long series of experimental studies was run to investigate the transformation mechanism of shock waves propagating in a liquid containing gas bubbles. As a result, a number of mathematical models were proposed to study analytically and numerically the main features of the interaction of shock waves with bubbly media [1–14]. The most interesting features of the process are due to the pressure nonequilibrium in the liquid and gas phases, the complicated character of absorption and reradiation of the energy of an incident wave by the two-phase medium [1–3]. These features are distinctly manifested both in the case of short shock waves (of arbitrary intensity) and in long waves with a steep front. If the relaxation process is relatively long, the initial volumetric concentration of the gas is low and bubble oscillations are essentially nonlinear. The term shock wave usually refers to the one whose positive phase is compared with the time of bubble collapse.

Fundamentally different effects were observed under propagation of shock waves in mixtures with a high (percentrange in tens) gas content and with short relaxation times, i.e., pressure in bubbles soon became comparable with average pressure in the liquid phase and their transport velocity to the average mass velocity of the liquid. These phenomena were also observed during the interaction of strong and long shock waves with gas bubbles, including the dissolution of bubbles. The results of experimental and numerical studies of the wave processes in such media [4–6] did not reveal the above features of the wave-medium interaction because they mainly occur under equilibrium pressure, velocity, and (sometimes) temperature of the two phases.

Some of these models should be considered from the viewpoint of the possibility of using the classical approaches to the description of complicated physical processes in bubbly media.

Lyakhov's model [5] applies the adiabatic approximation for both components of the bubbly medium on the assumption that there is no mutual penetration:

$$p_g = p_{g0}\left(\frac{\rho_g}{\rho_{go}}\right)^{\gamma}, \quad p_l = p_{l0} + \frac{\rho_l c_l^2}{n}\left[\left(\frac{\rho_l}{\rho_{l0}}\right)^n - 1\right]. \tag{5.1}$$

5 Shock Waves in Bubbly Media

If ρ_g and α_g, ρ_l and α_l are the density and the volume content of the gas and the liquid component of the liquid, respectively, and the density of the medium is defined as $\rho = \alpha_g \rho_g + \alpha_l \rho_l$, the equation of state of a bubbly system is as follows:

$$\rho = \frac{\rho_0}{\alpha_g (p/p_0)^{-1/\gamma} + \alpha_l P^{-1/n}},$$

where $P = n(p - p_0)/\rho_l c_l^2 + 1$. Using (5.1), one can easily obtain from formal considerations the expression for the speed of sound in a bubbly medium,

$$c^2 = \frac{[\alpha_g (p/p_0)^{-1/\gamma} + \alpha_l P^{-1/n}]^2}{\rho_0 [(\alpha_g/\rho_g c_g^2)(p/p_0)^{-(1+\gamma)/\gamma} + (\alpha_l/\rho_l c_l^2) P^{-(1+n)/n}]},$$

which is considerably simplified for an undisturbed medium:

$$c_0^2 = \frac{1}{\rho_0} \left(\frac{\alpha_g}{\rho_g c_g^2} + \frac{\alpha_l}{\rho_l c_l^2} \right)^{-1}.$$

The shock wave velocity U_{sh} and the mass velocity u are defined as

$$U_{\text{sh}}^2 = \frac{p - p_0}{\rho_0} \left\{ 1 - \alpha_g \left(\frac{p}{p_0} \right)^{-1/\gamma} - \alpha_l P^{-1/n} \right\}^{-1}$$

and

$$u^2 = \frac{p - p_0}{\rho_0} \left\{ 1 - \alpha_g \left(\frac{p}{p_0} \right)^{-1/\gamma} - \alpha_l P^{-1/n} \right\}.$$

One can see that c_0 decreases with increasing α_g due to the abrupt increase in the compressibility of the bubbly medium $(-(1/V)(\partial V/\partial p))$ at practically invariable density.

In the *Campbell–Pitcher* model [4], a small parameter μ (relation of the gas mass to the liquid mass) is used and it is assumed that the liquid is incompressible and the gas is subject to the Boyle law. The medium density ρ is defined as $(1 + \mu)/\rho = \mu/\rho_g + 1/\rho_l$, which results in the equation of state

$$\frac{p}{\rho} \left[1 - \frac{\rho}{(1 + \mu)\rho_l} \right] = \text{const} .$$

The speed of sound is

$$c^2 = \frac{p}{\rho} \left[1 - \frac{\rho}{(1 + \mu)\rho_l} \right]^{-1}.$$

Campbell and Pitcher [4] considered the problem on the propagation and collision of stationary shock waves in a bubbly mixture. The appropriate relations for average ρ, u, and p are written as

$$\rho_1 u_1 = \rho_2 u_2, \quad \rho_1 u_1^2 + p_1 = \rho_2 u_2^2 + p_2,$$

$$\frac{p_1}{\rho_1} + \frac{C + \mu C_{vg}}{1 + \mu} T_1 + \frac{u_1^2}{2} = \frac{p_2}{\rho_2} + \frac{C + \mu C_{vg}}{1 + \mu} T_2 + + \frac{u_2^2}{2},$$

where C and C_{vg} are the heat capacities of the liquid and the gas at constant volume. The equation of state for the mixture then acquires the form

$$\frac{p_1}{T_1}\left(\frac{1+\mu}{\rho_1}-\frac{1}{\rho_1}\right) = \frac{p_2}{T_2}\left(\frac{1+\mu}{\rho_2}-\frac{1}{\rho_1}\right).$$

Neglecting the heat capacity of the gas phase, Campbell and Pitcher obtained the relationships:

– For the temperature jump,

$$\Delta T = \frac{\bar{p}^2-1}{2\bar{p}}\frac{\mu BT_1}{mC};$$

– For the density and mass velocity jumps,

$$\bar{\rho} = \bar{p}\left(1+\frac{(\bar{p}-1)\rho_1}{(1+\mu)\rho_1}\right)^{-1} \quad \text{and} \quad \frac{u_2}{u_1} = \bar{\rho};$$

– For entropy,

$$\Delta S = \frac{\mu B}{(1+\mu)m}\left[\frac{\bar{p}^2-1}{2\bar{p}} - \ln \bar{p}\right],$$

where $\rho_2/\rho_1 = \bar{\rho}$, $p_2/p_1 = \bar{p}$, B is the gas constant, and m is the mean molecular weight of the gas. Propagation of the compression waves in water with very small (0.1–0.4 mm) bubbles at gas volumetric concentration of 10 % and higher was studied experimentally in a diaphragm shock tube. The experimental results confirmed high physical adequacy of the model with regard to the transformation of compression waves into shock waves and flattening of rarefaction waves.

Based on simple physical considerations on the character of interaction of compression waves with a bubbly medium, the following equation was proposed in [1]

$$\frac{\partial}{\partial x}(pu) = p\frac{dk}{dt},$$

which relates the distribution of the wave energy flux to the dynamics of potential energy in a unit volume of the bubbly system. This equation, together with the relation for volumetric concentration of the gas phase $k = n(4/3)\pi R^3$ and the equation of bubble oscillation, closes the system describing the waves in bubbly media for average pressure p and mass velocity u.

This simple approach was used for qualitative analysis of the transformation of a step-like wave propagating in a semi-infinite medium, at the boundary of which the pressure was taken to be a constant $p_b = 10$ MPa. The calculation results for the evolution of the pressure profile $p(t)$ are shown in Fig. 5.1 for two points placed at distances of $l = 1.35$ (a) and 7.5 cm (b) from the boundary.

Comparison with experimental data of [13] shows that the model proposed in [3] describes quite adequately the characteristic oscillating profile

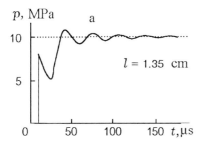 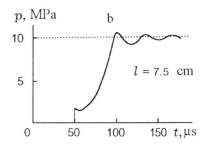

Fig. 5.1a,b. Evolution of the profile of a quasi-stationary shock wave

of the shock wave. However, comparison with the data of [1] shows that the system provides a general description of the initial transformation stage of the propagating wave.

The 1960s and 1970s were marked by increased attention to both experimental and theoretical studies of waves propagating in bubbly systems. Among the theoretical studies, we would like to mention primarily the models developed by Iordansky, Kogarko, and van Wijngaarden [7–10] for the media whose state changes dynamically under the oscillation of gas bubbles and is described by a certain subsystem considering the Rayleigh equation.

Extensive studies were performed on propagation of weak disturbances in bubbly liquids, their results were thoroughly reported by Nigmatullin [11] and Nakoryakov et al. [12]. Some of the studies dealt with mathematical analogs of the Korteweg–de Vries, Boussinesq, and Burgers models and were aimed at mapping the structure of the flows in the bubbly media.

Experimental studies were mainly focused on weak disturbances in bubbly media. As a rule, blast loads were primarily considered from the viewpoint of the possibilities of attenuating them by bubbly screens. The objectives were limited to determining the dynamics of shock wave amplitudes during their interaction with the screen, the efficiency of reducing the shock load and the possibility of control over its spectral distribution. But even in this statement, the data on the absorption of the shock wave energy were insufficient to explain even the character of the load behind the bubbly screen, not to mention the unexpected wave amplification, the considerable change of its duration, etc.

The experimental studies have shown that the damping of the incident shock wave is of secondary significance, while the mechanism of wave transformation and the reradiation of energy absorbed by the bubbly medium, as well as relaxation, dispersion, and dissipation effects accompanying the wave penetration into the bubbly medium are of primary interest. A detailed experimental study of the transformation mechanism of shock waves by a bubbly medium was completed, as well as the numerical experiment based on the mathematical models of [7–10]. The results of the studies are analyzed in this chapter. Let us start with certain features of the interaction of the gas layer with the shock wave, considering the layered structure as an extremely simplified model of a bubbly medium.

5.1.1 Shock-Gas Layer Interaction

Assume that a wave front propagates toward the x-axis and interacts with the gas layer, whose front and back walls at the initial time have the coordinates $x_0(0) = 0$ and $x_1(0) = h_0$, respectively. When a shock wave of amplitude p_* falls onto the layer, its front wall acquires the initial velocity $v_0 = 2(p_* - p_0)/\rho_0 c_0$ due to reflection and begins to move to the right, compressing the gas. The rarefaction wave propagating to the left (along the negative x-axis) defines the equation of motion of the wall that can be obtained from the invariance conditions of the function $G = x^0(w + v^2/2) = w + v^2/2$ along the characteristics propagating to the left with the velocity $c_0 - v$ [2]:

$$G_t - (c_0 - v)G_x = 0.$$

Here, w is the enthalpy and v is the mass velocity at the boundary of the layer from the side of the liquid. The equation relating the two has the form

$$\frac{dv}{dt} = -\frac{1}{c_0}\frac{dw}{dt}. \tag{5.2}$$

If the pressure in the gas is described by the adiabatic equation $p_1 = p_0[h_0/(x_1 - x_0)]^\gamma$, the wave profile is assigned as a triangle $p(t)$ with a positive phase τ, and the enthalpy can be written as $w = \rho_0^{-1}[p_1 - p(t)]$. The first integral of (5.2) will define the expression for the current velocity of the front wall of the gas layer:

$$\rho_0 c_0 \dot{x}_0 = \rho_0 c_0 v_0 - \left[p_* \frac{t}{\tau} + p_0\left(\frac{x_1 - x_0}{h_0}\right)^{-\gamma} - p_0\right]. \tag{5.3}$$

As the pressure p_1 in the gas increases, the back wall of the layer moves in the positive direction of the x-axis at a velocity $\dot{x}_1$ so that (5.3) can be written as

$$\rho_0 c_0 \dot{x}_1 = p_0\left(\frac{x_1 - x_0}{h_0}\right)^{-\gamma} - p_0. \tag{5.4}$$

A compression wave $p_1(t)$ is formed in the liquid to the right of the layer. The wave has a certain maximum amplitude that is determined by the minimum layer thickness h_* on the basis of the simultaneous solution of (5.3) and (5.4) and by the duration of the order of τ. The limiting value of the amplitude p_1 will correspond to the case when $\tau \to \infty$. With allowance for v_0, the rate of the layer thickness change $\dot{h} = \dot{x}_1 - \dot{x}_0$ can be written as

$$-\rho_0 c_0 \dot{h} = 2(p_* - p_0) - 2p_0\left[\left(\frac{h}{h_0}\right)^{-\gamma} - 1\right]. \tag{5.5}$$

It is clear that the layer will collapse until the velocities of the walls become equal, i.e., the condition $\dot{h} = 0$ is satisfied. One can see from (5.5) that at this moment the pressure in the gas becomes equal to $p_1 = p_*$, while $\dot{x}_0 = \dot{x}_1 = (p_* - p_0)/\rho_0 v c_0$, as follows from (5.3) and (5.4), and remains invariable. Thus, the stationary wave with a flat front will be recorded behind the layer. The rise time of the front, following (5.5), is close to the collapse time of an empty cavity $t_* = \rho_0 c_0 h_0/2p_*$ and depends linearly on the layer thickness h_0.

The equation for the layer in the variables $x = h/h_0$ and $t_1 = t/\tau$ for short waves has the form

$$\delta \ddot{x} = -\left(1 - \frac{t_1}{2}\right) + \frac{p_0}{p_*} x^{-\gamma} . \tag{5.6}$$

Here the dimensionless parameter $\delta = \rho_0 c_0 h_0 / 2 p_* \tau$ points to the dependence of the maximum pressure amplitude p_1 in the gas on h_0. The condition $\dot{x} = 0$ determines the relation $x_*(t_{1*})$. Calculations show that with increasing h_0 the maximum p_1 decreases and shifts toward higher values of t_{1*}, staying on the line governing the pressure decay behind the shock wave front. One can notice a typical spreading of the signal and the delay of the reradiation maximum, which depends on the time of the layer collapse only.

Analysis of the behavior of the gas layer under the interaction with shock waves provides an insight into wave transformation in bubbly media. A principal distinction of the process occurring in bubbly media from that in a gas layer consists in strong inertial and cumulative effects resulting from the collapse of spherical bubbles, as well as in transformation of the shock pulse as it passes through the bubbly medium.

It should be noted that if $\dot{x}_1 = 0$ (the layer is placed on a solid wall), the amplitude of the stationary wave is doubled after the compression of the layer, while the amplitude of the short wave increases by a factor of less than two. The inertial property of the bubble layer on the solid wall and the associated possible strong overcompression of the gas in bubbles can result in the intensification of the effect. In order to analyze the transformation of shock waves in detail we will study their structure in passing bubble layers of finite thickness.

5.1.2 Shock Waves in Bubble Layers

In analyzing the behavior of a single bubble in the shock wave field it is usually assumed that the energy intensity of the bubble, i.e., its ability to absorb an essential portion of the energy of the incident wave, is insignificant. Thus, it is assumed that the wave field in the liquid surrounding the bubble is independent from the state and dynamics of the bubble. This assumption is acceptable when studying single bubbles. But it is not valid for the case of the interaction of long or short shock waves with a group of bubbles, both due to the finiteness of the energy transferred by the wave over the duration of the interaction and the relatively high energy intensity of the bubbly area in general.

An appreciable portion of the energy transferred by the shock wave is spent for the work executed by the wave in such a system (the increase of the internal energy of gas in bubbles and the generation of radial liquid flows around bubbles during the collapse) already at low volumetric concentration of the gas. It is clear that some portion of the wave energy will not pass through the area and, consequently, the wave profile will be changed. One can increase the size of the area or volumetric concentration of gas in it, thus bringing the energy intensity of the medium to the value exceeding the energy of an incident short shock wave. Obviously, the shock pulse will not go through such a layer. We note that a similar effect is observed for long waves of constant amplitude behind the front, but only at the wave section corresponding to the time of bubble collapse in the given area.

5.1.2.1 Short Shock Waves

We define short waves as waves whose positive pressure phase is of the same duration as the time of bubble collapse. In the experiments, it corresponds to shock waves with an amplitude on the order of $p_{\max} \simeq 1$ MPa and the positive phase duration of ~ 100 μs for bubble radii of about 3–4 mm.

During the interaction with shock waves, air bubbles are compressed. This results in the increase of the internal energy of the gas and appearance of the radial velocity component, i.e., the kinetic energy of the added mass of the liquid surrounding the bubbles. As a result, an intense absorption of the energy of the incident shock wave is observed as the wave propagates in the bubbly medium. But simultaneous with the absorption of the shock wave energy, the wave is reradiated by the oscillating bubble layer. The processes occur simultaneously, but their maximum manifestations (the residue of the wave pulse after passing through a bubble layer of thickness l and the maximum reradiation of the energy of the incident wave absorbed by the layer) are essentially uncoupled in time [1,3]. This time interval is a function of the size of bubbles, and, as shown below, it also depends on their concentration in the mixture, the initial parameters of the shock wave, and the layer length. Practically, this time is determined as the time of "collective" oscillation of the bubble layer.

The process of shock wave transformation can be studied in detail only in the one-dimensional statement. For this purpose, we used an electric-discharge hydrodynamic shock tube, wherein a shock wave was produced by an explosion of a wire near the bottom (Fig. 5.2). The steady profile of the plane wave with stationary parameters was formed at a certain distance L_* from the explosion site. Beginning from $L > L_*$ pressure gauges 1 and 2 were mounted at a base of about 0.5 m from the working section to record the

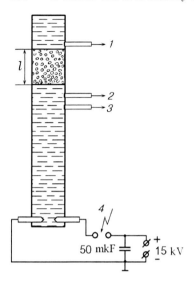

Fig. 5.2. Diagram of the setup for shock wave generation in a bubble layer: (1) and (2) pressure gauges, (3) triggering gauge, (4) ionizing high-voltage pulse

shape and parameters of the transmitted and incident waves. The pressure gauge 3 was designed to trigger an oscillograph. At a fixed distance from the gauge 2, a layer of calibrated gas bubbles made of thin rubber was fixed to a special lightweight frame, which practically did not affect the structure of the wave field.

The method of producing a two-phase layer that occupied the entire cross-section of the tube enabled modification of the gas concentration k_0 and the size of bubbles R_0, control over the length of the layer l, and provided reliable results. The upper edge of the tube was closed, if the reflection in the layer was studied at the solid wall, or a special device for wave absorption could be attached, thus the tube could be considered as infinitely long.

The passage of shock waves through the bubble layer was studied in detail for a wide range of parameters: $1 \leq l \leq 30$ cm, $0.004 \leq k_0 \leq 0.3$, $1 \leq p_* \leq 3$ MPa, and $50 \leq \tau \leq 100$ µs. Typical pressure oscillograms illustrating the initial absorption of a short shock wave by the layer are shown in Fig. 5.3 for various values of k_0 and l. The scale of signal amplification at the oscillograms with respect to the shock wave (Fig. 5.3a) incident on the layer varied within 2.5:1.

Oscillograms in Fig. 5.3b–c demonstrate the effect of splitting of the incident shock wave (Fig. 5.3a) into two waves [1]: a shock wave pulse (precursor) transformed by the layer and a reradiation of the bubble layer ($l = 1$ cm, $k_0 = 0.025$, $k_0 = 0.08$). The 3rd, 4th, and 5th oscillograms (b, c, and d, respectively) showed the same effect for a fixed concentration of bubbles ($k_0 = 0.08$) but varying layer length ($l = 1$, 2, and 3 cm). One can see that the steepness of the front of the residual shock pulse was practically preserved until it was completely absorbed, while the positive phase was abruptly reduced. The linear dimension of the pulse became comparable with

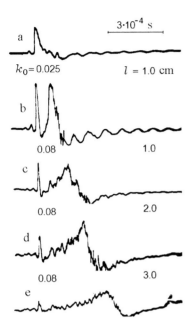

Fig. 5.3a–e. Splitting of a shock wave into a precursor and a main disturbance for different k_0 and l

the distance between the bubbles. Thus, the medium for this pulse ceased to be "continuous" and now "individual" bubbles not only absorbed the pulse but also dispersed it. The wave splitting was later recorded many times in experiments run by other authors [15].

The second wave is the reradiation of the incident wave energy absorbed by the layer. The energy dissipates during the interaction with bubbles and is transformed into the kinetic energy of radial oscillations and the internal energy of the compressed gas. Finally all the energy absorbed by the layer in the absence of other types of losses is accumulated in the compressed gas at the moment when bubble collapse terminates. Thus, the level of "collective" compression for the fixed medium and wave parameters will determine the degree of the dissipation of the incident wave energy and the reradiation amplitude, while the inertial character of the collapse process will determine the delay of the onset of the reradiation maximum.

The absorption of the wave energy and its reradiation by the bubbles of the layer are inseparably linked and simultaneous processes, whose extreme manifestations do not coincide in time in view of the sluggishness of the collapse processes. As bubbles collapse, pressure in them increases, while the pressure in the wave decreases due to the wave unsteadiness and "dissipative" processes. At a certain moment these pressures level off, and then the inertia of the liquid causes an overcompression of bubbles. Now the pressure in bubbles tends to suppress the velocity rise of the convergent radial flows around them. At the moment when the velocity of these lows achieves a maximum, the pressure in the transformed wave is minimum.

During splitting of a short shock wave, the residual pulse (precursor) and the reradiation wave, as a rule, are separated by a rarefaction region. This effect is explained by the fact that "dissipative" losses (for a rather high energy intensity of the layer) decrease pressure in the wave so quickly that it fails to compensate for the rarefaction arising in the medium due to the convergent radial flows around the bubbles. The rarefaction wave is also recorded by the gauge in front of the layer, which is particularly evident for the case of long shock waves. This interesting experimental fact is important for understanding the reflection of a shock wave from the bubble layer.

As k_0 or l increases, so does the collapse time of bubbles in the layers, while the radial velocity, the compression degree, and, consequently, the reradiation maximum decrease. It appears that there is such a value $l = l_k$ at which the layer behaves "collectively" – it completely absorbs the shock wave energy (the pressure gauges record only a precursor) and reradiates it as attenuating oscillation with their own characteristic frequency.

With further increasing of l, the process repeats, but now for the wave reradiated by the layer l_k, a new layer l_{k1} arises at which the first reradiation is completely absorbed and reradiated, and instead of it the gauge records a new precursor, etc. The maximum of the second reradiation is determined by the layer parameters and the first reradiation with a delay with respect to its precursor and is recorded with a delay relative to its precursor.

The effect of synchronous oscillation of bubbles in the layer is of principal importance, since it not only determines the distinct structure of the reradiated compression wave but also indicates the ability of the bubble system to "independent generation" of peculiar coherent structures in the medium. We note that the requirement of small size dispersion of the bubbles is not sufficient for explaining the coherence effect in this case. In the above experiment for a 3-cm layer (Fig. 5.3), the shock wave was completely absorbed, which evidenced the availability of a sharp pressure gradient along the layer that inevitably should lead to the considerable dispersion of the parameters of bubble oscillations in the layer, even in the case of equal initial size of bubbles, which was, however, not the fact.

Figure 5.4 shows the oscillograms of the wave field under the interaction of the shock wave with bubble layers (up to 30 cm long) for volumetric concentrations of the gas phase $k_0 = 0.04$, 0.06, 0.08, 0.10, and 0.15.

One can easily notice that as the layer thickness increased, wave packets with distinct periodic structure were formed with a certain regularity of the decrease of amplitudes and the constant frequency of succession. Despite the considerable difference in the medium parameters, one could trace the general character of the absorption of an initial wave, the passage of successive reradiations through the layer, and the formation of the packet. Some features of the transformation were seen by the example of oscillograms in Fig. 5.4b (the volumetric concentration of the gas phase was $k_0 = 0.06$):

5.1 Wave Field Structure 163

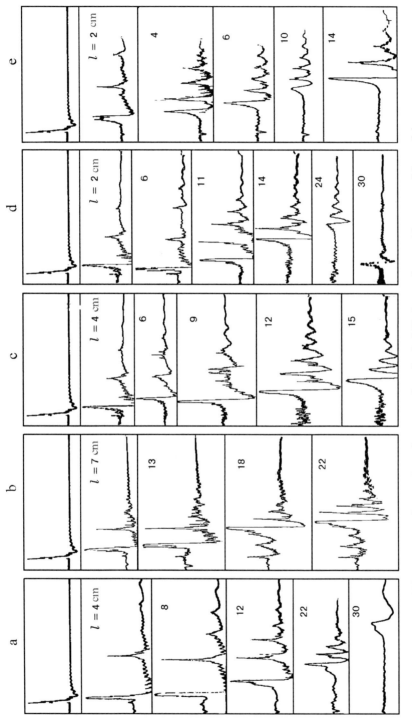

Fig. 5.4a–e. Generation of wave packets by bubble layers for a wide range of k_0 and l

164   5 Shock Waves in Bubbly Media

1. The wave energy was absorbed immediately behind the front, it changed the amplitude and the duration of the pulse and "splitted" it but did not change the front steepness.
2. The "split" pulse retained its position in time, while the maximum of the wave reradiated by the medium shifts to the right in the oscillograms (see the top-down sequence). The character of the reradiation suggested synchronous collapse of bubbles, i.e., the layer behaved as a coherent structure. The experimental data confirmed the existence of a series of precursors and indicated the distinct periodic structure of the resulting signal.
3. The length of the coherent layer l_k was limited and was determined for each concentration and initial parameters of the shock wave. Figure 5.5 shows the dependence of the delay times T_{del} on the reradiation maximum for four values of k_0: 0.04, 0.08, 0.15, and 0.30.

As the reference point for T_{del}, we took the moment of complete absorption of the initial shock wave by the layer whose length $l_{k,0}(k_0)$ was marked on the diagram as a cross-point of the dependence $T_{\text{del}}(l)$ with the l-axis. This approach made it possible to analyze the evolution of the resulting signal by the example of the propagation of an already relatively weak reradiation wave with steep front and periodic structure. The main feature of the transformation of bubbles by the bubble layer was distinctly shown by the pressure oscillograms (see Fig. 5.3): the recording moment of the first precursor corresponded to the propagation velocity c_0 in the pure liquid, the reradiation was shifted with respect to it, and the shift (Fig. 5.5) increased linearly with increasing the layer length.

The increasing delay in recording the reradiation maximum was associated with a principal question of the formation and propagation mechanism of the so-called long-wave disturbance. The point is that the shift is often used to justify the conclusion that the low-frequency component of the signal (reradiation of the layer), unlike its high-frequency part (the precursor) propagates at practically equilibrium speed of sound

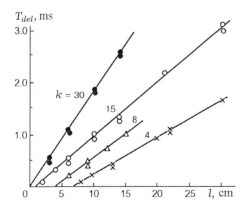

Fig. 5.5. Maximum radiation delay T_{del} of the layer as a function of its length l for various volumetric concentrations of the gas phase (in percent)

$c_*^2 = \gamma p_0/\rho_0 k_0$ [1] characteristic of a long-wave disturbance in the two-phase medium [5]. The reradiation amplitudes in the experiment were rather small, the fronts were sloping, and the estimate for c_* seemed realistic, at least for high concentrations and large l. In this case, if the delay time T_{del} was indeed related to the low *propagation velocity* of the main part of the resulting signal, the value $\zeta = c_*^2 k_0$ should be constant $(\gamma p_0/\rho_0)$ for the same gas and hydrostatic pressure and should not depend on k_0. But in the range of k_0 shown in Fig. 5.5 the value of ζ changed approximately two-fold, which raised doubts about the conclusion on the nature of the delay and pointed to a certain contradiction in the direct "wave analysis" of the experimental data.

Another contradiction arose in the numerical analysis of the process within the mathematical model of a two-phase medium [7–10] under the assumption of the incompressibility of the liquid component. In this case, $c_0 \to \infty$ and no wave structure of the "precursor–reradiation" type should exist. However, the data of [1] indicated the possibility of existence of the two-wave structure even in this case. This suggests that the contradiction consisted in the interpretation of the mechanism of the delay T_{del}. The example of the plane gas layer, as well as the analysis of the oscillograms of the wave field (Fig. 5.4) and the data presented in Fig. 5.5, showed that the value T_{del} was determined by the collapse time of bubbles in the layer or the collapse of the layer itself.

Thus, the delay in recording of the reradiation maximum was the result of the delay in its generation, rather than the low propagation velocity. In this case, both the initial shock wave and the disturbances that formed the resulting reradiation propagated in the medium with the same velocity c_0 typical of the liquid component. The total delay was the superposition of the successive delays of the wave reradiation (Fig. 5.4a), while the width of the front of the resulting reradiation was determined by the time interval from its maximum to the nearest precursor.

4. As a result of these processes, a reradiation as a wave packet was formed behind the bubble layer. The structure of the packets for different k_0 and l is shown in Fig. 5.4. The duration of the packets exceeded that of the positive phase of the incident shock wave by an order of magnitude and the frequency ν depended on the layer length l at fixed k_0.

The frequency characteristics of the wave packet depending on the layer length and gas concentration within a wide range of their values proved (Fig. 5.6) that $\nu(l)$ at $k_0 = \text{const}$ had a maximum ν_*, which with increasing k_0 was shifted to the region of lower values l and its absolute value decreased. Thus, at $k_0 = 0.04$ the packet frequency was $\nu_* = 4$ kHz at the distance of 22 cm, and at $k_0 = 0.23$ the maximum frequency decreased to $\nu_* = 1.4$ kHz while the layer length was $l = 3.5$ cm.

In principle, all frequency characteristics of the process have a distinct maximum at a certain layer length l_k. The experiments have shown that beginning from this moment ν_1 became equal to the frequency of succession of the other pulses ν_n. A kind of resonance occurred, which manifested itself in the fact that the reradiation of the "collective" layer was totally absorbed by the next layer and was radiated with the equal frequency (characteristic of a single bubble at low oscillation amplitudes) that depend on k_0 and the "upward" length of the layer l_k. At l_k we had a distinct formed wave packet with a constant succession frequency and the appropriate decline of amplitudes.

Here we should dwell on the feedback, the most important fact in the entire process of forming the packet of reradiations. Let us turn back to Fig. 5.4b ($k_0 = 0.06$). If the pressure after 14 cm of the layer were taken as initial, i.e., the boundary value for the next 8 cm (the pressure was recorded without this 8-cm addition), it would have been impossible to get the oscillograms in the form shown in the last photo. The point is that as the additional layers of the medium were involved, a redistribution of the absorbed energy near the layer interface occurred, due to which the oscillation velocity of bubbles and the reradiation frequency decreased. Naturally, the oscillations remained nonlinear, which affected the onset of the reradiation. Analysis of the experimental data made it possible to state that despite the discrete character of the formation of "collective" layers, the transformation of the waves had unconditionally continuous character.

The maximum was observed for the dependence of the frequency of the reradiation series in the packet on the concentration for any fixed value, which with increasing k_0 shifted to small values of l.

Usually the studies of shock wave propagation in a bubbly media deal with the shock wave attenuation and what remains after its absorption is considered as the acting factor. However, experimental studies of wave transformation showed that maximum loads behind the bubble layer are primarily determined by the reradiation of the bubbly medium already at relatively thin layer rather than by the passing shock wave.

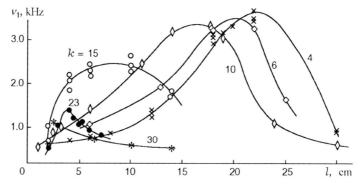

Fig. 5.6. General frequency characteristics of the reradiation of the bubble layer (k_0 is in percent)

Figures 5.7 and 5.8 confirm the validity of the above conclusions. Figure 5.7 illustrates the relative change of the shock wave amplitude during propagation: for the concentration $k_0 = 0.06$ the amplitude reduced to less 10% of the initial amplitude already at the layer length $l \simeq 3$ cm. For a denser bubbly medium, the effect was even more pronounced: for $k_0 = 0.15$ only a weak trace as a precursor remained after the shock wave passed through the 2-cm layer, while the maximum of the reradiation amplitude $p_1/p_{\max}$ was approximately $0.3\,p_*$ (Fig. 5.8). The experiments also revealed a weaker (with respect to the absorption of the shock wave amplitude) decline of the reradiation maximum with increasing thickness of the bubbly screen.

Comparison of Figs. 5.7 and 5.8 shows that the reradiation of the medium itself soon became the main acting factor in the interaction between a shock wave and a bubbly medium and the data presented in the figures reflected the exponential character of the attenuation.

Since the oscillation period of a single cavity is inversely proportional to the square root of the wave amplitude, if the pressure behind the front is constant or has a more complicated character for the waves with the pressure gradient behind the front, one can assume that the increase of the amplitude

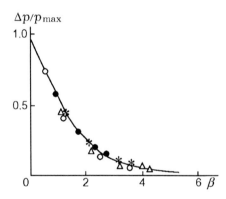

Fig. 5.7. Attenuation of the shock wave amplitude as a function of $\beta = l\sqrt{3k_0}/R_0$: (*) $k_0 = 0.004$, ($\bullet$) 0.02, ($\circ$) 0.06, ($\triangle$) 0.08

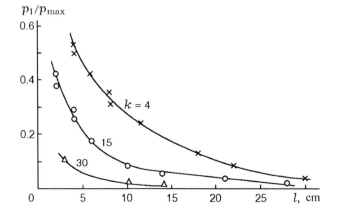

Fig. 5.8. Attenuation of the amplitude reradiated by layer

will result in the increase in the oscillation frequency, and, consequently, the frequency of the wave series in the packet. Such a check was performed for several gas concentrations at $l \simeq 6$ cm and two shock wave amplitudes of 1 and 2.5 MPa, respectively. The results summed up in Table 5.1 showed that by increasing the pressure one could increase the frequency by several times.

Using the data of Fig. 5.4e, we estimated the change of the pulse and the reradiation wave energy depending on the layer length. It appeared that at $k_0 = 0.15$ the relative impulse remained practically the same, while the relative energy per unit area changed exponentially. In this case, the concentration influenced essentially the value of the reradiation energy: at $k_0 = 0.04$ and $l = 12$ cm for the same initial shock wave (Fig. 5.4) the relative energy during reradiation was approximately 30% against 3% for $k_0 = 0.15$ and a practically identical pulse.

The increase of the amplitude of the initial shock wave for the above pressure ranges practically did not influence the relative reradiation energy: for $k_0 = 0.05$, $l = 6$ cm and $p_{\max} = 2.5$ MPa we obtained $\bar{E} \simeq 75\%$, the same was obtained for $p_{\max} = 1$ MPa.

In [1,3], on the basis of processing of the experimental data, it was shown that the amplitude p_* of the transformed shock wave for different parameters of the bubble layer changes in a similar way, if their combination $\beta = \sqrt{3k_0}\, l/R_0 = \Omega l/c_*$ is constant. Here Ω is the eigenfrequency of the bubble oscillations. It turned out that p_* decreased almost instantaneously over the layer length l and complied with the theoretical dependence $p_*(\beta)$ obtained on the assumption of incompressibility of the liquid component of the medium. Indeed, the wave front passed a 1-cm thick bubble layer in about 7 μs regardless of the concentration k_0 (see Figs. 5.3 and 5.4). Over the time the bubble radius remained practically unchanged, the velocity of the radial flows remained close to zero, and the shock wave amplitude (for example, for $k_0 = 0.08$) decreased by a factor of 2.5.

Physically, the effect is obvious and is explained by the pressure decrease in the wave front due to its interaction with the rarefaction waves

Table 5.1. Comparison of changes in shock wave frequencies with changes in volumetic concentration of the gas

k_0	$\nu_{n,1}$, kHz ($p_m = 1$ MPa)	$\nu_{n,2}$, kHz ($p_m = 2.5$ MPa)	$\nu_{n,2}/\nu_{n,1}$
0.20	1.87	2.87	1.59
0.15	2.25	3.125	1.39
0.10	2.9	3.5	1.52
0.08	1.875	3.625	1.93
0.06	1.375	3.75	2.73
0.05	1.1	3.8	3.45
0.04	0.95	3.75	3.95
0.025	0.75	3.5	4.67

from the free surfaces of gas bubbles. It is interesting to analyze the possibility of describing this effect within the assumption on invariant initial values of $k(0)$ and $\dot{k}(0)$ over the time it takes for the front to pass the thickness of the "collective" layer, which is equivalent to the concept of infinite velocity of the front propagation. Analysis of the mathematical model of the propagation of waves in bubbly media [2] confirmed that the main features of the wave transformation are associated with the variation of the function $\partial^2 k/\partial t^2$, which at the initial moment is determined by the pressure jump in the front, whereas $k \simeq k_0$, $\partial k/\partial t \simeq 0$. This and other effects will be analyzed below within the framework of two-phase mathematical models.

5.1.2.2 Long Shock Waves, Reflection From Interface

Experimental studies of long waves in bubbly media were carried out following the classical Glass's scheme: a hydrodynamic shock tube with a layer of bubbles placed on the solid wall (the tube bottom) and a shock wave produced by the rupture of a diaphragm separating a high-pressure (gas) and a low-pressure (liquid with a bubble layer) chamber. The simplicity of the arrangement was due to the fact that the amplitude of the incident wave propagating in the liquid after the rupture of the diaphragm was equal, to a high accuracy, to the initial pressure in gas. Naturally, this arrangement led to additional difficulties in analyzing pressure oscillograms, since they recorded not only an incident and reflected (from the tube edge) wave, but also a rarefaction wave resulting from the reflection of the wave system from the free surface of the liquid column near the diaphragm (Fig. 5.9, oscillograms on the left: the lower line is for the signal reflected from the bubble boundary, the upper line records pressure in the bubble layer).

The propagation of long waves, whose positive phase was much longer than the bubble collapse time, was studied to reveal the reflection mechanism of the shock wave from the front boundary of the medium. The interaction mechanism of a long wave with a bubbly medium (Fig. 5.9) was similar to

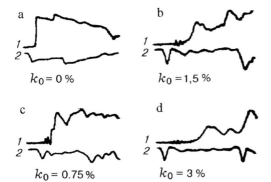

Fig. 5.9a–d. Dynamics of reflection of a shock wave from the interface with a bubble layer

the case of short waves: the wave was transformed with absorption of the energy of the head, the formation of a precursor, and reradiation.

The records of pressure by the gauge placed at a distance L ahead of the boundary of the bubbly medium (lower oscillograms) were of the most interest. They showed the development of tensile stresses and their increase with increasing k_0. Indeed, taking into account the significant difference in the acoustic impedance ρc_0 of the adjacent media at the location of the gauge, one could expect that after $t_* = 2L/c_0$, the recording of the front of the wave incident on the interface had been started and a wave reflected from the front boundary of bubbles could appear as a rarefaction wave.

Figure 5.9 shows that after a certain time the gauge ahead of the medium recorded the appearance of the rarefaction wave. However, the interval differed significantly from the time t_* required for the front to pass the distance from the gauge to the boundary and back at velocity c_0 and was equal to 180 μs instead of anticipated 20 μs.

As a result, a narrow pressure peak and a rather long-duration zone of reduced pressure were recorded in the reflected wave. It was obvious that the width of the peak was determined by the time the particles took to acquire the additional mass velocity due to the collapse of bubbles (additional as compared with the velocity of the incident compression wave behind the front). In fact, during the collapse the radial velocity of bubble walls increased, so does the average velocity of liquid particles, and when it reached $\approx 2\Delta p/\rho_0 c_0$, one could suggest that the reflection took place. No similar effects were observed for short waves. This means that their reflection (in the used scale range of recording parameters of the medium and waves) were either not allowed, or did not happen because of relatively low contribution of radial bubble oscillations in short shock waves. However, in principle, they should exist and should be dynamically related to the oscillation of bubbles.

5.1.3 Two-Phase Model of a Bubbly Liquid: Three Estimates of Wave Effects

In the early 1960s, a pressure-nonequilibrium two-phase mathematical model was published independently by Iordansky (1960) and Kogarko (1961) in Russia and by van Wijngaarden (1964) in the Netherlands to describe wave processes occurring in liquids containing gas or gas–vapor bubbles.

The Iordansky–Kogarko–van Wijndaarden (IKvanW) model has become a revolutionary achievement in the mechanics of multiphase media. It is based on a simple physical model:

a) The conservation laws are written for the averaged pressure p, density ρ, and mass velocity v;
b) The state of the medium is described by a dynamic subsystem including the Rayleigh equation for the bubbles radius R and the expression relating the average density ρ and the volumetric concentration of the gas (gas–vapor) phase k.

The basic equations of the model have the form

$$\frac{d\rho}{dt} + \rho \, \text{div} \, v = 0,$$

$$\frac{dv}{dt} + \frac{1}{\rho}\nabla p = 0,$$

$$\rho = (1 - k_b)\rho_l, \quad k_b = k_0(R/R_0)^3, \quad (5.7)$$

and

$$R\frac{d^2R}{dt^2} + \frac{3}{2}\left(\frac{dR}{dt}\right)^2 = \frac{1}{\rho_l}(p_g - p).$$

This system should be supplemented by the equation of state for liquids (Tait's equations) and gas in the bubbles: in a simplest case this can be an adiabat of the $pR^{3\gamma} = \text{const}$ type.

Within the acoustic approximation the IKvanW model was reduced to so called "pk-model" [1–3]:

$$\Delta p - c_l^{-2}\frac{\partial^2 p}{\partial t^2} = -\rho_0 k_0 \frac{\partial^2 k}{\partial t^2} \quad (5.8)$$

and

$$\frac{\partial^2 k}{\partial t^2} = \frac{3k^{1/3}}{\rho_0 R_0^2}(p_0 k^{-\gamma} - p) + \frac{1}{6k}\left(\frac{\partial k}{\partial t}\right)^2.$$

Here $k = k_b/k_0(R/R_0)^3$.

In principle, now we can neglect bubbles and consider the medium as homogeneous but having some special properties described by the above subsystem of equations of state. Let us estimate the possibility of application of pk-model (system (5.8)) to the description of wave effects observed experimentally.

5.1.3.1 Precursor

A precursor is a residual pulse of the incident shock wave that has preserved the steepness of its front and has an oscillating profile. The wave packet results from the radiation of the bubble layer absorbing the energy of the incident wave. It appears that the degree of compression of "collective" bubbles in a layer for constant parameters of the medium and shock waves determines the dissipation of the wave energy and the amplitude of the layer radiation, while the inertial character of the bubble collapse determines the delay of the reradiation maximum. In this case, the absorption of the energy of the incident wave and its reradiation by the bubble layer are continuous and concurrent processes, whose maxima do not coincide in time for the above reason.

According to experimental data, one can assume that as a precursor forms, bubbles are radially accelerated ($k \simeq 1$, $\partial k/\partial t \simeq 0$) and, consequently, the equation for the concentration k becomes

$$\frac{\partial^2 k}{\partial t^2} \simeq \frac{3}{\rho_0 R_0^2}(p_0 - p).$$

which makes it possible to obtain from (5.8) the Klein–Gordon equation that describes the formation of a precursor

$$\triangle p - c_0^{-2}\frac{\partial^2 p}{\partial t^2} = \alpha^2 (p - p_0).$$

The equation involves a *similarity parameter*

$$\alpha^2 = 3k_0/R_0^2,$$

which depends on the characteristics of the bubbly medium.

The solution of the Klein–Gordon equation has the form

$$p(x, c_0 t) - p_0 = p_{\text{bond}}(c_0 t - x) - \alpha x \int_0^{c_0 t - x} p_{\text{bond}}(\tau) \frac{I_1(\alpha\sqrt{(c_0 t - \tau)^2 - x^2})}{\sqrt{(c_0 t - \tau)^2 - x^2}} \, d\tau,$$

where $p_{\text{bond}}(t)$ is the pressure at the boundary of the bubbly medium and I_1 is the Bessel function.

Comparison with experimental data (Fig. 5.7) shows that the parameter

$$\beta = \alpha \cdot l \quad \text{or} \quad \beta = \sqrt{3k_o} \cdot \frac{l}{R_o}$$

(l is the length of the bubble layer) is a *similarity criterion of the damping amplitude of a shock wave in the bubbly medium*.

5.1.3.2 Dispersion relationship

To estimate the conformity of the IKvanW model to the real process one can use the determination of the propagation velocity of small disturbances as a criterion [3]. As an example we can refer to the dispersion phenomenon of the sound speed characteristic of multiphase media that was studied experimentally by Fox et al. [17] for bubbles of a certain size.

Let there be a discrete size distribution of bubbles for which in the system describing the state of the medium (system (5.8)) the expression for density can be given more precisely by

$$\rho = \left(\rho_0 + \frac{p - p_0}{c_0^2}\right)(1 + k_b)^{-1},$$

where $k_b = \Sigma k_{bj}$, $k_{bj} = k_{bj0}(R_j/R_{j0})^3$, and the equation for concentration is rewritten for $k_{bj}/k_{bj0} = k_j = (R_j/R_{j0})^3$.

If small disturbances of the density and the concentration of each kind of bubbles are presented as

$$\rho' = \rho - \rho_0 = \frac{p'}{c_0^2} - \rho_0 k_b \quad \text{and} \quad k_{bj} = k_{bj0}\left[1 + 3\left(\frac{R_j}{R_{j0}} - 1\right)\right],$$

the initial system (5.8) for the one-dimensional problem will become
$$c_0^{-2} p_{tt} - p_{xx} - \rho_0 \Sigma k_{bj0}(k_j)_{tt} = 0 \tag{5.9}$$

and
$$(k_j')_{tt} + \Omega_j^2 k_j' = -\frac{\Omega_j^2}{\gamma p_0} p',$$

where $\Omega_j^2 = 3\gamma p_0 / \rho_0 R_{j0}^2$ is the eigenfrequency of a bubble of the jth kind, $k_j = k_{bj}/k_{bj0}$, $k_j' = k_{bj}/k_{bj0} - 1$, $p' = p - p_0$.

We seek the solution in the form $p = A e^{i(\omega t - mx)}$ and $k_{bj} = B_j e^{i(\omega t - mx)}$. Substitution of the relations into the linearized system (5.9) after simple transformations leads to the following relation for the phase speed of sound $c_{\text{ph}} = \omega/m$ (here m is the wave number):

$$\frac{c_0^2}{c_{\text{ph}}^2} = 1 + \frac{c_0^2}{c_{\text{eq}}^2} \sum_{j=1}^{N} k_{bj0} \frac{1}{1 - (\omega^2/\Omega_j^2)},$$

where c_{eq} is the equilibrium speed of sound in the two-phase medium, according to Lyakhov [5].

For $N \to \infty$, we obtain the dispersion relation for the continuous size distribution of bubbles [1]:

$$\frac{1}{c_{\text{ph}}^2} = \frac{1}{c_0^2} + \frac{1}{c_{\text{eq}}^2} \int_0^\infty \frac{k(R) dR}{1 - (\omega/\Omega)^2}. \tag{5.10}$$

In this case,
$$\int_0^\infty k(R) \, dR = 1, \quad c_{\text{eq}}^2 = \frac{\gamma p_0}{\rho_0 k_0 (1 - k_0)}, \quad \text{and} \quad \Omega^2 = \frac{3\gamma p_0}{\rho_0 R_0^2},$$

where $k(R)$ is the concentration ratio of bubbles of radius R.

The integral in (5.10) can be transformed as

$$\int_0^\infty \frac{k(R) dR}{1 - (\omega/\Omega)^2} = \frac{1}{2} \left(\int_0^\infty \frac{k(R) dR}{1 - (\omega/a) R} + \int_0^{-\infty} \frac{k(-R_1) d(-R_1)}{1 - (\omega/a) R_1} \right),$$

on the condition that the function $k(R)$ acquires the form

$$\int_0^\infty \frac{k(R) dR}{1 - (\omega/\Omega)^2} = \frac{1}{2\omega/a} \int_{-\infty}^\infty \frac{k(R) dR}{a/\omega - R}.$$

Now we can rewrite (5.10) as

$$\frac{1}{c_{\text{ph}}^2} = \frac{1}{c_0^2} + \frac{1}{c_{\text{eq}}^2} \frac{1}{2\omega/a} \int_{-\infty}^\infty \frac{k(R) dR}{a/\omega - R},$$

where $a = \sqrt{3\gamma p_0/\rho_0}$.

We present the experimental data of Fox et al. [15] for the bubble concentration range as

$$k(R) = q\left(\frac{R}{b}\right)^2 \left(1 + \left(\frac{R}{b}\right)^4\right)^{-1},$$

where b is an arbitrary scale factor, which is selected based on some physical considerations and q is found from the obvious conditions

$$\int_{-\infty}^{\infty} k(R)\mathrm{d}R = 2 \quad \text{or} \quad \int_{-\infty}^{\infty} k(R)\mathrm{d}R = qb \int_{-\infty}^{\infty} \frac{z^2 \mathrm{d}z}{1+z^4}.$$

Since the critical points of the subintegral function are $(1+i)/\sqrt{2}$ and $(-1+i)/\sqrt{2}$, following the theorem on deductions [16],

$$2 = qb2\pi i \left[\frac{((1+i)/\sqrt{2})^2}{4((1+i)/\sqrt{2})^3} + \frac{((-1+i)/\sqrt{2})^2}{4((-1+i)/\sqrt{2})^3}\right] = \frac{qb\pi}{\sqrt{2}},$$

where $q = \frac{2\sqrt{2}}{\pi b}$. Then the integral in the expression for the phase velocity can be transformed as follows:

$$I = qb^2 \int_{-\infty}^{\infty} \frac{R^2 \mathrm{d}R}{(a/\omega - R)(b^4 + R^4)} =$$

$$= qb^2 2\pi i \left[\frac{(b(1+i)/\sqrt{2})^2}{4(b(1+i)/\sqrt{2})^3(a/\omega - b(1-i)/\sqrt{2})} + \right.$$

$$\left. + \frac{(b(-1+i)/\sqrt{2})^2}{4(b(-1+i)/\sqrt{2})^3(a/\omega - b(-1+i)/\sqrt{2})} - \frac{1}{2}\frac{(a/\omega)^2}{b^4 + (a/\omega)^4}\right] =$$

$$= \frac{\sqrt{2}}{2}\pi qb\frac{a}{\omega}\frac{(a/\omega)^2 - b^2}{(a/\omega)^4 + b^4}.$$

Thus, the dispersion relation for the continuous size distribution of bubbles becomes

$$c_{\mathrm{ph}}^{-2} = c_0^{-2} + c_{\mathrm{eq}}^{-2}\left[1 - \left(\frac{\omega}{\Omega(b)}\right)^2\right]\left[1 + \left(\frac{\omega}{\Omega(b)}\right)^4\right]^{-1}. \quad (5.11)$$

Here $\Omega(b)$ is the scale factor selected, for example, from the condition of coincidence of theoretical and experimental data at a point of the dispersion curve $c_{\mathrm{ph}} = c_0$.

Figure 5.10 presents the calculation results for the dispersion curve using (5.11) and the experimental data of [17] obtained for the following conditions: $p_0 = 0.1$ MPa, $\gamma = 1.4$, $k_0 = 0.00025$, 0.0002, and 0.00015 (curves 1–3, respectively). We note that the results of [17] suggested that the authors

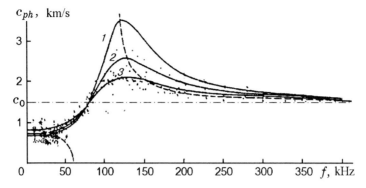

Fig. 5.10. Dispersion curves for different volumetric concentration of bubbles: (1) $k_0 = 2.5 \cdot 10^{-4}$, (2) $2 \cdot 10^{-4}$, (3) $1.5 \cdot 10^{-4}$

failed to record the expected resonance peculiarity because of the scatter in the experimental data.

The calculations were performed for the above three values of concentrations as the average for the three continuous size distributions to enable better correlation with the experiment. The value of $\Omega(b)$ was calculated at the point $c_{ph} = c_0$, which corresponded to the frequency $f = 80$ kHz. The dashed line is the dispersion curve for the bubbles with a radius $R = 0.055$ mm, which corresponded to the highest "partial" concentration $k_0 = 0.00015$ among experimental values.

One can readily see that for $f \to \infty$ the phase velocity converged towards the "frozen" value, which was higher than c_0, because the bubbles in this case "transformed" into solid particles. The evidently good correspondence of the experimental data with the model proved that the experimentally observed "acoustically transparent window" in the region of resonance frequencies was explained by the existence of the initial spectrum of bubble concentrations, i.e., their size dispersion rather than dissipative losses.

5.1.3.3 Pressure Waves, Incompressible Liquid Component

For monodisperse gas phase, application of the complete two-phase model (5.8) to the analysis of the structure of shock waves in a bubbly medium during the separation of a precursor [1] and rarefaction waves in a real liquid [19–21] showed that the model allows further essential simplifications. Some of them are physically well justified. For example, we can neglect:

a) The compressibility of the liquid component as compared to that of the gas phase, which determines this feature of the bubbly medium as a whole,
b) The nonlinear terms in the conservation laws for average values, but the nonlinear term in the pulsation equation (in the subsystem of the equations of state) should be retained.

Other simplifications are less obvious and require theoretical justifications. However, if their use provides numerical or analytical estimates, which at least do not contradict the experimental data, the appropriate approximation can be considered valid. The main approximation is associated with the introduction of a new function $\zeta = p - p_g$ (instead of pressure) and a new spatial variable $\eta = r\alpha k^{1/6}$ containing the initial parameters of the bubble system as a similarity parameter $\alpha = \sqrt{3k_0/R_0^2}$ and the current value for concentration $k^{1/6} = y^{1/2} = (R/R_0)^{1/2}$, which can be considered as an allowance for the microstructure of the medium in determining the average wave field in it.

Under these conditions the system (5.8) is transformed as (see [3]) follows:

$$\frac{d^2\zeta}{d\eta^2} + \frac{\nu}{\eta}\left(\frac{d\zeta}{d\eta}\right) - \zeta = 0, \tag{5.12}$$

and

$$\frac{\partial^2 k}{\partial t^2} = -\frac{3k^{1/3}}{\rho_1 R_0^2}\zeta + \frac{1}{6k}\left(\frac{\partial k}{\partial t}\right)^2,$$

where $\nu = 0, 1$, and 2 specifies the flow symmetry type. The major result of this approach to the description of the processes in bubbly media is that the first equation of the system (5.12) yields analytical relations between the average pressure p in the medium and the volumetric concentration k for plane ($\nu = 0$), cylindrical ($\nu = 1$), and spherical ($\nu = 2$) bubble clusters:

$$\zeta = Ae^{-\eta} + Be^{\eta} \quad (\nu = 0),$$
$$\zeta = AI_0(\eta) + BK_0(\eta) \quad (\nu = 1),$$
$$\zeta = \eta^{-1/2}[AI_{1/2}(\eta) + BK_{1/2}(\eta)] \quad (\nu = 2).$$

Here I_0, K_0, $I_{1/2}$, and $K_{1/2}$ are modified Bessel functions, and the coefficients A and B are determined from the boundary conditions for each specific problem statement.

The dynamics of the pressure field structure behind the bubble layers of finite thickness and in the semi-infinite bubble space was studied numerically within this model. At the boundary of bubbles, pressure $p_b(t)$ was set to simulate the profile of a shock wave "incident" on the boundary. The numerical analysis proved that the "incident" wave behind the layer split into a precursor and a radiation [1] in the case of incompressible liquid component as well, which confirmed that the delay in the pressure maximum was associated with the time of bubble collapse in the layer.

Pressure distribution in the bubble half-space is of interest because all disturbances should "propagate" instantaneously (in the case of incompressible liquid component). In [1], the pressure distribution in the medium was calculated for different times for the shock wave incident on the boundary and having a triangular profile with a maximum amplitude of 2 MPa. The duration of the positive pressure phase was 100 μs (at $k_0 = 0.002$ and $R_0 = 0.4$ cm).

The calculations proved that *a compression wave propagates* in the medium with a variable velocity and with an amplitude exponentially vanishing with time (Fig. 5.11).

Some time later the bubbles continued oscillating and a compression wave arose again in the initial layers, which propagated inside the layer: a periodic system of waves with a decreasing amplitude initiated in the fixed point of the space. The above examples show that the IKvanW model yields numerical results that are adequate to the experimental data on the development of wave processes in bubbly media.

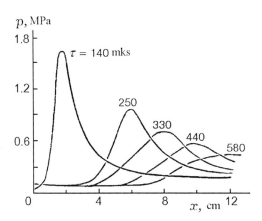

Fig. 5.11. Profile of the "traveling" compression wave in the half-space of a bubbly liquid with an incompressible liquid component

5.1.4 Amplification, Collision, and Focusing of Shock Waves

For several decades, experimental, theoretical, and numerical studies of the amplification and focusing of shock waves in various media were traditionally focused on the generation of high pressures in homogeneous and multiphase systems. As to the multiphase systems, one should mention the studies of the disintegration mechanisms of kidney stones by convergent shock waves in liquids, the initiation of explosive processes, and the problem of an acoustic "laser."

Among the studies of the disintegration mechanisms we would like to mention the surveys by Gronig and Sturtevant, experimental data and numerical models for convergent cylindrical shock waves in a homogeneous medium by Takayama, Watanabe, Nagoya, Stuka, Kuwahara, Isuzukawa, Horiuchi et al. [22–35] (1989–1993), and the studies of the role of cavitation effects in the disintegration mechanism [21].

Publication of the studies of the so-called acoustic "laser" [36, 37] dealing with the possible methods of excitation of coherent acoustic radiation and its amplification refreshed attention to the peculiar properties of bubbly media and their ability to amplify shock waves and form a system of

178   5 Shock Waves in Bubbly Media

layers with coherent properties in the medium [1,2]. The cavitation cluster arising in the heterogeneous liquid under the effect of tensile stresses and the liquid with bubbles containing explosive gas mixture can be considered as *hydrodynamic and physical analogs of pumping in laser systems*, respectively. Even under the interaction with a weak shock wave the cavitation cluster radiates a series of compression pulses with the frequency of collective oscillations of bubbles [1,2], while the interaction of waves in a bubbly liquid can result in a strong pressure pulse. These effects can underlie the principles of creation of *acoustic analogs of laser systems*. Therefore, the problem of adequate numerical modeling of such processes and the possibility of estimation of the level of amplification of the wave field as a result of such interactions are not only of fundamental but also of applied significance.

5.1.4.1 Shock Wave Amplification by a Cavitation Cluster Near the Solid Wall

Discovery of the possibility to generate high dynamic loads near a solid wall [1] by an oscillating cavitation (bubble) cluster allowed observation of unexpected effects. Under sudden recovery of the hydrostatic pressure on the boundary of the cavitation cluster, the cumulative character of the bubble collapse and strong inertial effects lead to the generation of a series of strong pressure pulses on the wall surface [1].

This effect was found in numerical calculations of the layer dynamics at a wall carried out within the model (5.12). Let us consider the statement of the problem in detail. At a solid wall there is a layer of thickness l formed by uniformly distributed cavitation bubbles with an initial radius R_0 and volumetric concentration k_0 formed as a result of pressure drop from the hydrostatic value p_{hs} to p_0, for example, due to the emergence of the wall from the liquid. At the time $t = 0$ hydrostatic pressure p_{hs} is instantaneously "restored" and is kept constant on the layer boundary. The problem is solved with the following boundary conditions:

$$\zeta = p_{hs} - p_0 k^{-\gamma} \quad \text{at} \quad x = l \quad \left(\beta = \sqrt{\frac{3k_0}{R_0^2}} l\right)$$

and

$$\frac{\partial \zeta}{\partial \eta} = 0, \quad \text{at} \quad x = 0,$$

where the latter is due to the symmetry.

The calculations were performed for a wide range of rarefaction waves in the layer $p_{hs}/p_0 = 300, 200, 100$, and 70 and for $\beta = 1, 2, 3$, and 4. The characteristic pressure profile at the solid wave is shown in Fig. 5.12. One can see that in spite of the low pressure at the boundary of the cavitation zone ($p_{hs} = 0.1$ MPa), a series of strong pressure pulses with amplitudes of several megapascals arose on the wall. This was confirmed by the conclusions of Manen [20] that the collapse of cavitation bubbles on the plane results not only in the erosion damage of the surface but also in intense pressure pulses generated by a bubbly medium on the wall surface.

Calculations (Fig. 5.12) showed that for the change in the amplitude of hydrostatic pressure below 0.1 MPa over a range of initial pressures in a bubbly layer $p_0 = (0.33\text{--}10) \cdot 10^3$ Pa and the range of the similarity parameter $\beta = 1\text{--}6$, the loads on the wall varied from $10 p_{\text{hs}}$ to $80 p_{\text{hs}}$ and had a maximum at $\beta \simeq 3$. In this case, an effect of acoustic dissipation was observed in the two-phase medium with an incompressible liquid component (Fig. 5.12), i.e., the attenuation of the radiation pulse amplitudes was a result of the acoustic losses, bubble oscillations in the layer.

The cause of this effect is obvious: the bubbly medium preserved its main properties, abnormal compressibility, and nonlinearity. Comparison of the above characteristics for water and bubbly medium supports this conclusion:

– The equations of state:

$$\frac{p_{\text{liq}} + B}{\rho_{\text{liq}}^n} = \text{const}, \quad \rho_{\text{bl}} = \left(\rho_{\text{liq},0} + \frac{p - p_0}{c_{\text{liq}}^2} \right)(1 - k)$$

– The speed of sound:

$$c_{\text{liq}}^2 = \frac{n(p + B)}{\rho_{\text{liq}}}, \quad \frac{1}{c_{\text{bl}}^2} = \frac{1 - k}{c_{\text{liq}}^2} + \frac{k \rho_{\text{liq}}}{\gamma p} \simeq \frac{k \rho_{\text{liq}}}{\gamma p}$$

– Compressibility $\varepsilon = -\frac{1}{V}\frac{dV}{dp}$:

$$\varepsilon_s^{\text{liq}} = \frac{1}{n(p + B)} = \frac{1}{\rho_{\text{liq}} c_{\text{liq}}^2}, \quad \varepsilon_s^{\text{bl}} = \frac{k}{\gamma p} + \frac{1 - k}{\rho_{\text{liq}} c_{\text{liq}}^2} \simeq \frac{1}{\rho_{\text{liq}} c_{\text{bl}}^2}$$

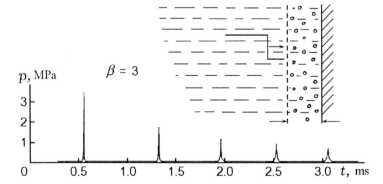

Fig. 5.12. The effect of shock wave amplification during the collapse of a cavitation layer on the solid wall (at $p_{\text{hs}}/p_0 = 100$)

180   5 Shock Waves in Bubbly Media

- Nonlinearity ($\frac{B}{A} = \frac{\partial(1/\varepsilon)}{\partial p}|_s - 1$):

$$\frac{B}{A}|_{bl} = \gamma n \frac{nk(p+B)^2 + \gamma(1+k)p^2}{[nk(p+B) + \gamma(1-k)p]^2} - 1,$$

$$\frac{B}{A}|_{gas} = \gamma - 1, \quad \frac{B}{A}|_{liq} = n - 1, \quad \frac{B}{A}|_{bl} \simeq \frac{\gamma}{k} - 1,$$

here $k = \frac{k_0}{1-k_0}\rho_{liq}\beta^3$, $\beta = \frac{R}{R_0}$ and $B = 321.4$ MPa, $n = 7$, following Ridah (see Chap. 1, [3]).

In conclusion one can note the experimental results of Reisman et al. [38] on the dynamics and acoustics of cloud cavitation formed by the periodic breakup and collapse of a sheet or vortex cavity. It was found that several types of shock waves are formed in a collapsing cloud which were associated, in particular, with the coherent collapse of a well-defined and separate cloud when it is convected into a region of higher pressure. This type of global structure causes the largest impulsive pressures and radiated noise. It was concluded that the ubiquity and severity of these propagating shock wave structures provides a new perspective on the mechanisms reponsible for noise and damage in cavitating flows involving clouds of bubbles. Besides it would appear that shock wave dynamics rather than the collapse dynamics of single bubbles determine the damage and noise in many cavitating flows.

5.1.4.2 Collision of Stationary Shock Waves. Statement of the Problem

Here we consider a general statement of the problem on the interaction of shock waves that allows solving a wider spectrum of interaction problem [39].

Let us assume that at the time $t = 0$, pressure p_b increases abruptly and is kept constant at two opposite boundaries placed at a distance L from each other. At $t > 0$ shock waves start moving towards each other in the bubbly medium and separate into precursors and main disturbances in the form of shock waves with an oscillating front.

The two-phase mathematical model that enables the description of the formation, propagation, and interaction of waves involves:

- The laws of conservation for average p, ρ, and u:

$$\frac{\partial \rho}{\partial t} = -\rho^2 \frac{\partial u}{\partial s} \quad \text{and} \quad \frac{\partial u}{\partial t} = -\frac{\partial p}{\partial s}$$

- The Rayleigh equation:

$$\frac{\partial \beta}{\partial t} = S \quad \text{and} \quad \beta \frac{\partial S}{\partial t} + \frac{3}{2}S^2 = C_1 \frac{T}{\beta^3} - \frac{C_2}{\beta} - C_3 \frac{S}{\beta} - p$$

- The equation for temperature:
$$\frac{\partial T}{\partial t} = \delta(\gamma - 1)\mathrm{Nu}\frac{\beta^3(1-T)}{T} - 3(\gamma-1)\frac{TS}{\beta} \qquad (5.13)$$

- The parameters and the equation of state of a liquid:
$$p = 1 + \frac{\rho_0 c_0^2}{n p_0}\left[\left(\frac{\rho}{1-k}\right)^n - 1\right], \quad k = \frac{k_0}{1-k_0}\rho\beta^3,$$

$$\mathrm{Pe} = C_4(\gamma-1)\frac{\beta|S|}{|1-T|}, \quad \mathrm{Nu} = \begin{cases} \sqrt{\mathrm{Pe}}, & \mathrm{Pe} > 100, \\ 10, & \mathrm{Pe} \le 100. \end{cases}$$

$$C_1 = \frac{\rho_{g0} T_0 B}{p_0 M}, \quad C_2 = \frac{2\sigma}{R_0 p_0}, \quad C_3 = \frac{4\mu}{R_0 \sqrt{p_0 \rho_0}}$$

and

$$C_4 = \frac{12 R_0 \sqrt{p_0/\rho_0}}{\nu}.$$

Here s is the Lagrangian variable. The coefficient δ is used in the equation for energy to account for the heat exchange and to correlate its intensity. The necessity of the latter is associated with the approximate nature of the relations used in the system for the Nusselt and Peclet numbers proposed by Nigmatullin, which is quite applicable to weak waves.

Since the velocities of main disturbances are much lower than those of the precursors, the latter collide first in the flow symmetry plane, reflect from it, and interact with the main waves, somewhat distorting their profile before the collision.

The dynamics of the collision of stationary waves will be considered below by the example of reactive bubbly media. Here we note only that the amplification of the wave amplitude under the collision is explained by the inertial properties of the "liquid–bubble" system, the abnormal compressibility of the bubbly medium, and the dynamic character of the collision. The data on the amplification of shock waves under collision (the analog of reflection from a solid wall in the case of the equal amplitudes) in the bubbly medium are presented in Fig. 5.13 and Table 5.2 as the dependencies of p_ref on k_0 (the amplitude of the incident wave is 1 MPa, $R_0 = 0.1$ cm). The calculations were performed for various L sufficient to shape a stationary wave structure before the reflection. The scatter in the data was due to the differences in the phases of interaction of the reflected precursors with the main incident wave.

One can see that the volumetric concentration was crucial for the amplification: the increase in k_0 resulted in the monotonic growth of the maximum pressure in the collision plane (on a "solid wall"), which can be approximated by the following dependence (Fig. 5.13, dotted curve):

$$p_\mathrm{ref}/p_\mathrm{b} \simeq 2 + 24.5 k_0^{1/4}.$$

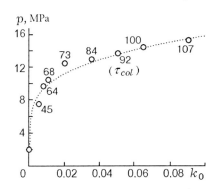

Fig. 5.13. Shock wave amplification during collision: the amplitudes of incident waves $p_{1,r} = 1$ MPa; collision time τ_{col}, μs

Table 5.2. Pressure of the reflected wave as a function of volumetric concentration

k_0, %	p_{ref}, MPa		
	$L = 10$ cm	$L = 8$ cm	$L = 7$ cm
0	2	–	–
0.5	7.5	–	–
0.75	9.7	–	–
1.0	10.7	10.5	–
2.0	12.4	13.0	–
3.5	12.8	–	–
5.0	13.7	–	–
6.5	14.4	–	–
9.0	–	–	15.3

The collision of stationary shock waves in a bubbly medium is an essentially dynamic process that takes a certain time τ_{col} (shown in Fig. 5.13 in microseconds). The higher the concentration, the longer the process of pressure growth in the collision plane, which is apparently determined by the steepness of the incident wave front.

5.1.4.3 Focusing of Shock Waves (Cylindrical Symmetry)

Collision effects are naturally expected to be significantly amplified by shock wave focusing. This problem was not considered before for bubbly media, though it is obvious that the features of the formation of the fine structure of the wave field are essentially the same, while the focusing of the precursor, its reflection from the symmetry axis, and interaction with the main incident wave can complicate the general pattern of the process.

The system of governing equations for describing the cylindrical focusing of waves in a two-phase medium consists of the subsystem of gas dynamic equations for average p, ρ, and u:

$$\frac{\partial u}{\partial t} = -\frac{1}{\rho_0}\left(\frac{x(r,t)}{r}\right)^2 \frac{\partial p}{\partial r},$$

$$\frac{\partial x}{\partial t} = u, \qquad (5.14)$$

$$\frac{1}{\rho} = \frac{1}{\rho_0}\left(\frac{x(r,t)}{r}\right)^2 \frac{\partial x}{\partial r},$$

and the subsystem of kinetic equations presented in the system (5.13).

Figure 5.14 presents the numerical results for the focusing of a stationary shock wave with a front steepness of 1.5 μs in a one-phase liquid. By analogy with the reflection, one can expect that, as compared to a pure liquid in which the shock wave amplitude increases by a factor of 20, the focusing in a bubbly system will yield a much stronger effect. The converging one-dimensional cylindrical shock waves were calculated using the system (5.14) for a bubbly medium at similar initial and boundary conditions mentioned in the previous problem (focusing radius $r = 5$ cm).

The entire pattern of focusing on the shock wave axis in the bubbly medium is shown in Fig. 5.15. Here we should first note the focusing of a precursor (with time interval of 20–40 μs) has a reflection that will naturally affect the profile of the main wave reaching the focus by about 100 μs. It is also noteworthy that the wave amplitude in the focus will increase by an order of magnitude in bubbly media as compared to pure liquids.

Analysis of the calculation results for a wide range of k_0 showed that the pressure dynamics in the wave front $p_{\max}(r)$ as the wave focused on the axis is approximated by the Lorenz profile (L):

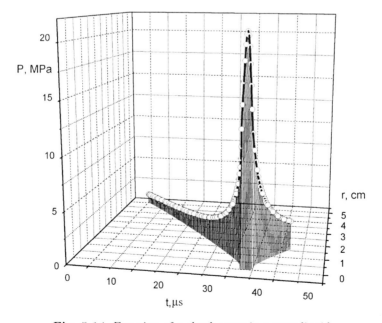

Fig. 5.14. Focusing of a shock wave in a pure liquid

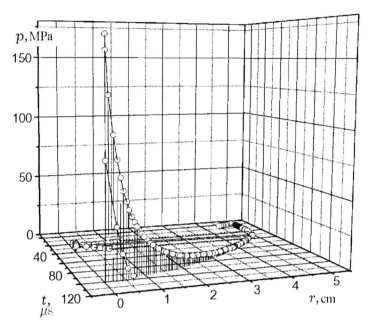

Fig. 5.15. Shock wave amplification under focusing: $k_0 = 0.01$, $R_0 = 0.2$ cm, $p_{sh} = 1$ MPa, and $r = 5$ cm

$$p_{max}(r) = \frac{2ab/\pi}{b^2 + 4(r-c)^2},$$

whose parameters for different values of k_0 are summarized as follows:

k_0	0.01	0.02	0.03	0.05	0.07
a	4 192	7 191	8 565	10 069	10 242
b	0.795	0.435	0.305	0.194	0.141
c	−0.413	−0.319	−0.245	−0.202	−0.171

A typical example of the dynamics of focusing a shock wave amplitude is shown in Fig. 5.16 for the volumetric concentration $k_0 = 0.03$. The dotted line corresponds to the Lorenzian approximation. The agreement was satisfactory. It follows from the example that the wave amplification occurred in the vicinity of the focus.

Figure 5.17 shows the effect of concentration k_0 on the maximum pressure in the focus (on the boundary $p_{sh} = 1$ MPa) in nonreactive bubbly media. One can notice the explicit oscillating character of the calculation resulted (dots) due to the effect of reflected precursors. The approximation of this dependence (solid line) can be presented as a second-order polynomial,

$$p = p_{foc}/p_{sh} \simeq 18.6 + 17\,18 k_0 - 95\,52 k_0^2,$$

describing the focusing result with a sufficient accuracy.

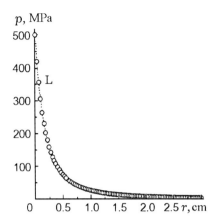

Fig. 5.16. Dynamics of the shock wave amplitude under focusing: $k_0 = 0.03$, $R_0 = 0.1$ cm, $p_b = 1$ MPa, $r = 5$ cm.

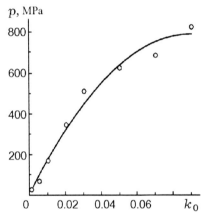

Fig. 5.17. Pressure distribution in the focus versus k_0

5.2 Generation of Radiation by Free Bubble Systems

The problems of high-pressure pulses in liquids and gases have been attracting much attention for a long time. Often these problems are not limited by the requirements of energy cumulation, but involve the problem of directional radiation. Among numerous publications on this subject we wish to single out the studies of one-dimensional shock waves in axial symmetry, in which experimental and theoretical aspects are associated with circular sources of shock waves and conical flows. The axial focusing of a pressure wave (generated by a high-voltage pulse on the surface of a toroidal discharger) in air was studied experimentally in [40]. Theoretical studies within the framework of the Chester–Chisnell–Witham (CCW) model [41–43] for gas showed that the effects of unbound cumulation also occur during focusing of circular and conical shock waves [44–46]. In [46], it was noted that the velocity of the Mach wave (neglecting the dissipative processes) increases with respect to the incident wave by a factor of α^{-1}, where α is the cone half-angle.

186   5 Shock Waves in Bubbly Media

Experimental and theoretical analysis of the topological transformation of the nonuniform shock wave front under the front cumulation was done in [47]. Nonregular reflection of a circular shock wave in gas from the symmetry axis and the solid wall was studied experimentally in [48–50]. Barkhudarov et al. [50] noticed the formation of a quasi-spherical convergent shock wave, which resulted in the amplification of the cumulative effect in the local region. The nonregular reflection of an axisymmetrical shock wave (generated in the atmosphere) from a liquid surface was observed in the experiments on modeling of a surface pointwise explosion [51]. In this study, the peculiarities of the wave field structure were considered at underwater high-voltage explosion of a circular conductor generating a toroidal shock wave in water. Upon focusing the wave interacted with the expanding toroidal cavity containing the explosion products. A bubbly cavitation zone was formed behind the front of the resulting rarefaction wave in the center of the torus.

The structure of the pressure field formed under the explosion of circular and spiral DC charges in air and water was studied in [51–54]. The numerical results for the focusing of a toroidal shock wave for different Mach numbers and geometrical parameters of the scheme were presented in [55]. However, these studies were concerned with continuous one-phase media only. Bubbly media can also be effective, because as we showed, they can amplify shock waves by orders of magnitude. Below we will consider the so-called free bubbly systems in two statements, which amplify the outer disturbances and enable certain concentration of the radiation in a given direction [56, 57].

5.2.1 Toroidal Bubble Cloud, Mach Disks

5.2.1.1 Statement of the problem

At a moment $t = 0$, a piston generates a pressure jump at the butt end of a cylindrical shock tube of radius r_{st} filled with water. The shock tube contains a toroidal bubble cluster, whose center is placed on the z-axis of the shock tube at a distance l_{cl} from its left boundary. The plane of the base circle of the torus (further the torus plane) of radius R_{tor} ($R_{tor} < r_{st}$) is perpendicular to the shock tube axis, the torus section radius is R_{circ} (Fig. 5.18).

The volumetric concentration of the gas phase in the cluster is k_0. Gas bubbles at the initial moment of time have the same radius R_b and are uniformly distributed in the toroidal cluster. At $t > 0$ the external shock wave propagating along the positive z-axis interacts with the toroidal bubble cloud, envelops it, and refracts into the cluster in the zone of contact with the front. The refracted wave propagates inside the cluster and is transformed by the bubble system. The difference in the propagation velocities of the external shock wave in the liquid (1.5 km/s) and in the bubbly medium (hundreds of meters per second) leads to the formation of a shock wave with a concave front inside the torus. The focusing results in the amplification of the wave

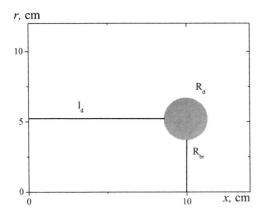

Fig. 5.18. Layout of a toroidal bubble cluster (the torus radius is hatched, z is the symmetry axis)

inside the cluster to the level governed by the system parameters and the radius of the torus section R_{circ}. The shock wave amplified by the cluster is radiated in the ambient liquid as a toroidal quasi-stationary shock wave with oscillating profile. The transformation and focusing of the refracted wave in the cluster were calculated using the modified IKvanW model [58], which represents a physically heterogeneous medium as a uniform medium with peculiar properties described by the Rayleigh equation.

The solution region has rectangular form where $0 \le z \le z_{\max}$ and $0 \le r \le r_{\text{st}}$ in cylindrical coordinates. The boundary conditions on the plane $(z = 0)$ describe a stationary shock wave with the amplitude P_{sh} by specifying the axial velocity component on the assumption of zero radial component. On the axis, $r = 0$ and the symmetry conditions are specified. The calculations were performed for $k_0 = 0.001$–0.1, $R_0 = 0.01$–0.4 cm, and the values of the amplitudes of the shock wave interacting with the cluster $P_{\text{sh}} = 3$–10 MPa. The condition set on the boundary $r = r_{\max}$ (shock tube wall) excluded shock wave reflection. The exit of the wave from the region (at $z = z_{\max}$) was governed by the equality to zero of the second derivatives of all functions in the axial direction. To solve the gas dynamic system of equations an explicit scheme with directed differences and a split scheme described in [59,60] were adapted to the problem. To calculate the subsystem for the state of the bubbly medium, the implicit fourth-order Runge–Kutta–Merson scheme was used.

5.2.1.2 Dynamics of Wave Field Structure

A typical pattern of isobars for different times of focusing of a shock wave generated by a toroidal bubble cloud is shown in Fig. 5.19a–c. The value of pressure can be estimated from the scale of distribution of the relative pressure p (in the units of hydrostatic pressure $p_0 = 0.1$ MPa, Fig. 5.19c).

Each moment is presented by two photos: a general view (on the right) and a close-up (on the left) of the region limited by the torus plane $z = 10$ cm, its

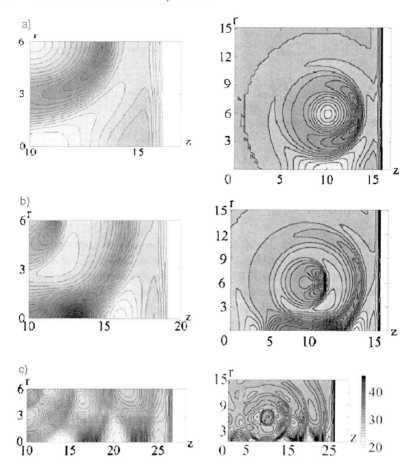

Fig. 5.19a–c. Pressure fields as a system of isobars presented for the times $t = 110$ μs (**a**), $t = 130$ μs (**b**), and $t = 180$ μs (**c**) (two shots for each, the reflection zone scale is magnified in the left shots)

radius R_{tor}, and the front of the shock wave interacting with the torus (in the right photos). The calculations were performed for the following parameters: $p_{\text{sh}} = 3$ MPa, $r_{\text{st}} = 20$ cm, $z_{\text{max}} = 40$ cm, $l_{\text{cl}} = 10$ cm, $R_{\text{tor}} = 6$ cm, $R_{\text{circ}} = 1$ cm, $k_0 = 0.01$, and $R_b = 0.1$ cm.

Figure 5.19a shows the wave field before axial focusing of a shock wave radiated by the torus. It should be noted that the pressure distribution along the front was essentially nonuniform and the axial reflection of the precursors running ahead of the wave was already nonregular. Naturally, axial reflection of the toroidal wave became irregular at the initial stage already (Fig. 5.19b), which was clearly demonstrated by the form of the first isobar of the incident wave. In the same photo, there is the second maximum of the oscillating shock wave generated by the cluster. The third pair of photos (Fig. 5.19c) demonstrates the formation of two Mach discs on the axis. The complicated wave structure, including the system of rarefaction waves, the precursors and the fading sequence of pressure peaks of the oscillating shock wave, slightly "obscured" the detailed pattern of the region of irregular reflection and requires higher resolution.

The results are presented in Fig. 5.20a–d for four values of the volume content of the gas phase and the times corresponding to the appearance of the irregular reflection at fixed values of other parameters. One can see that for this type of waves the width of the zone of axisymmetrical irregular reflection (the Mach disc) was finite and was equal to 4–5 cm. A high-pressure zone limited by the system of closed isobars, which can be defined as the Mach disc core, was distinctly visible in the disc.

In Figure 5.20a–c, the core center was slightly shifted along the axis to the right with respect to the coordinate $z = 15$ cm; in Fig. 5.20d the core zone along the axis was placed approximately in the range $z = 12$–15 cm. The isobars ahead of the core belonged to the fore front of the appropriate maximum of the focusing shock wave; behind it, the isobars of its trailing edge that represent the structure of the reflected wave. A distinctive feature of the reflection process was that due to the oscillation structure of the focusing wave, its subsequent maxima with the system of isobars with increasing pressure interacted with the reflected wave and pressed it to the axis. This suggests that this effect resulted in the appearance of the zone of relative rarefaction behind the Mach disc (see Fig. 5.19c). The numerical results for the growth of

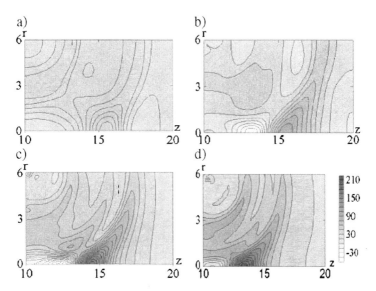

Fig. 5.20a–d. The structure of isobar distribution in the Mach discs for volumetric concentrations of the gas phase in the torus: (a) $k_0 = 0.1$, (b) $k_0 = 0.05$, (c) $k_0 = 0.01$, (d) $k_0 = 0.001$

the Mach disc radius R_{mach} with increasing distance from the torodial plane, Z_{mach}, show that it increased monotonely for the entire range of k_0.

At first sight the unexpected effects arose as one of the geometrical parameters of the torus, the radius of its section R_{circ}, changed at fixed radius of the base circle R_{tor} (Fig. 5.21a–d). The results are presented for $R_{circ} = 0.5$, 2, 4, and 6 cm at fixed $p_{sh} = 3$ MPa, $R_{tor} = 6$ cm, $k_0 = 0.01$, and $R_b = 0.1$ cm. The flow topology remained practically unchanged for the first three values of R_{circ}. Calculations showed that with increasing R_{circ} there was a considerable increase of the wave amplitude in the Mach disc core: $p = 81.7$ at $R_{circ} = 0.5$ cm, $p = 99.4$ at $R_{circ} = 1$ cm, $p = 166$ at $R_{circ} = 2$ cm, $p = 258$ at $R_{circ} = 3$ cm, $p = 386$ at $R_{circ} = 4$ cm, $p = 568$ at $R_{circ} = 5$ cm, and $p = 859$ at $R_{circ} = 6$ cm.

In the fourth case, $R_{circ} = 6$ cm, the outer torus boundary closed in a point on the axis. It turned out that the dynamics of the pressure field in the liquid was modified considerably: for this source configuration the front of the radiated shock wave converging to the axis was a concave surface with the pressure gradient directed toward the symmetry axis. Despite the fact that the pressure was minimum near the axis at the converging front, the flow cumulation finally led to the formation of a strong solitary wave in the near zone with the amplitude exceeding that of the wave interacting with the torus by almost a factor of 30 (Fig. 5.21d).

Analysis of the wave field structure showed that, as the Mach disc propagated along the axis, the dynamics of pressure on the axis was a nonmonotonic function of the distance to the torus (Fig. 5.22, curves 1–4). At fixed geometrical parameters of the torus and the initial radius of bubbles, the pressure

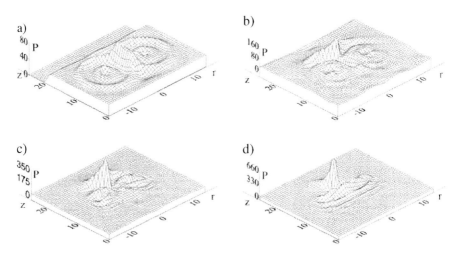

Fig. 5.21a–d. The radius of the section of a bubble torus vs. the evolution of the first solitary wave generated by the torus in the liquid: (**a**) $R_{circ} = 0.5$ cm, (**b**) $R_{circ} = 2$ cm, (**c**) $R_{circ} = 4$ cm, (**d**) $R_{circ} = 6$ cm.

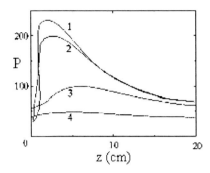

Fig. 5.22. Change of the maximum pressure amplitude in the Mach disc core with increasing distance to the torus plane for fixed $R_{\text{circ}} = 1$ cm, $R_{\text{tor}} = 6$ cm: (1) $k_0 = 0.1$, (2) $k_0 = 0.05$, (3) $k_0 = 0.01$, (4) $k_0 = 0.001$

distribution had a distinct maximum. The higher the maximum, the greater the gas phase concentration. The maximum was formed in the near zone of the torus plane and had a relatively abrupt pressure growth in the Mach disc core (the amplitude of the wave was amplified by a factor of 6–7). The character of the subsequent behavior of pressure had the tendency to approach asymptotics, for which pressures in the Mach disc cores practically leveled out over the range $k_0 = 0.01$–0.1 at a distance of 20 cm from the torus, but were still higher (2–2.5 times) than the amplitude of the wave interacting with the torus.

As was shown in Fig. 5.22, for fixed medium parameters the change of the torus volume at the expense of the radius influences appreciably the pressure in the region of the Mach disc core. A similar effect can be achieved by retaining the torus volume for the appropriate combinations of its dimensions. Calculations showed that as the torus radius decreased two-fold from 8 to 4 cm, the relative pressure increased by 40 units, and under further decrease from 4 to 2 cm, by 200. Thus, for equal radii the amplitude of the generated wave was maximum.

The statement with the toroidal bubbly cluster was first proposed and studied in [56]. The objective of the study was to access the efficiency of this method for the generation of strong shock waves in a liquid and for their directional radiation prior to proceeding to the experimental arrangement. Both tasks were practically fulfilled: a considerable amplification of the incident shock wave and the directedness of the radiation (at least in the near zone, 4–5 R_{tor}) at the expense of the Mach discs were achieved.

The numerical simulation of the problem on the interaction of a shock wave with a "free" bubbly system showed that:

– The convergence of the toroidal wave reradiated by the bubble cluster to the symmetry center was the classical cumulative process;
– The reflection from the symmetry axis was irregular, a Mach disc of finite thickness occurred with a characteristic core (pressure in the core was maximum);
– The shock wave generated by the tore in the liquid had an oscillating profile characteristic of a bubbly source with amplitude-damping maxima;

- Focusing of the shock wave resulted in the subsequent formation of a chain of Mach discs on the axis.

5.2.2 Spherical Bubbly Clusters: SW Cumulation with a Pressure Gradient Along Front

One of the problems concerned with the so-called acoustic "laser" is that of reactive media capable of absorbing the external disturbance, amplify and reradiate it as an acoustic pulse. Since the "free" systems may be of principal importance for experimental studies, we consider the statements focused on analysis of the interaction of shock waves with bubble clusters. The wave processes in such free systems are distinguished by a wide spectrum of time and space parameters, associated with the generation of shock waves with the amplitudes of tens of megapascals, and governed by a great number of parameters whose effect can hardly be analyzed in physical experiments.

Therefore, the dynamics of the state of complicated reactive systems and the development of wave processes in them should be simulated numerically. The numerical analysis of the interaction of a plane stationary shock wave with a nonreactive spherical cluster within the framework of a two-phase mathematical model of a bubbly liquid shows that the difference in the wave propagation velocities in the cluster and the ambient liquid, as well as the selection of a real cluster geometry, lead to unexpected effects, which will be analyzed below.

5.2.2.1 Statement of the Problem

A jump of mass velocity is arranged at a time $t = 0$ at the butt end of a cylindrical shock tube of radius r_{st} filled with water. A center of a spherical bubble cluster of radius R_{cl} with the volumetric concentration of the gas phase k_0 is placed on the z-axis of the tube at a distance l_{cl} from the boundary. Gas bubbles in the cluster have the same initial radius R_0. At $t > 0$ the shock wave propagating along the positive z-axis collides with a spherical bubble cloud, envelops it, and refracts into the cluster in the contact zone. We note that the equilibrium speed of sound in the cluster c_b depends significantly on the volumetric concentration and, for example, for $k_0 = 0.01$ is on the order of 10^2, which is much below the propagation velocity of the wave in the liquid.

Here, as before, the wave processes in the bubbly medium are described by the modified IKW system [58], including the conservation laws written for a cylindrical region, and the subsystem involves

- The Rayleigh equation

$$\beta \frac{ds}{dt} + \frac{3}{2}s^2 = C_1 \frac{T}{\beta^3} - \frac{C_2}{\beta} - C_3 \frac{s}{\beta} - p, \quad s = \frac{d\beta}{dt},$$

- The equation of temperature

$$\frac{dT}{dt} = (\gamma - 1) \cdot Nu \cdot \frac{\beta^3(1-T)}{T} - 3(\gamma-1) \cdot \frac{Ts}{\beta},$$

- The equations of state of a liquid and a bubbly medium

$$p = 1 + \frac{\rho_0 c_0^2}{n p_0}\left[\left(\frac{\rho}{1-k}\right)^n - 1\right], \quad k = \frac{k_0}{1-k_0}\rho\beta^3.$$

- Here

$$\mathrm{Nu} = \sqrt{\mathrm{Pe}} \quad \text{if} \quad \mathrm{Pe} > 100, \quad \mathrm{Nu} = 10 \quad \text{if} \quad \mathrm{Pe} < 100,$$

$$\mathrm{Pe} = C_4(\gamma-1)\frac{\beta|s|}{|1-T|},$$

$$p = \frac{p_l}{p_0}, \quad \rho = \frac{\bar{\rho}}{\rho_0}, \quad t = \tau \frac{1}{R_0}\sqrt{\frac{p_0}{\rho_0}},$$

$$C_1 = \frac{\rho_{g0} T_0 B}{p_0 M}, \quad C_2 = \frac{2\sigma}{R_0 p_0}, \quad C_3 = \frac{4\mu}{R_0\sqrt{p_0\rho_0}}, \quad \text{and} \quad C_4 = \frac{48}{C_3}.$$

5.2.2.2 Focusing of Shock Wave in a Cluster

When a cluster is enveloped by an incident shock wave the cluster is excited with a delay in different points of its surface due to the finite wave propagation velocity. In view of relatively low propagation velocity and the cluster geometry, the shock wave formed in the cluster through the reradiation of the refracted wave absorbed by bubbles will differ principally from the one-dimensional case.

Figure 5.23 presents the schemes of interaction of a bubble cluster (1 is the cluster boundary) with a shock wave (2 is the incident shock wave front and 3 is the system of isobars) for the time $t = 110$ µs and the relative pressure distribution (in the units of hydrostatic pressure $p_0 = 0.1$ MPa) in the space $[r, z]$ (4 is the shock wave front in the cluster, 5 is the undisturbed zone). The calculation was performed for the amplitude of an incident shock wave $p_{sh} = 3$ MPa, $k_0 = 0.01$, $R_0 = 0.1$ cm, $R_{cl} = 4.5$ cm, $l_{cl} = 10$ cm. The tube radius was $r_{st} = 15$ cm, its length was $L = 40$ cm. One can readily see that the shock wave front 4 in the cluster was concave, while pressure along the front had a strong gradient. This fact was due to the unsteady formation of shock waves in the bubbly medium and the difference of the process phases that were distributed along the front at a certain time, which determined the appearance of the gradient. At this stage a rarefaction wave arose in the inner region of the cluster due to the pressure drop behind the front of the refracted wave caused by the absorption by bubbles. The wave propagated in the liquid surrounding the cluster.

194   5 Shock Waves in Bubbly Media

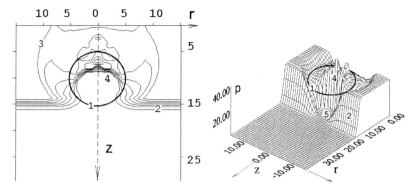

Fig. 5.23. The interaction of a shock wave with a bubble cluster and the pressure distribution $p(r,z)$ for $t = 110$ μs

Calculations (Fig. 5.23, $p(r,z)$) proved that, although by 110 μs the front of the incident wave 2 had completed enveloping the cluster, a part of the bubbly zone 5 (restricted by the curvilinear front of the wave generated by bubbles near the "far" boundary of the cluster, $z = 10$–14.4 cm) remained unperturbed.

By the time $t = 140$ μs the initial stage of the wave amplification stage was observed (see Fig. 5.24): the focusing zone in the cluster was formed, one can distinctly see the pressure gradient along the curvilinear front 4 in the bubbly system. The focusing led to the formation of a strong shock wave with the amplitude p_{foc} of over 30 MPa (Fig. 5.25, $t = 160$ μs) at the cluster–liquid interface near the coordinate $z = 13.5$ cm. Then, the cluster generated a shock wave in the ambient liquid. The wave 1 was of the "bore" type (Fig. 5.26) with the front envelope as a parabola, a pressure maximum on the axis, and rather abrupt decline along the "branches" ($t = 180$ μs).

It is necessary to mention the "nonclassical" type of the focusing accompanied by the absorption of the shock wave in the cluster by gas bubbles and their subsequent reradiation, let alone the strong pressure gradient along the front. Thus, at each time step a new wave arises in the cluster, each with a less curved front. The stability of the wave focusing wave in the cluster

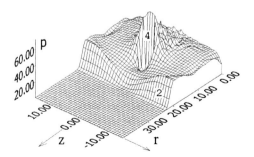

Fig. 5.24. The final stage of shock wave focusing in a bubbly cluster ($t = 140$ μs)

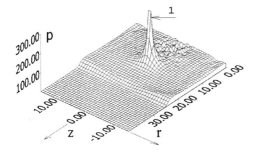

Fig. 5.25. Generation of a strong acoustic pulse 1 by a cluster ($t = 160$ μs)

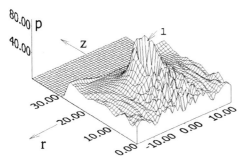

Fig. 5.26. Profile of the shock wave 1 radiated into the liquid by a bubbly cluster ($t = 180$ μs)

was tested by introducing a special disturbance, a liquid sphere of radius $r_{dr} = 0.5\text{–}1.0$ cm was placed in the cluster with the center on the z-axis at a distance of $l_{dr} < l_{cl} = r_{dr}$. Calculations showed that the disturbance introduced by the sphere in the pressure field was insignificant and soon vanished without any effect on the final result. Analysis of the calculation results proved that the pressure amplitude in the focus depended on several parameters (the volumetric concentration k_0 of the gas phase, the radius of the cluster R_{cl} and the bubble R_0). These dependences can be approximated by the relationships (1)–(3):

$$\frac{p_{foc}}{p_{sh}} \simeq 1 + 1.56 \cdot 10^3 \cdot k_0 - 1.6 \cdot 10^4 \cdot k_0^2 ; \tag{1}$$

$$\frac{p_{foc}}{p_{sh}} \simeq 1 + 2.6 \cdot R_{cl} + 0.11 \cdot R_{cl}^2 ; \tag{2}$$

$$\frac{p_{foc}}{p_{sh}} \simeq 17.9 - 8.5 \cdot R_0 - 135.6 \cdot R_0^2 . \tag{3}$$

The dependences (1)–(3) were obtained for $p_{sh} = 3$ MPa, in (1) $R_{cl} = 3$ cm, $R_0 = 0.1$ cm, in (2) $k_0 = 0.01$, $R_0 = 0.2$ cm, and in (3) $k_0 = 0.01$, $R_{cl} = 5$ cm. It is noteworthy that the dependence of pressure in the focus p_{foc} on the amplitude of the incident shock wave p_{sh} increased monotonically and was approximated by

$$\frac{p_{foc}}{p_0} \simeq 62 \cdot \left(\frac{p_{sh}}{p_0}\right)^{0.21} - 30 . \tag{4}$$

The dependences (1)–(4) were obtained for the ranges

$$k_0 = 0\text{–}0.05, \quad R_{\text{cl}} = 1\text{–}5 \text{ cm},$$

$$R_0 = 0.05\text{–}0.3 \text{ cm}, \quad \text{and} \quad p_{\text{sh}} = 0.4\text{–}12 \text{ MPa},$$

respectively.

Regarding the production principles for the sources of a strong acoustic pulse, one can conclude that the nonreactive spherical bubble cluster excited by the shock wave is an reactive medium capable of absorbing, amplifying, and reradiating the external disturbance as a strong acoustic signal. To control the location of the focus field with respect to the "cluster–liquid" interface and, consequently, to eliminate the absorption of the cluster-generated bubble by the cluster, one should properly select the volumetric concentration of the gas phase, k_0.

5.3 Reactive Bubble Media, Waves of Bubble Detonation

We define bubble detonation as a quasi-stationary self-sustained regime for the formation and propagation of a solitary wave or a wave packet in reactive media [61].

Historically, discovery of this process stretched for almost a quarter of a century. The investigations started with the study of the dynamics of solitary bubbles containing reactive explosive gaseous mixtures by Kedrinskii [62]. Only 20 years later a regime of the type of a wave "rolling" with a constant velocity was detected by Hasegawa and Fujiwara during an investigation of the propagation of the external shock wave along the chain of bubbles [63]. Finally, the normal statement of the problem on the transformation of waves in reactive bubbly media was studied by Sychev and Pinaev [61].

The formation of a quasi-stationary solitary wave in bubble detonation depends on the character of the chemical reaction in the closed microvolume of a single gas bubble collapsing under the effect of an incident shock wave. It is known that the reaction can proceed in various regimes. However, over the last decade the studies of the formation and interaction of bubble detonation waves [64–72] and different approaches to the determination of its propagation velocity [68–70] were lacking detailed analysis of the dynamics of a single bubble. It was assumed a priori that a reactive gas mixture in bubbles exploded when they were compressed to a certain degree. The mechanism of the wave amplification in reactive bubbly media was not considered.

The experiments run by Barbone et al. [70] repeated and refined already known facts concerning special features of the dynamics of a reactive single bubble, proving once again that the interaction of the single bubble with a shock wave, being a kind of "elementary" act of the complicated process of wave transformation and energy release by the exploding bubbles, which is far from being completely investigated.

5.3.1 Shock Waves in Reactive Bubbly Systems

5.3.1.1 Experimental Arrangements

Hasegawa and Fujiwara [63] studied experimentally the propagation of a shock wave in a liquid column containing a vertically located chain of bubbles with a reactive gas mixture. The parameters of the shock wave and the bubbles were selected so that the first bubble located in the path of the shock wave could completely absorb the wave energy and collapse to the ignition temperature of the mixture. As a result of the chemical reaction, the bubble exploded and radiated a secondary shock wave, whose interaction with the following bubble repeated the previous stage. This effect was called a *bubble detonation*.

It is obvious that the propagation velocity of such a "rolling" wave is governed by the collapse time of bubbles in the chain to the moment of detonation initiation in the gas mixture rather than by the equilibrium velocity in a "conventionally homogeneous" two-phase medium.

Pinaev and Sychev were the first to perform detailed experimental studies of the structure of shock waves in reactive bubbly systems (where the bubbles occupied the whole shock tube cross-section) [61, 72, 73]. They revealed the existence of a self-sustained regime of wave generation as a solitary wave packet with a propagation velocity D higher than that of shock waves in nonreactive bubble systems with the same volumetric concentration. They found the existence of lower and upper limits of the volumetric concentration of bubbles ($> 0.5\,\%$ and $< 8\,\%$), beyond which detonation did not occur. Experiments with acetylene–oxygen mixture ($C_2H_2 + 2.5O_2$) showed that for bubbles 3–4 mm in diameter and with volumetric concentration $k \simeq 6\,\%$ the detonation process had the following parameters: the pressure in the wave packet can vary within 15–40 MPa, the length of the initiation zone was 6–7 cm, the luminescence time was 2–3 µs, the duration of the bubble detonation wave was 100–200 µs, and the wave velocity was about 560 m/s.

Later Beilikh and Gulhan [74] and Lee, Scarinci et al. [75] confirmed the main conclusions of Pinaev and Sychev [61, 72].

The problem of shock ignition thresholds was also considered by Kang and Butler [76]. They have suggested a mathematical model which taken into account energy balances for the liquid and gas phase, as well as the heat transfer between the gas and surrounding liquid. The gas-phase thermodynamic properties and chemical reaction rates were assembled by incorporating a real-gas version of the CHEMKIN chemical kinetics package in which the Nobel–Abel equation of state is used. The liquid energy equation is solved by the integral method to yield an ordinary differential equation for interface temperature. Some typical numerical results and shock ignition thresholds, or critical shock pressures necessary to ignite the gas-phase mixture, under various conditions for argon-diluted hydrogen/oxygen bubbles in water and glycerin are presented. Comparison with experimental data indicates that the model can predict the process of bubble collapse and ignition accurately, especially if ignition occurs during the first cycle. Calculations show that initial bubble radius, temperature, pressure, and mixture composition have strong influences on bubble collapse and ignition. The ignition threshold decreases with increasing initial radius and temperature and decreasing initial gas pressure. Ignition is also favored by stoichiometric or fuel-rich mixture diluted by inert gas.

5.3.1.2 Liquid HE, Combustible Mixtures: Mechanisms of Detonation Ignition

A comparison of the research results for the regimes of bubble detonation with the data on detonation initiation in liquid HE and explosions in containers with fuel enabled conclusions on the adequacy of the mechanisms governing these processes [21, 58, 66, 77, 78]. Studies of liquid HE and explosions in volatile and ignitable liquids were performed in parallel over several decades without using the models of the mechanics of multiphase systems. Nevertheless, the models of *hot spots and fractional impact* [79–85], *vapor explosions, and explosions of volume-detonating systems* [70, 86] involved the nonuniformity of the media, the data on the physics of cavitation phenomena, and the energy release under the effect of external pulsed loads.

Often a detailed study of the role of a certain mechanism responsible for the process evolution, its justification, and experimental justification meet serious difficulties. Therefore, thorough and detailed numerical modeling of the processes is required. The cavitation processes in volatile and ignitable liquids, arising at sudden depressurization of containers and collecting bins and often leading to catastrophic breakdowns, might be of principal importance. These processes can be associated with the so-called steam explosions. The above effects were considered by many authors, among them Barbone et al. [69], Hill and Sturtevant [86], Chaves et al. [87]. Two mechanisms of catastrophic breakdowns were usually considered [70]:

a) Possible ejection of liquids in the form of two-phase cavitating jets and their subsequent spreading with the formation of a gas-droplet cloud, the so-called volume detonation charge;
b) Propagation of a rarefaction wave inside the liquid fuel after partial depressurization of the tank resulting in rapid boiling-type processes: the pressure inside the tank could abruptly increase by several times.

In [21, 77, 78], a fundamentally new mechanism for the generation of high pressures in compressed combustible liquid systems was proposed. The essence of the mechanism is as follows: numerous bubbles containing an air–fuel vapor mixture are formed when filling a container, the bubbles can ignite under adiabatic compression. If the container moving at a high velocity comes up against an obstacle, the impact produces a system of compression waves in the liquid that can initiate a bubble detonation wave. On the contrary, under depressurization of the container a system of rarefaction waves appears in the liquid, whose interaction, as shown below, can lead to unexpected effects.

The influence of inhomogeneities on the initiation mechanism of liquid explosives was studied long ago. Thus, Bowden and Ioffe [79] pointed to the thermal nature of the shock initiation of explosions of liquid HE and emphasized that small gas bubbles, which are usually available in liquids, are potential heat sources of explosion (hot spots model). Andreev [80] and Johansson [81] considered small particles or gas phase of HE in bubbles as the most plausible reason for detonation initiation.

Within the same model Dubovik and Bobolev [82] studied the effect of droplets or HE vapors in the gas atmosphere of the bubble on the development of the "hot spots" mechanism by estimating the value of the adiabatic induction period t_i and the critical size d_0 such that the droplet manages to warm up during the collapse of the bubble:

$$d_0 \leq \pi(\xi_0 \tau_R)^{1/2} \quad \text{and} \quad t_i = \frac{c_p B T_*^2}{A Q E_a} \exp(E_a / B T_*).$$

Here T_* is the temperature of the liquid drop, E_a and A are the activation energy and the preexponential factor for the monomolecular reaction, Q is the heat effect, B is the universal gas constant, c_p is the specific heat capacity, τ_R is the time of the bubble collapse, and ξ_0 is the coefficient of heat conductivity. The estimate is valid only if T_* is valid over t_i.

The mechanism of fractional impact when the detonation process was initiated by a sequence of shock waves was considered as well. The first wave was reflected from the free surface and initiates cavitation, then the second wave compressed bubbles by forming a system of "hot spots" and initiated detonation.

Campbell et al. [83] proved experimentally that the mechanism of shock initiation was determined by the interaction of the wave with microinhomogeneities and led to the formation of "hot spots" and further amplification of the propagating wave until the energy released behind the front was sufficient to initiate detonation.

Field et al. [84] studied the interaction of strong shock waves (amplitude of the order of 3 GPa) with a single cavity and an array of bubbles. They related light flashes observed in reactive emulsive systems under bubble collapse to the formation of a cumulative jet and its impact on the far wall of the bubble. Roberts and Field [85] showed that the inhomogeneities of diameter 1.5 mm introduced in the sample near the impact point were the main initiators of the reaction.

The studies produced the following conclusions:

1. The shock sensitivity of liquid HE depends on quantitative characteristics and the character of microinhomogeneities of different origin;
2. Detonation in liquid HE occurs only in the presence (origination) of the inhomogeneities.

Finally, a comparison of the hot spots model and detonating bubbles leads to the conclusion that the mechanisms governing the above processes and the bubble detonation can be regarded as identical. Let us consider some scenarios of the development of wave processes and wave amplification in bubbly media, which can justify the validity of this statement.

5.3.1.3 Model of Instantaneous Release of Energy

Wave processes in bubbly liquids are described by a pressure-nonequilibrium two-phase mathematical model. Let us apply it to a reactive medium and add the multiplier δ in the Rayleigh equation [69]:

$$R\frac{d^2 R}{dt^2} + \frac{3}{2}\left(\frac{dR}{dt}\right)^2 = \rho_l^{-1}(\delta p_g - p) \ .$$

The multiplier is equal to unity until the gas mixture in a bubble is adiabatically compressed to the temperature of initiation of chemical reaction. At this moment *adiabatic explosion at a constant volume* causes an abrupt pressure increase in bubbles to the value determined by the expression

$$p = \rho_*(\gamma_* - 1)Q_{\text{expl}} \ ,$$

where Q_{expl} is the explosion heat. The expression follows from the condition that the total energy released during the reaction is transformed into the internal energy of its products. In this case, the value of adiabatic index changes ($\gamma = 1.25$) and the process continues to develop under new conditions without any changes in the mathematical model.

Calculations within such a simple kinetics system show that the parameters and the structure of the forming bubble detonation wave (Fig. 5.27) are adequate to the real process.

In [58, 69], the following approximation formula was proposed for the propagation velocity of the bubble detonation wave:

$$D_{\text{bubl}} \simeq \lambda \sqrt{\frac{\rho_*(\gamma_* - 1)Q_{\text{expl}}}{\rho_0 k_0}} \ ,$$

where $*$ denotes detonation products, λ is an empirical coefficient that is selected assuming that the velocity coincides with experimental data at a certain point. We used the notion of the time of "collective" collapse t_{col} of

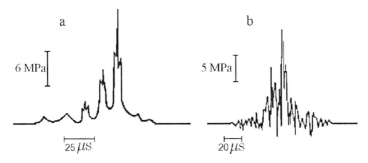

Fig. 5.27a,b. Comparison of calculated (**a**) and experimental (**b**) pressure oscillograms in the bubble detonation wave

a bubble layer of thickness l introduced in [1] on the basis of analysis of numerical estimates:

$$t_{col} \simeq 0.9 R_0 \sqrt{\frac{\rho_0}{p_{sh}}} \exp\left(\frac{l}{2}\sqrt{\frac{3k_0}{R_0^2}}\right).$$

The function $D_{bubl}(k_0)$ was obtained for the mixture $C_2H_2 + 2.5 O_2$ and an explosion heat of $Q = 15.2$ MJ/m^3 at $\gamma_* = 1.15$, $\lambda^2 \simeq 0.16$, and $\rho_* = \rho_{g0}(R_0/R_*)^3$, where R_* was the bubble radius at the moment of explosion of the mixture.

We note that the reactive bubbly system can be considered as a specific type of condensed HE and used the classical relation for the amplitude p_{bubl} and velocity D_{bubl}

$$p_{bubl} = \frac{(\rho - \rho_0)\rho_0}{\rho} D_{bubl}^2,$$

where $\rho = \rho_{liq}(1 - k_0 k^*)$, $\rho_0 = \rho_{liq}(1 - k_0)$, and $k^* = (R_*/R_0)^3$. Then finally we obtained

$$p_{bubl} = \rho_{liq} k_0 (1 - k^*) D_{bubl}^2.$$

The result corresponded to the experimentally established weak dependence of the detonation wave amplitude on concentration.

5.3.1.4 Bubble Detonation Waves and "Hot Spots" Mechanism

The above system supplemented by the Todes kinetic model (bimolecular kinetics) [88]

$$\frac{dn}{dt} = A\sqrt{T} e^{-E_a/BT}(a - n)^2,$$

can be reduced to the following form (Lagrangian coordinates, dimensionless characteristics):

$$\frac{\partial \rho}{\partial t} = -\rho^2 \frac{\partial u}{\partial s}, \quad \frac{\partial u}{\partial t} = -\frac{\partial p}{\partial s},$$

$$\beta\frac{\partial S}{\partial t} + \frac{3}{2}S^2 = C_1\frac{T}{\beta^3} - C_2\frac{S}{\beta} - p, \quad \frac{\partial \beta}{\partial t} = S,$$

$$\frac{\partial T}{\partial t} = \eta\frac{\partial N}{\partial t} + \delta(\gamma-1)\mathrm{Nu}\frac{\beta^3(1-T)}{T} - 3(\gamma-1)\frac{TS}{\beta},$$

$$\frac{\partial N}{\partial t} = \zeta\frac{1}{\beta^3}\sqrt{T}\mathrm{e}^{-\alpha/T}(1-N)^2,$$

$$p = 1 + \frac{\rho_0 c_0^2}{n p_0}\left[\left(\frac{\rho}{1-k}\right)^n - 1\right], \quad k = \frac{k_0}{1-k_0}\rho\beta^3,$$

$$\mathrm{Pe} = C_3(\gamma-1)\frac{\beta|S|}{|1-T|}, \quad \mathrm{Nu} = \begin{cases} \sqrt{\mathrm{Pe}}, & \mathrm{Pe} > 100, \\ 10, & \mathrm{Pe} \leq 100, \end{cases}$$

$$C_1 = \frac{\rho_{g0} T_0 B}{p_0 M}, \quad C_2 = \frac{4\mu}{R_0\sqrt{p_0 \rho_0}} \quad \text{and} \quad C_3 = \frac{12 R_0 \sqrt{p_0/\rho_0}}{\nu}.$$

Here ρ is the ratio of the density to the initial value, $N = n/a$ is the relative fraction of the reacted components of the gas mixture, a is the initial concentration of the component, n is the number of molecules formed in a unit volume during the reaction, T is the temperature of the gas mixture taken with respect to the initial T_0, $\beta = R/R_0$, α is the relative activation energy, ζ is a constant depending on the initial parameters and the mixture composition, $\eta = Q/c_m T_0$ is the specific reaction heat, E_a is the activation energy, and B is the gas constant.

The coefficient δ in the temperature equation is used to trigger the heat transfer and to correlate its intensity. The necessity of the coefficient is caused by the approximate nature of the relationships used in the system for the Nusselt and Peclet numbers proposed by Nigmatullin, which is quite acceptable for the case of weak waves.

It was shown that as a wave propagated through a bubbly medium it attenuated due to losses for imparting kinetic energy to the liquid component and for increasing the internal energy of the gas phase in bubbles under compression. If the bubbles were filled with an explosive gas mixture, a reaction with high energy release could occur under adiabatic heating. In this case, compression waves are radiated into the ambient liquid, which compensates for the loss of energy of the incident wave until a self-sustained regime is established in the medium.

The Todes kinetics model describes adequately the formation and propagation of bubble detonation, but overestimates the value of the amplitude, which can be explained by inadequate description of the kinetics itself. Calculations were performed for ($\alpha = 51.5$, $\zeta/R_0 = 2.7 \cdot 10^9$, $\eta = 70$).

In analyzing possible scenarios of the development of wave processes, a question naturally arises on the threshold values of the wave intensity required to initiate detonation in reactive bubbly media. A partial answer to this question is obtained in calculating the collision of weak shock waves with an amplitude of 0.4 MPa (Fig. 5.28). It was shown that such waves could not initiate detonation, however, their collision initiated chemical reaction in the gas phase: at $t = 361$ µs there was an abrupt temperature rise T in the detonation products. In the vicinity of the collision plane, as a result of the explosion, the average pressure p in bubbly medium increased up to 8 MPa over 2 µs and, thus, the wave reflected after the collision became the deto-

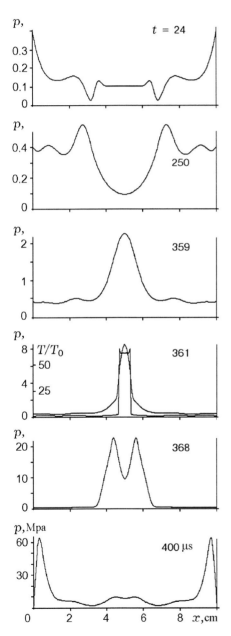

Fig. 5.28. Collision of weak shock waves and generation of a bubble detonation wave

nation wave. One can see that it was formed in the course of propagation to the periphery, attaining 60 MPa at a distance of 5 cm from the center, as if the bubble layer explodes from inside.

An unexpected effect was found in the numerical analysis of the collision of rarefaction waves (Fig. 5.29) [21, 77] aimed at simulation of both the

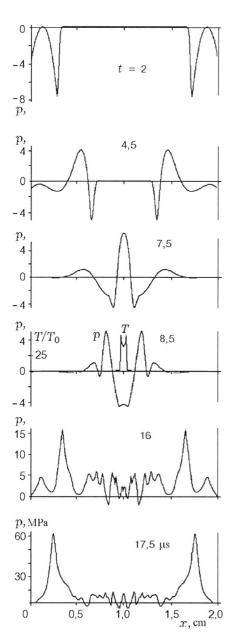

Fig. 5.29. Generation of a bubble detonation wave by the colliding rarefaction waves

probable development of the wave process in a container under partial depressurization and the natural process of pulse loading of a liquid HE sample with a free boundary. In the latter case, the compression wave generated by the shock interacted with the free surface from which a rarefaction wave should propagate in the sample.

The calculation (Fig. 5.29) showed *a possibility of the initiation of a bubble detonation wave at the sample center resulting from the collision of rarefaction waves*. This effect is based on the transformation of the rarefaction wave as it propagates through the medium with microinhomogeneities and excites the cavitation in it: an oscillation with intense positive phase arises behind the rarefaction wave front. The oscillation under collision is eventually responsible for the heating of the gas mixture to the ignition temperature. We note that intensities of the positive phase of a solitary rarefaction wave are, as a rule, insufficient to do this. Following the calculation results, the wave collision (Fig. 5.29) occurred within 8.5 µs after the beginning of the propagation at the center of a 2-cm thick sample. In this case, the reflected waves became bubble detonation waves, whose amplitude near the free surface achieved practically 40 MPa after 17.5 µs. Here we observed the explosion of a sample from inside. However, the loading conditions were "paradoxical" in terms of the initiation of the explosive process: loading by impulsive tensile stresses was accompanied by the development of cavitation behind the rarefaction wave front.

5.3.1.5 SW Focusing in a Reactive Bubble System

Heterogeneous structures in real liquids play a principal role in initiating explosions (see Barbone et al. [70], Field et al. (1992) [84], Kedrinskii, Vshivkov et al. [39, 64, 66, 77], etc.). Undoubtedly, the bubbly media with reactive gas phase attract attention as a probable source of powerful acoustic radiation, which can be of help in solving the above problem of the so-called *acoustic laser*.

Bubble clusters containing explosive gas mixtures can be considered as *physical analogs of pumping in laser systems*. Even under interaction with a weak shock wave (or under wave collision) such a cluster generates a strong pressure pulse in the form of a bubble detonation wave [65], which can be considered as one of the principles for creating *acoustic analogs of laser systems*. Obviously, the focusing in reactive bubbly media intensifies considerably the amplification of bubbly media. For the calculations we use the known system of governing equations for the cylindrical symmetry that is composed of two subsystems:

– *Conservation laws for average p, ρ, and u:*

$$\frac{\partial u}{\partial t} = -\frac{1}{\rho_0}\left(\frac{x(r,t)}{r}\right)^2 \frac{\partial p}{\partial r}, \quad \frac{\partial x}{\partial t} = u, \quad \frac{1}{\rho} = \frac{1}{\rho_0}\left(\frac{x(r,t)}{r}\right)^2 \frac{\partial x}{\partial r}$$

206   5 Shock Waves in Bubbly Media

- A set of kinetic equations describing the dynamic state of the medium, the heat exchange, and the kinetics of chemical reactions.

The formation of the self-sustained wave regime requires a certain time and a distance that depends on k_0. Although in the above examples the linear dimensions of the considered regions were smaller than the distance, the amplification of the divergent waves was quite sufficient to speak about the bubble detonation as an initiation mechanism of explosive processes in containers with fuel and as a powerful source of a hydroacoustic signal. How does the focusing influence extremely weak amplitudes of the focused wave?

The focusing of a short shock wave was calculated using the Todes kinetics for the amplitude $p_b = 5$ MPa and the duration of the positive and negative pressure phases $t_+ = 5$ μs and $t_- = 15$ μs, respectively. Such a shock wave profile simulated the profile of a wave generated by underwater explosion. The concentration of the reactive gas phase in the bubbly medium was intentionally taken to be lower than the critical value, $k_0 = 0.001$. The wave was focused from the distance of $r = 10$ cm. The calculations showed that at first the wave attenuated and its amplitude decreased to the value at which the incident wave was unable to initiate the detonation regime. Near the focus the wave amplitude increased up to 15 MPa ($t = 65$μs), which was sufficient to compress bubbles to the ignition temperature of the mixture and initiate the chemical reaction. The detonation wave excited in the focus further propagates to the periphery. As a result of bubble explosion, the pressure increased to 27.5 MPa ($t = 200$ μs) and remained practically invariant until the bubble detonation wave reached the outer boundary of the region. The problem statement with detonation excited at the center of a cylindrical sample could apparently be considered as one of the models of hydroacoustic sources.

When the volumetric concentration of the gas phase was increased by an order of magnitude, $k_0 = 10^{-2}$, $R_0 = 0.2$ cm, the wave pattern changed dramatically. As the amplitude of the generating wave decreased (the parameters of positive and negative phases were unchanged), the incident shock wave could be separated during focusing by initiating the detonation at the boundary approximately at the moment when the initiating wave achieved the focus line.

Figure 5.30 shows the pressure distribution for different times, $p_b = 3$ MPa, and the focusing radius $r = 5$ cm. One can see that at $t = 26$ μs (with a delay for the bubble collapse to the ignition temperature of the mixture) bubble detonation was excited at the periphery (the temperature jump shown at the right in Fig. 5.30), the shock wave initiating it reached the focus after 32 μs: the wave amplitude increased by a factor of 5. However, the bubble concentration in this example was much higher and the intensity of the focused wave was insufficient to excite the detonation regime in the focus.

The bubble detonation wave moving from the periphery focused and formed simultaneously. After $t = 60$ μs its amplitude (at a distance 2 cm

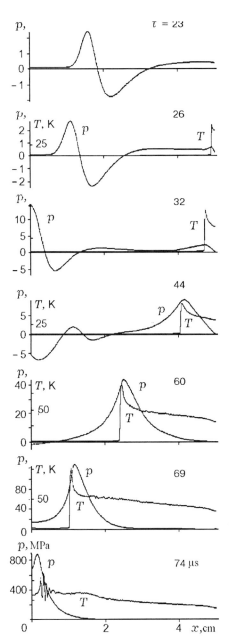

Fig. 5.30. Initiation and focusing of a bubble detonation wave: $k_0 = 10^{-2}$, $R_0 = 0.1$ cm, $p_b = 3$ MPa

from the axis) reached 40 MPa. The calculation showed that near the axis the detonation wave amplitude achieved an unrealistic value on the order of 10^3 MPa. Undoubtedly, the appearance of such stresses inside a container could result in catastrophic consequences.

It should be noted that high pressures arising during the focusing were a consequence of the inertial effects accompanying bubble oscillation and the overcompression of detonation products. The temperature of the products achieved $(1.5–2) \cdot 10^4$ K, which required application of a more realistic kinetics.

5.3.1.6 Modern Kinetics, Phase Transitions

The bubble detonation model is formulated with due regard to the liquid compressibility and viscosity, the availability of an induction period of chemical reaction, the shift of chemical equilibrium, and intensive interphase heat and mass exchange (see Chap. 4). The model includes the modern kinetics (Eqs. (4.18–4.21)) and the equations of the reactive bubble dynamics, taking into account the evaporation of a liquid phase and the IKvanW model of a two-phase medium [89, 90]. Only this approach enables simulation of the bubble detonation in a system where an oxidizer and a fuel are originally in different phases. It is clear that the detonation process can occur only due to phase transition (evaporation of the liquid from the bubbles) when an explosive gas mixture is created inside the bubbles. The detonation is initiated by a shock wave $P_\infty = $ const. The system of equations is solved using the explicit finite difference scheme and applying the Neumann–Richtmyer linear and quadratic artificial viscosities. The formation and propagation of bubble detonation waves are calculated using the given model in the System I [$H_2+\lambda O_2$ (gas phase) $-(1-\alpha)H_2O+\alpha$ glycerine (liquid phase)] at $T_0 = 300$ K and the system II [H_2 (gas phase) $-O_2$ (liquid phase)] at $T_0 = 87$ K. Here λ is the stoichiometric factor and α is the volumetric concentration of the glycerin solution.

It was shown in Chap. 4 that the injection moment of microdroplets scarcely affects the bubble dynamics and gas parameters. Therefore, it is supposed that the injection occurs during the first bubble oscillation at the stage of bubble compression when $R/R_0 = 0.3$. This choice is based on the experimental results of Sychev and Pinaev [61], who observed the instability of the bubble surface at the last stages of bubble compression. Figure 5.31 shows the typical structure of the bubble detonation wave upon attainment of a constant propagation regime in System I ($\lambda = 0.5$, $\alpha = 0$). The calculation was performed for the liquid evaporation ($M_L/M_0 = 1$, $D_0 = 5$ μm and $k_0 = 0.01$). The dependence of the velocity of bubble detonation wave W on the initial concentration of the gas phase k_0 with and without evaporation was analyzed. The computational results for System I without evaporation and the experimental data of [73] are presented in Fig. 5.32. One can see that the model was in good agreement with the experiment. For $k_0 < 0.0025$ and $k_0 > 0.1$, the detonation wave did not separate from the initiating wave and no "solitary" wave was formed. The former was caused by the limited amount of specific energy released in the medium due to the low concentration of bubbles, the latter was due to considerable damping of the detonation wave.

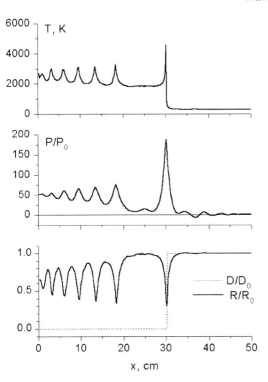

Fig. 5.31. The structure of a stationary bubble detonation wave in System I ($t = 330$ μs).

Analysis of the dependence of the bubble detonation velocity W on the composition of the hydrogen–oxygen mixture in the gas phase for various concentrations of bubbles showed that the variation of λ practically did not affect W. The dependence had a weakly expressed maximum corresponding to the stoichiometry ($\lambda = 0.5$). The diameter and total mass of the evaporating microdroplets were varied in the calculations. The evaporation in System I occurred at high gas temperature (about 3000 K), whereas the velocity of microdroplet evaporation was high enough. Thus, the evaporation of microdroplets with diameters of up to 20 μm was practically instantaneous. Microdroplets with diameters of over 20 μm were not considered, since they had no time to evaporate before the moment of maximum bubble compression and their total volume by that time was comparable with the bubble volume.

An unexpected result was obtained: the increase in the evaporated liquid mass led to an increase (though moderate) of the wave velocity. It was likely that increasing gas pressure in bubbles completely compensated for energy losses for the liquid evaporation reducing the wave velocity. With increasing liquid mass in System II, the detonation wave velocity increased monotonically, and the pressure passed through the maximum at the ratio $M_{\rm L}/M_0 = 8$, which corresponded to the stoichiometric mixture in gas. As

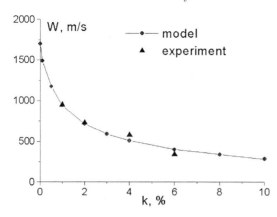

Fig. 5.32. The detonation velocity W vs. the gas phase concentration in System I: $2H_2+O_2$ (gas phase) – 50%-aqueous glycerin solution (liquid phase)

opposed to System I, where the velocity and detonation pressure decreased with increasing initial diameter of microdroplets, because the evaporation in this system integrally led to an increase in the internal energy of gas in a bubble; increasing heat release completely compensated for all energy losses for evaporation. With smaller diameters of microdroplets, the evaporation proceeded as faster rate, and thus the amount of chemical energy contributed to the solitary wave increased.

5.3.2 Shock Tube with Changing Cross Sections

5.3.2.1 Statement of the Problem

Shock tubes with sudden changes of the cross section area and a one-phase liquid waveguide can be used to implement some approaches to SASER (shock amplification by systems with energy release) problems, which we analyze numerically [91] below. In this configuration, the amplification effect should be a consequence of the two-dimensional cumulation of a shock wave after it leaves the annular channel and reflects irregularly from the shock tube axis (Fig. 5.33).

As was shown in [21,77], both cavitating and bubble systems with a non-reactive gas phase and explosive gas mixtures can be regarded as reactive

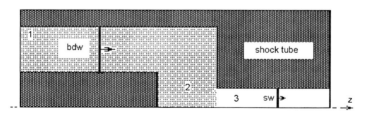

Fig. 5.33. Shock tube geometry with suddenly expanding and contracting channel: (1) annular channel; (2) working section; (3) waveguide

media capable of absorbing external disturbance energy, amplifying the disturbance, and then reradiating it as a power acoustic pulse. Studies of these systems in designing hydroacoustic analogs of laser systems (hydroacoustic "lasers" or SASER [89]) involve the problem on the transmission of an acoustic pulse generated by the system into a liquid with the least losses.

Sudden changes in the cross section area of a shock tube of radius R_{ST} (Fig. 5.33) are produced by changing the shock-tube profile (transition from section 1 to section 2) or/and by inserting a coaxial rigid cylinder (radius r_{cyl} and length L_{cyl}) forming an annular channel 1 filled with a two-phase mixture similarly to section 2. The condition $L_{SW} \leq L_{cyl}$ must be satisfied, where L_{SW} is the characteristic distance at which the steady-wave regime initiated at the left butt end of the shock tube is established. Owing to the inner geometry of the ST the pressure field in the section 2 can be modified by varying the distance L between the butt end and the wall. And in the result of reflection and focusing in the reactive bubbly medium (section 2) the shock wave can be intensified immediately in the vicinity of the "waveguide 3 – section 2" interface.

5.3.2.2 Two-Phase Model

To describe wave processes in bubbly hydrodynamic shock tubes we use the physical-mathematical model based on the modified IKvanW model [2, 7–10, 58], which for a reacting gas phase is supplemented by the equations of the type of the Todes kinetics of chemical reactions [88], a more general kinetics [90] or a simple condition of instantaneous adiabatic explosion at constant volume. Analysis of different approaches to the description of wave processes in reactive media containing bubbles filled with explosive gas mixtures showed that the above simplified statement is sufficient [69]. The statement suggests that when gas in bubbles is heated to the ignition temperature, the reaction occurs instantaneously after the appropriate degree of compression of bubbles (R_0/R^*) is achieved. An adiabatic explosion occurs, the bubble volume remains unchanged, and the pressure p_g in the bubble increases instantaneously to $p_* = \rho_*(\gamma_* - 1)Q_{expl}$ (Q_{expl} is the explosion heat and ρ_* is the density of the detonation products). In this case, the coefficient η and the adiabatic exponent γ change instantaneously. The adiabatic exponent becomes equal to that of detonation products γ_*. Thereafter the process develops under new "initial" conditions without changing the mathematical model.

The governing system of equations [64, 66, 69] within this physical model is determined by the conservation laws for average density, pressure, and velocity, and the subsystem describing the state of the medium and including the equation of state of the liquid phase (Tait's equation), the adiabat for the gas phase, and the Rayleigh equation for the dynamics of "average solitary" bubbles. The collective effect of the bubbles on the field characteristics is taken into account using the average pressure in the bubbly medium,

212   5 Shock Waves in Bubbly Media

which replaces the pressure at infinity in the right-hand part of the Rayleigh equation.

In the calculations water was considered as a reactive phase, for the case of a reactive bubbly medium the gas mixture was $2H_2+O_2$. At the time $t = 0$ a constant pressure $p(0)$ was specified on the left boundary of the shock tube. The numerical scheme was tested by the example of the one-dimensional problem on the formation of wave structures in the shock tube of constant cross section filled with a liquid containing gas bubbles. Characteristic structures of steady shock waves in the nonreactive and reactive media are presented in Figs. 5.34a and 5.34b, respectively, horizontal dashed lines mark the values of pressure jumps $p(0) = 10$ atm specified at the tube butt end and initiating the wave process. In both cases, the gas-phase volume fraction was $k_0 = 0.01$ and the bubbles were of the same size ($R_0 = 0.1$ cm).

For the indicated parameters, the establishment of a steady-state regime, for example, for a bubble detonation wave (with the maximum amplitude of about 150 atm), was recorded at a distance of $x \approx 15$ cm from the left wall of the shock tube. A stationary wave in the nonreactive medium (Fig. 5.34a) and the "tail" of a detonation wave (Fig. 5.34b) running in the bubble system with detonation products (already nonreactive system) had the classical oscillation structure. A system of precursors was generated ahead of the detonation leader and the shock-wave front.

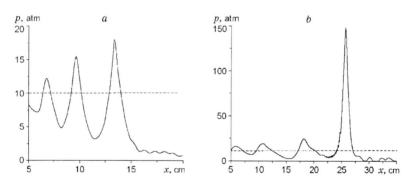

Fig. 5.34a,b. Characteristic profiles of stationary shock waves in nonreactive (a) and reactive (b) media

5.3.2.3 Bubbly Detonation Wave: Expansion, Focusing and Reflection

The shock tube shown in Fig. 5.33 consisted of three sections: a circular section 1, a section 2 bound from the right by a rigid wall and the waveguide 3 of radius r_{out}. The bubbly medium occupied the sections 1 and 2. Considering the complexity of the wave pattern expected in the shock tube, whose profile had abrupt cross section jumps, the characteristic peculiarities of the wave field structure for the two statements in parallel was analyzed preliminarily: the reflection from the wall in the shock tube channel without a rod and the expansion of the bubble detonation behind the butt end of the coaxial rod in a "semi-infinite" shock tube filled with a reactive bubbly medium.

Expansion and Focusing of a Bubbly Detonation Wave. This statement deals with analysis of the wave structure of the pressure field formed in the shock tube under the expansion of a detonation wave in the bubbly medium behind the edge of a coaxial rod. It is obvious that as the wave comes out of the circular channel 1 in the section 2 (Fig. 5.33), a rarefaction region arises behind the rod edge (Fig. 5.35a, on the left, marked by shading). Evidently, at a certain distance from the rod edge the unloading is not intense and a sufficient amount of energy is preserved in the wave to compress bubbles to the ignition temperature of the mixture. In this case, the energy radiated by exploding bubbles will compensate for the wave losses and the wave can be amplified due to the focusing, which is confirmed by the calculation results. The characteristic pattern of these effects is shown in Figs. 5.35a and 5.35b. In Fig. 5.35a, the rarefaction zone, in which the bubble detonation wave does not occur at this stage of the process, is seen in the immediate proximity to the rod edge (the zone between 15 and 17.5 cm on the axis). But since the reactive mixture is preserved in the bubbles within the zone, the subsequent oscillations of bubbles can result in detonation initiation.

The focus zone is distinct in both figures. As follows from Fig. 5.35b, the wave reflection from the axis becomes irregular at a certain distance from the rod edge. In the reactive bubbly medium ($k_0 = 0.01$, $R_0 = 0.1$ cm), a Mach disc is formed, whose characteristic width is over 1 cm for a relatively small focusing radius $r_{\rm cyl}/R_{\rm ST} = 0.5$ cm. The Mach structure shifts along the axis and for the case shown in Fig. 5.35b is at a distance of 3 cm from the rod

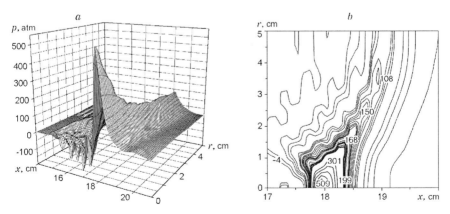

Fig. 5.35a,b. Focusing of a bubble detonation wave (**a**) and the Mach disc structure (**b**)

edge. The disc is a core with high pressure in the center and rather abrupt pressure gradient to the periphery. The core is the zone near the symmetry axis limited by the system of closed isobars.

Reflection of a Detonation Wave From the Wall With a Hole. As for an axisymmetrical problem on the formation and propagation of a bubble detonation wave in a round tube (Fig. 5.33) with one section jump as a solid wall with a hole. In this arrangement the bubbly detonation wave is initiated along the whole tube section, the coordinate of the wall and the bubbly medium–liquid interface coincides with the waveguide outfall. Analysis of the calculation results for the fields shows that the bubble detonation wave reflected from the wall has a spatial circular form and is focused on the symmetry axis of the shock tube (but already in the nonreactive system with burnt mixture products). The pressure amplitude of the focus is up to 300 atm, while the wave amplitude at the entry to the waveguide exceeds 200 atm with practically uniform pressure distribution along its radius. The pressure on the wall by this moment decreases to the amplitude of the incident wave. The reflected wave focus shifts to the left in the nonreactive medium. The shock wave generated in the waveguide is essentially one-dimensional. The axial symmetry is noticeable only in the thin layer near the waveguide walls.

5.3.2.4 Shock Tube with Two Section Jumps

Calculations show that in a general statement one can use the focusing effect and the wall effect for the amplification of the wave radiated in the waveguide, for example, by selecting the parameter L (the distance between the rod edge and the wall). Thus, a Mach structure can be formed near the contact boundary (below $L = 4$ cm). As the second governing parameter one can use the waveguide radius r_{out}, whose value should be adjusted to the Mach disc to ensure a one-dimensional configuration of the wave field in the waveguide channel. For example, if the waveguide radius is equal to the rod radius ($r_{out} = r_{cyl} = 2.5$ cm, $r_{cyl}/R_{ST} = 0.5$), the pulse amplitude in the waveguide is 200 atm. As the waveguide radius r_{out} decreases to the size of the Mach stem (about 1 cm), the wave amplitude increases to 500 atm.

Calculation results presented as characteristic pressure fields are shown in Figs. 5.36–5.38 for the times 215, 225, and 235 µs, respectively. They confirm the above suggestion on the possibility of a significant amplification of the wave generated in the waveguide by a reactive bubbly medium under the interaction of the bubble detonation wave in the shock tube with cross section jumps. At $t = 215$ µs the pressure amplitude was over 600 atm (Fig. 5.36) in the center of the focus with the coordinate $x = 16.5$ cm. After 10 µs (Fig. 5.37) the focus shifted for 2 cm, while the amplitude increased to 800 atm (due to the effect of the wall). Pressure distribution in the waveguide by the time $t = 235$ µs was presented in Fig. 5.38. The wave crest at this moment was at a distance of 1 cm from the contact boundary and had a maximum amplitude

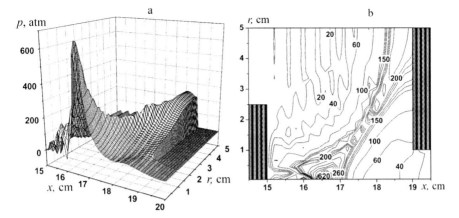

Fig. 5.36. Wave field structure in the channel 2, focusing of the wave reflected from the wall

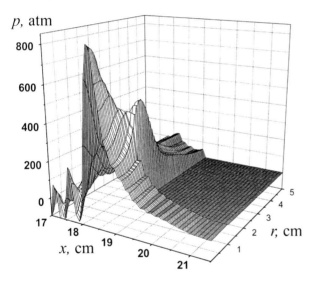

Fig. 5.37. Amplification of a bubble detonation wave, wall effect

of about 600 atm on the channel axis. As approaching the waveguide wall the pressure slightly decreased, approximately to 500 atm at a distance of 0.9 cm from the axis (at $r_{\text{out}} = 1$ cm). An abrupt pressure decay was observed at the remaining "base" of about 1 mm in the "boundary" channel zone.

These results suggest that the proposed configuration of the shock tube with section jumps and a bubbly liquid with exploding gas mixture can be used as one of the SASER models. The effect of amplification in such a configuration is a result of two-dimensional cumulation of the shock wave after it exits the circular channel, under the reflection from the wall, and the formation of a Mach disc. Geometrical characteristics of the shock tube make it possible (within certain limits) to regulate the amplification coefficient and

216   5 Shock Waves in Bubbly Media

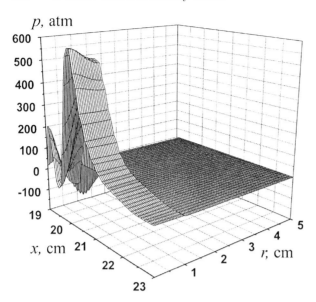

Fig. 5.38. Solitary shock wave in the waveguide

the coordinates of the focus pit. In particular, the wave can be focused in the vicinity of the second jump of the section, the solid wall (near the transition to the waveguide through the contact boundary). When the waveguide radius is equal to the Mach stem radius, the radiated wave has a maximum amplitude.

In conclusion, note the recent experimental and theoretical investigations of wave processes in passive and reactive bubbly media [92–95]. The refraction of a bubble detonation wave from a chemically active bubble media into a liquid, its structure and evolution as well as reflection from the butt-end of a shock tube were studied experimentally by Sychev [92,93] to evaluate the effects of poly-dispersion of bubble media, gas-bubble size and to understand the energy-dissipation mechanisms. Mathematical and numerical modeling of two-phase compressible flows with micro-inertial were carried out by Gavrilyuk and Saurel [94] which using the Hamilton principle of stationary action have developed the new model for compressible multiphase mixtures with full coupling between micro- and macroscale motion and validated for computation of oscillating shock waves as well as for the multidimensional interaction of a shock wave with a large bubble. Teshukov [95] has considered a kinetic approach based on the approximate calculation of the fluid flow potential and formulation of Hamilton's equations for generalized coordinates and momenta of bubbles to describe processes of collective interaction of gas bubbles moving in an inviscous incompressible fluid. Kinetic equations governing the evolution of the distribution function of bubbles (similar to Vlasov equation) were derived.

References

1. V.K. Kedrinskii: *Propagation of Disturbances in Liquid with Gas Bubbles*, Zh. Prikl. Mekh. i Tekh. Fiz. **9**, 4, pp. 29–34 (1968)
2. V.K. Kedrinskii: *Shock Waves in a Liquid Containing Gas Bubbles*, Fiz. Goren. Vzryva **21**, 5, pp. 14–25 (1980)
3. V.K. Kendriskii: Propagation of disturbances in a liquid containing gas bubbles. Ph.D. Thesis (Lavrentyev Institute of Hydrodynamics, Novosibirsk, 1968)
4. I.J. Campbell, A.S. Pitcher: *Shock Waves in a Liquid Containing Gas Bubbles*, Proc. Roy. Soc. A **243**, 1235, pp. 534–545 (1958)
5. G.M. Lyakhov: *Shock Waves in Multi-Component Media*, Izv. Akad. Nauk SSSR. Mekh. i Mashinostr. **1**, 1, pp. 46–49 (1959)
6. B.P. Parkin, F.R. Gilmore, G.L. Brode: Shock waves in bubbly water, Memorandum RM-2795-PR (Abridged), Oct. (1961)
7. S.V. Iordansky: *On Motion Equations of Liquid with Gas Bubbles*, Zh. Prikl. Mekh. i Tekh. Fiz. **1**, 3, pp. 102–110 (1960)
8. B.S. Kogarko: *On a Model of Cavitating Liquid*, Dokl. Akad. Nauk SSSR **137**, 6, pp. 1331–1333 (1961)
9. B.S. Kogarko: *One-Dimentional Transient Fluid Motion and Resulting Generation and Development of Cavitation*, Dokl. Akad. Nauk SSSR **155**, 6, pp. 779–782 (1964)
10. L. van Wijngaarden: *On the Equations of Motion for mixtures of Liquid and Gas Bubbles*, J. Fluid Mech. **33**, pp. 465–474 (1968)
11. R.I. Nigmatullin: *Fundamentals of Mechanics of Heterogeneous Media* (Nauka, Moscow, 1987)
12. V.E. Nakoryakov, B.G. Pokusaev, I.R. Scheiber: *Wave Dynamics of Gas and Vapor-Liquid Media* (Energoatomizdat, Moscow, 1990)
13. L. Nordzii: *Shock Waves in Bubble-Liquid Mixtures*, Phys. Communications **3**, 1 (1971)
14. L. Nordzii, L. van Wijngaarden: *Relaxation Effects Caused by Relative Motion on Shock Waves in Gas-Liquid/Liquid Mixture*, J. Fluid Mech. **66**, 1, pp. 115–145 (1974)
15. V.E. Nakoryakov, B.G. Pokusaev et al: Experimental investigation of shock waves in a liquid with gas bubbles. In: S. Kutateladze, V. Nakoryakov (eds.) Wave Processes in Two-Phase Systems, (Institute of Thermophysics, Novosibirsk, 1975)
16. V.G. Gasenko, V.E. Nakoryakov, I.R. Schreiber: *The Two-Wave Model of Disturbance Propagation in Liquid with Gas Bubbles*, Zh. Prikl. Mekh. i Tekh. Fiz. **20**, 6, pp. 119–127 (1979)
17. F. Fox, S. Kerly, G. Larson: *Phase Velocity and Absorption Measurements in Water Containing Air Bubbles*, Probl. Sovr. Fiz. **27**, 3, pp. 534–539 (1955)
18. M.A. Lavrentiev, B.V. Shabat: *Methods of the Complex Variable Theory*, (Nauka, Moscow, 1965)
19. V.K. Kedrinskii: *Dynamics of Cavitation Zone at Underwater Explosion Nearby free Surface*, Zh. Prikl. Mekh. i Tekh. Fiz. **16**, 5, pp. 68–78 (1975)
20. I.D. Manen: *Bent Trailing Edges of Propeller Blades of high Powered Single Screw Ships*, Int. Shipbuilding Progress **10** (1963)
21. V.K. Kedrinskii: *The Role of Cavitation Effects in the Mechanisms of Destruction and Explosive Processes*, J. Shock Waves. **7**, 2, pp. 63–76 (1997)

22. H. Gronig: Past, present and future of shock focusing research. In: K. Takayama (ed.) Proc. Intern. Workshop on Shock Wave Focusing (Sendai, Japan, 1989) pp. 1–38
23. B. Sturtevant: The physics of shock wave focusing in the context of extracorporeal shock wave lithotripsy. In: R. Brun, L.Z. Dumitrescu (eds.) *Proc. 19th Intern. Symp. on Shock Waves.* vol. 4 (Springer, Berlin, Heidelberg, New York, 1995) pp. 81–86
24. D. Book, R. Lohner: Quatre foil instability of imploding cylindrical shock. In: K. Takayama (ed.) Proc. Intern. Workshop on Shock Wave Focusing (Sendai, Japan, 1989) pp. 193–206
25. K. Takayama: High pressure generation by shock wave focusing in ellipsoidal cavity. In: K. Takayama (ed.) Proc. Intern. Workshop on Shock Wave Focusing (Sendai, Japan, 1989) pp. 1–37
26. M. Watanabe et al: Shock wave focusing in a vertical annular shock tube. In: R. Brun, L.Z. Dumitrescu (eds.) Proc. 19th Intern. Symp. on Shock Waves, vol. 4, (Springer, Berlin Heidelberg New York, 1995) pp. 99–104
27. H. Nagoya et al.: Underwater shock wave propagation and focusing in inhomogeneous media. In: R. Brun, L.Z. Dumitrescu (eds.) Proc. 19th Intern. Symp. on Shock Waves, vol. 4 (Springer, Berlin Heidelberg New York, 1995) pp. 439–444
28. C. Stuka et al.: Nonlinear transmission of focused shock waves in nondegassed water. In: R. Brun, L.Z. Dumitrescu (eds.) Proc. 19th Intern. Symp. on Shock Waves. vol. 4, (Springer, Berlin, Heidelberg, New York, 1995) pp. 445–448
29. M. Kuwahara et al.: The problems of focused shock waves effect on biological tissues. In: K. Takayama (ed.) Proc. Intern. Workshop on Shock Wave Focusing, (Sendai, Japan, 1989) pp. 41–48
30. K. Isuzukawa, M. Horiuchi: Experimental and numerical studies of blast wave focusing in water. In: K. Takayama (ed.) Proc. Intern. Workshop on Shock Wave Focusing, (Sendai, Japan, 1989) pp. 347–350
31. K. Fujiwara et al.: New methods for generating cylindrical imploding shock. In: R. Brun, L.Z. Dumitrescu (eds.) Proc. 19th Intern. Symp. on Shock Waves. vol. 4 (Springer, Berlin, Heidelberg, New York, 1995) pp. 81–86
32. F. Demmig et al.: Experiments and model computation of cylindrical shock waves with time-resolved deformation and fragmentation. In: R. Brun, L.Z. Dumitrescu (eds.) Proc. 19th Intern. Symp. on Shock Waves. vol. 4 (Springer, Berlin, Heidelberg, New York, 1995) pp. 87–92
33. T. Hiroe et al.: A numerical study of explosive-driven cylindrical imploding shocks in solids. In: R. Brun, L.Z. Dumitrescu (eds.) Proc. 19th Intern. Symp. on Shock Waves. vol. 4 (Springer, Berlin, Heidelberg, New York, 1995) pp. 267–272
34. S. Itoh et al.: Converging underwater shock waves for metal processing. In: R. Brun, L.Z. Dumitrescu (eds.) Proc. 19th Intern. Symp. on Shock Waves. vol. 4 (Springer, Berlin, Heidelberg, New York, 1995) pp. 288–294
35. R. Neemeh: Propagation and stability of converging cylindrical shocks in narrow cylindrical chamber. In: R. Brun, L.Z. Dumitrescu (eds.) Proc. 19th Intern. Symp. on Shock Waves. vol. 4 (Springer, Berlin, Heidelberg, New York, 1995) pp. 273–278
36. I.V. Volkov, S.T. Zavtrak, I.S. Kuten: Rev. E. **56**, 1, pp. 1097–1101 (1997)
37. S.T. Zavtrak, I.V. Volkov: J. Acoust. Soc. Am. **102**, 1, pp. 204–206 (1997)
38. G.E. Reisman, Y.C. Wang, C.E. Brennen: J Fluid Mech. **355**, 255–283 (1998)

39. V.K. Kedrinskii, V.A. Vshivkov, G.I. Dudnikova, Yu.I. Shokin: *Amplification of Shock Waves at Collision and Focusing in Bubbly Media*, Dokl. Akad. Nauk. **361**, 1, pp. 400–404 (1998)
40. N.K. Berezhetskaya, E.F. Bolshakov, S.K. Golubev et al.: Zh. Exp. Teor. Fiz. **87**, 6 (12), p. 1926 (1984)
41. G.B. Whitham: Fluid Mech. **22**, 2, p. 146 (1957)
42. W. Chester: Phil. Mag. **45**, 7, p. 1293 (1954)
43. R.F. Chisnell: J. Fluid Mech. **22**, 2, p. 286 (1957)
44. I.V. Sokolov: Zh. Exp. Teor. Fiz. **91**, 4(10), p.1331 (1986)
45. I.V. Sokolov: Thermofiz. Vys. Temp. **26**, 3, p. 560 (1988)
46. I.V. Sokolov: Izv. Akad. Nauk SSSR, Mekh. Zhidk. Gaza **4**, p. 148 (1989)
47. E.M. Barkhudarov, I.F. Kossyi, M.O. Mdivnishvili, I.V. Sokolov, M.I. Taktakishvili: Izv. Akad. Nauk SSSR, Mekh. Zhidk. Gaza **164**, 2, p. 164 (1988)
48. E.M. Barkhudarov, M.O. Mdivnishvili, I.V. Sokolov et al.: Piz'ma Zh. Exp. Teor. Fiz. **52**, 7, p. 990 (1990)
49. E.M. Barkhudarov, M.O. Mdivnishvili, I.V. Sokolov, M.I. Taktakishvili, V.E. Terekhin: Izv. Akad. Nauk SSSR, Mekh. Zhidk. Gaza **183**, 5, p. 183 (1990)
50. E.M. Barkhudarov, M.O. Mdivnishvili, I.V. Sokolov et al.: Shock Waves **3**, 4, p. 273 (1994)
51. V.K. Kedrinskii: *Hydrodynamics of Explosion*, 1st ed. (SB RAS Publishing House, Novosibirsk 2000)
52. V.K. Kedrinskii: *Features of Shock Wave Structure at Underwater Explosions of Spiral Charges*, Zh. Prikl. Mekh. i Tekh. Fiz. **21**, 5, p. 51 (1980)
53. A.V. Pinaev, V.K. Kedrinskii, V.T. Kuzavov: *On Shock Wave Focusing at Coil Charge Explosion in Air*, Fiz. Goren. Vzryva **37**, 4, p. 51 (2001)
54. A.V. Pinaev, V.K. Kedrinskii, V.T. Kuzavov: *Shock Wave Structure in Near Field of Spatial Charges at Explosion in Air*, Zh. Prikl. Mekh. i Tekh. Fiz. **41**, 5, pp. 81–90 (2000)
55. Z. Jiang Z., K. Takayama: Computer and Fluids **27**, 5–6 (1998)
56. V.K. Kedrinskii, V.A. Vshivkov, G.I. Dudnikova, Yu.I. Shokin, G.G. Lazareva: *Focusing of an Oscillating Shock Wave Emitted by a Toroidal Bubble Cloud*, Zh. Exp. Teor. Fiz. **125**, 5, pp. 1138–1145 (2004)
57. V.K. Kedrinskii, Yu.I. Shokin, V.A. Vshivkov, G.I. Dudnikova, G.G. Lazareva: *Shock Wave Generation in a Liquid by Spherical Bubbly Clusters*, Dokl. RAN, **381**, 6, pp. 773–777 2001
58. V.K. Kedrinskii: The Iordansky-Kogarko-van Wijngaarden model: shock and rarefaction wave interactions in bubbly media. In: A. Biesheuvel, G. van Heust (eds.) Fascination of Fluid Dynamics, (Reprint from Applied Scientific Research **58**, 1–4 (1997/1998) pp. 115–130)
59. V.M. Kovenya, A.S. Lebedev: Zh. Vych. Mat. Mat. Fiz. **34**, 6, p. 886 (1994)
60. V.M. Kovenya: Vych. Tekhnol. **7**, 2, p. 59 (1992)
61. A.I. Sychev, A.V. Pinaev: *Self-Sustaining Detonation in a Liquid with Bubbles of Reactive Gas*, Zh. Prikl. Mekh. i Tekh. Fiz. **27**, 1, pp. 133–138 (1986)
62. Kedrinskii V.K.: Shock-wave compression of a gas cavity in water. MA Thesis. (Leningrad Polytechnic Institute, Leningrad, 1961)
63. T. Hasegawa, T. Fujiwara: Detonation in oxyhydrogen bubbled liquids. In: Proc. 19th Intern. Symp. on Combustion (Haifa, 1982) pp. 675–683
64. V.K. Kedrinskii, F.N. Zamaraev: Wave amplification in chemically reactive bubble media (plenary lecture). In: W. Yong (ed.) Proc. 17th Intern. Symp.

on Shock Tubes and Waves (17-21 July, Lehigh Univ., Penns., USA, 1989) pp. 51–62.
65. A.V. Trotsyuk, P.A. Fomin: *Bubbly Detonation Model*, Fiz. Goren. Vzryva **28**, 4, pp. 129–136 (1992)
66. V.K. Kedrinskii, V.A. Vshivkov, G.I. Dudnikova, Yu.I. Shokin: *Shock Wave Interaction in Reactive Bubble Media*, Dokl. RAN **349**, 2, pp. 185–188 (1996)
67. V.Sh. Shagapov, N.K. Vakhitova: Waves in a bubbly liquid in the presence of chemical reactions in gas medium. In: V.K. Kedrinskii (ed.) Proc. XI Intern. Symp. On Nonlinear Acoustics , Ed. by Kedrinskii V.K (Novosibirsk, 1987) pp. 56–58
68. V.Yu. Lyapidevskii: *On Velocity of Bubbly Detonation*, Fiz. Goren. Vzryva **26**, 4, pp. 138–140 (1990)
69. V.K. Kedrinskii, Ch. Mader: Accidential detonation in bubbly liquids. In: H. Gronig (ed.) Proc. 16th Intern. Symp. on Shock Tube and Waves (1987) pp. 371–376
70. R. Barbone, D. Frost, A. Makris, J. Nerenberg: Explosive boiling of a depressurized volatile liquid. In: S. Morioko, L. van Wijngaarden (eds.) Proc. IUTAM Symp. on Waves in Liquid–Gas and Liquid–Vapor Two-Phase Systems (Kluwer Acad. Publ., Kyoto, 1995) pp. 315–324
71. A.A. Vasiliev, V.K. Kedrinskii, S.P. Taratuta: *Dynamics of Single Bubble with Chemical Reactive Gas*, Fiz. Goren. Vzryva **34**, 2, pp. 121–124 (1998)
72. A.V. Pinaev, A.I. Sychev: *Structure and Properties of Detonation in Systems Liquid-Gas Bubbles*, Fiz. Goren. Vzryva **22**, 3, pp. 109–118 (1986)
73. A.V. Pinaev, A.I. Sychev: *Discovery and Study of Self-Sustaining Regimes of Detonation in Systems Liquid Fuel-Bubbles of Oxidizer*, Dokl. Akad. Nauk SSSR **290**, 3, pp. 611–615 (1986)
74. A.E. Beylich, A. Gulhan: Waves in reactive bubbly liquids. In: G.E. Meier, P. A, Thompson (eds.) Proc. IUTAM Symp on Adiabatic Waves in Liquid–Vapor Systems (Springer, Berlin, Heidelberg, New York, 1990) pp. 39–48
75. T. Scarinci, X. Bassin, J. Lee, D. Frost: Propagation of a reactive wave in a bubbly liquid. In: K. Takayama (ed.) Proc. 18th ISSW, Sendai, Japan, July 1991, vol. 1, (Springer Verlag, Sendai, 1992) pp. 481–484
76. J. Kang, P.B. Butler: Combustion Science and Technology **90**, (1–4) (1993), pp. 173–192
77. V.K. Kedrinskii: Bubbly cavitation in intense rarefaction waves and its effects (plenary lecture). In: B. Sturtevant, J.E. Shepherd, H.G. Hornung (eds.) Proc. 20th Intern. Symp. on Shock Waves, Pasadena, CalTech., July 1995 (World Scientific, New York, 1996)
78. V.K. Kedrinskii, V.A. Vshivkov, G.I. Dudnikova, Yu.I. Shokin: *Role Cavitation in Disintegration Mechanics and Large Scale Explosion Processes*, Vych. Tekhnol. **2**, 2, pp.63–77 (1997)
79. F.P. Bowden, A.D. Yoffe: *Fast Reactions in Solid* (London: Butterworths Sci. Publ., 1958)
80. K.K. Andreev: *Some Consideration on the Mechanism of Detonation Initiation in Explosive*, Proc. Roy. Soc. London. A **246**, pp.257–267 (1958)
81. C.H. Johansson: *The Initiation of Liquid Explosives by Shock and the Importance of Liquid Break-Up*, Proc. Roy. Soc. London. A **246**, pp. 269–283 (1958)
82. A.V. Dubovik, V.K. Bobolev: *Shock Sensitivity of Explosive Systems* (Nauka, Moscow, 1978)

83. A.W. Campbell, W.C. Davis, J.R. Travis: *Shock Initiation of Detonation in Liquid Explosives*, Phys. Fluids **4**, 4, p. 498 (1961)
84. J.E. Field et al: *Hot-Spot Ignition Mechanisms for Explosives and Propellants*, Phil. Trans. Roy. Soc. London **339**, pp. 269–283 (1992)
85. P. Roberts, J. Field: Simulated of fragment attack on cased munition. In: Proc. 10th Intern. Det. Symp. (Boston, 1993)
86. L. Hill, B. Sturtevant: An experimental study of evaporation waves in a superheated liquid. In: G. Meier, P. Thompson (eds.) Proc. IUTAM Symp. on Adiabatic Waves in Liquid–Vapor Systems, (Springer, Berlin, Heidelberg, New York, 1990) pp. 25–37
87. H. Chaves, H. Lang, G. Meier, H. Speckmann: Flow in real fluids. In: Lecture Notes in Physics (Springer, Berlin Heidelberg New York, 1985)
88. O.M. Todes: J. Phys. Chem. **4**, 1 (1933)
89. V.K. Kedrinskii, Yu.I. Shokin, V.A. Vshivkov, G.I. Dudnikova: Shock amplification by bubbly systems with energy release (SABSER). In: Proc. 6th Japan–Russian Joint Symposium on Computational Fluid Dynamics, (Nagoya, 1998) pp. 58–61
90. V.K. Kedrinskii, P.A. Fomin, S.P. Taratuta, A.A. Vasiliev: Phase transition role in bubble detonation problems. In: Proc. 22nd Int. Symp. on Shock Waves (London, 1999) pp. 223–228
91. V.K. Kedrinskii, I.V. Maslov, S.P. Taratuta: J Appl. Mech. Tech. Phys. **43**, 2 (2002), pp. 101–109
92. A.I. Sychev: *Structure of Wave Field in Reactive Bubbly Systems in a Shock Tubes with Jump of Cross-Sections*, Combustion Explosion and Shock Waves **36**, 3 (2000)
93. A.I. Sychev: Combustion Explosion and Shock Waves **38**, 2 (2002)
94. S. Gavrilyuk, R. Saurel: J. Comput. Phys. **175**, 1 (2002), pp. 326–360
95. V. Teshukov: J. Appl. Mech. Tech. Phys. **41**, 5 (2000), pp. 879–886

6 Problems of Cavitative Destruction

6.1 Dynamics of Liquid State in Pulsed Rarefaction Waves

Fracture of a liquid in intense rarefaction waves produced by explosive loading near the free surface is a new field in the hydrodynamics of explosions. Only a few of the relevant concepts, such as critical tensile stress and tensile strength in liquids, have been used for a century. The first studies were static in nature and the first results in this field were obtained in the middle of the 1950s by Berthelot [1]. The first dynamic studies using underwater explosions near the free surface were performed by Hilliard in 1919. In a recent study by Wilson et al. [2] the strength of a liquid was estimated by measuring the velocity of a dome of sprays produced by shallow underwater explosion. The critical tensile stress, p_*, was determined from the amplitude of the wave, p, at which the velocity of the free surface $v = (2p - p_*)/\rho_0 c_0$ became zero. The resulting value of p_* was 0.85 MPa for settled tap water and 1.5 MPa for deionized and vacuum-evaporated water. Carlson and Henry [3] used high-speed loading of a thin liquid layer by a pulse electron beam and obtained the value $p_* = 60$ MPa, which may be explained by the very short loading time.

Practically all of the experiments dealt with visible discontinuities in the liquid, which developed on the cavitation nuclei under the effect of intense cavitation waves. These discontinuities (spalls) were observed by the author in experiments with underwater explosions of linear HE charges (two-dimensional problem) near the free surface (Fig. 6.1) [4].

It was noted that spalls formed only in a narrow layer near the free surface, in spite of the fact that a strong cavitation expanded over a volume that was orders of magnitude larger. The spalls were strongly cavitating layers and their structure resembled foam, which then rapidly broke into separate fragments and drops, forming the spray dome [5].

The fracture of a liquid under explosive loading involves a series of essentially nonlinear phenomena, which can be defined as *the inversion of the two-phase state of the medium*, i.e., the transformation of the cavitating liquid into a gas–droplet system. The inversion comprises the following stages:

- Formation and development of bubble clusters
- Unbound growth of the cavitation nuclei to a foamy structure

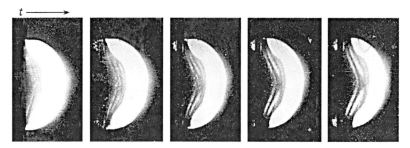

Fig. 6.1. High-speed photography of spallation in two-dimensional experiments (the free surface with the spalls is located vertically, the HE charge is on the left beyond the frame)

- Disintegration of the foamy structure into cavitating fragments
- Transition into the droplet state and its evolution

Each of these stages is self-contained and forms an integral part of the process. Therefore, the mechanisms corresponding to each stage are of great importance. In this chapter we consider basic results for each stage, including experimental data, experimental methods, and physical and mathematical models of the processes [6].

6.1.1 Real Liquid State (Nucleation Problems)

In physical acoustics, bubble cavitation has been studied for decades. Fundamental statements were primarily focused on the state of a real liquid in terms of its homogeneity and on the mechanism of bubble cluster formation and strength. The two-phase state of a real liquid should be taken into account in view of the intense character of cavitation. A mathematical model describing the cavitating liquids should be constructed to analyze the structure and the parameters of the wave field and the limiting values of the tensile stress allowed by a cavitating liquid.

In contrast to the fracture in solids, pulse loading in liquids does not involve a stage at which fracture centers develop. The macroscopic structure of the liquid is such that even when the liquid is carefully purified by distillation or deionization there are always microinhomogeneities, which act as cavitation nuclei. These can be microbubbles of free gas, solid particles, or their conglomerates (Fig. 6.2a). Figure 6.2b shows the probable size distribution of microinhomogeneities based on the results of Hammitt et al. (dots) [7] averaged over a wide spread of experimental data, Strasberg's results (crosses) [8], and those obtained in the author's laboratory [9]. Determination of the nature of these microinhomogeneities, their parameters, density, and size spectra are the basic problems of the analysis of the state of a real liquid.

The most reliable results are provided by a combination of light scattering and electromagnetic shock tubes. A typical experimental setup is shown

6.1 Dynamics of Liquid State in Pulsed Rarefaction Waves 225

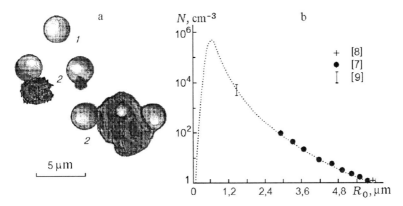

Fig. 6.2a,b. The structure of microinhomogeneities (a) and the spectrum of cavitation nuclei (b): (1) free gas bubbles and (2) combination structures

schematically in Fig. 6.3: a pulse magnetic field was generated by an electromagnetic source in a narrow gap between a membrane and a plane helical coil on which a high-voltage capacitor was discharged [9]. The features of the sources were discussed in Chap. 2. Here we note only that the parameters of the high-voltage circuit were chosen to ensure that the discharge was aperiodic and to eliminate pressure fluctuation in the shock wave generated in the liquid by a moving membrane driven by the magnetic field. This method

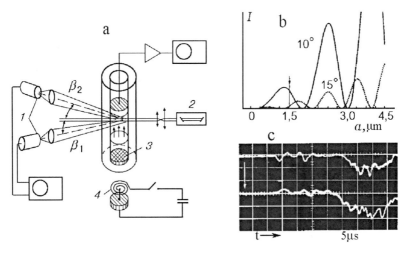

Fig. 6.3a–c. Diagram of electromagnetic shock tube (a); (1) photomultipliers, (2) an He–Ne laser, (3) a membrane, (4) a plane helical coil in the high-voltage discharging circuit; the curve of the intensity of scattered light versus the size of microinhomogeneities and the viewing angles (b); dynamics of the scattering indicatrix behind the shock-wave and rarefaction-wave fronts (c)

could produce shock waves with amplitudes of up to 10 MPa and durations of the positive phase of about 3–5 μs.

Distilled water was placed in the transparent working section of the shock tube. The light source was a He–Ne laser whose beam of diameter 1.5 mm was transmitted at a depth of 3 mm from the free surface of the liquid. The scattered light was collected by photomultipliers whose position relative to the direction of the laser beam was adjusted to a specific problem [9].

The signal from photomultipliers was fed into AD converters and a PC. A variable-capacitance transducer registered the displacement of the free surface during reflection of the shock wave. The angular distribution of the intensity of scattered light (the so-called *scattering indicatrix*) had characteristic maxima, which corresponded to concrete sizes of microinhomogeneities (Fig. 6.3b). This fact is clearly illustrated by Fig. 6.4 (curve 1) for the angle $> 32°$ where the magnification was increased by an order of magnitude. The figure also presents the calculation results obtained using the known relationships [10]

$$I = I_0|\alpha|^2 \frac{16\pi^4}{\lambda^4} v^2 (1 - \sin^2 \beta \cos^2 \varphi) f^2(q) ,$$

where $f(q) = 3(\sin q - q \cos q)/q^3$, $q = 4\pi a \sin(\beta/2)/\lambda$, $v = 4\pi a^3/3$, a is the radius of the particle, $\alpha = 3(m^2 - 1)/4\pi(m^2 + 2)$, $\lambda = \lambda_0/n$, n and m are the indices of refraction of the medium and the particle with respect to the medium, the angle β is counted from the direction of propagation of the light beam to the vector $\mathbf{R}$ (direction toward the photomultiplier), φ is the angle between the direction of the vector of the electric field $\mathbf{E}$ and the projection $\mathbf{R}$ onto the plane perpendicular to the wave vector $\mathbf{k}$. Following [11], the resulting expression for $m > 1$ can be used to determine the intensity of scattered light on free gas microbubbles if the light is polarized at a right angle to the observation plane (Fig. 6.3).

The light scattering technique was used in [12] to study microcracks within 200–2000 Å in the experiments on pulse loading of Plexiglas specimens. The

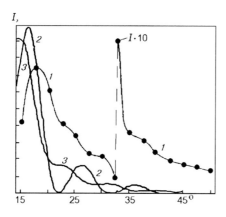

Fig. 6.4. Experimental scattering indicatrix for distilled water (1), calculated indicatrices for particles of radius 1.5 μm (2), and for a mixture of particles of radii 1.3, 1.5, and 1.7 μm (3)

scattering indicatrix was close to the Rayleigh value and consisted of one petal. The dynamics of submicrocracks was reconstructed from the change of the relation of intensities measured in two fixed directions within the zero maximum. In the case of microbubbles, the indicatrix had several petals, whose intensity, number, and position depended on the size of scattering particles. Thus, their dynamics could be reconstructed from the number of petals recorded by the photomultiplier at fixed observation angle and the requirements to the measurement accuracy of the intensity could be relaxed. Application of this method yielded a number of important results on the state of microinhomogeneities in distilled water.

Static measurements reported in [9] revealed stable existence of microinhomogeneities and proved that, in addition to the zero maximum, the scattering indicatrix had a clear maximum at $\beta = 17.5°$ and two smooth maxima at $\beta = 28°$ and $37°$. Their position unambiguously determined the radius of scattering particles as $a \approx 1.5$ μm. One could assume that the indicatrix petals were smoothened due to the size dispersion of the microparticles. To check this assumption, the integral indicatrix was calculated for the mixture of particles with radii 1.3, 1.5, and 1.7 μm (Fig. 6.4, curve 3). The calculations confirmed smoothening of the second and the third maxima to the same extent as in the experiments.

Static experiments (unperturbed liquid) with distilled water and a wavelength $\lambda = 0.63$ μm showed that the radii of the nuclei were approximately 1.5 ± 0.2 μm and were nearly monodisperse. We note that this result might be associated with a certain selectivity of the recording system, which would limit its capability of determining the true distribution.

The distribution and sizes of microparticles in distilled water were also measured using the Malvern Instrument M6.10: the distribution maximum in fresh water was about 4 μm and was 0.85 μm (with a magnetic mixer) in settled water (after 10–12 hours of settling). The experimental results of [7] for settled water were generalized in [13] to the simple relation $\sqrt{N_i}V_i = C$, where i is the bubble species, N_i is the number of bubbles of a given species, V_i is their volume, and $N \simeq 10^{-9}$.

Obviously this relation does not describe the entire distribution, which from physical considerations should have a maximum and asymptotically approached zero as the volume of bubbles approached zero and at infinity. A distribution with these properties (see Fig. 6.2b, curve) can be presented as [14]

$$N_i = N_0 \frac{(V_i/V_*)^2}{1 + (V_i/V_*)^4} . \tag{6.1}$$

This distribution involves two unknown parameters: the total number of particles per unit volume N_0 and the normalization parameter V_*, which can be taken as the volume of a bubble of radius R_* corresponding to the distribution maximum. If the above experimental value $R \simeq 0.85$ μm is taken as R_* and the "tail" of the distribution (6.1) fits the data of [7,8] for $R_i \geq 3$ μm, the total density of microinhomogeneities is $N \simeq 1.5 \cdot 10^5$ cm^{-3}, which is consistent with the experimental data, 10^5–10^6 cm^{-3}, obtained by measuring the tracks of diffraction spots from light scattered at inhomogeneities of any nature (Fig. 6.5).

Of principal importance is the question on the nature of inhomogeneities in a liquid. An original technique combining the shock tube and the scattering indicatrix method was proposed in [9]. The technique made it possible to solve the problem unambiguously. The idea was based on the essential dependence of the intensity of scattered light on the size of inhomogeneities. We selected two observation angles $\beta = 10°$ and $15°$. The intensity distribution of the light scattered at these angles is presented in Fig. 6.3b. The arrow marks the initial size of the inhomogeneities (1.5 μm).

If the inhomogeneities were the microbubbles of free gas, when the shock wave passed along the liquid the cavitation nuclei would collapse and the intensity of scattered light would vary in different ways for the selected registration angles (Fig. 6.3b), i.e., increase with respect to the background for $\beta = 10°$ and decrease for $\beta = 15°$ [9].

In the experiment, a laser beam 4 mm in diameter was focused along the tube diameter (Fig. 6.3) at a depth of 6 mm beneath the free surface of water. The oscillograms shown in Fig. 6.3c ($\beta = 10°$ for beam 1 and $\beta = 15°$ for beam 2) had three characteristic sections: oscillations due to the particle dynamics in the shock wave, reconstruction of unperturbed state, and the section of increasing intensity of oscillations and subsequent decline. The incident shock wave consisted of three attenuating pressure pulses with the amplitude of 6–7 MPa and a passage period of about 3 μs. The intensity fluctuations of scattered light coincided with the period of passage of the pulses and characterized the dynamics of the average radius of the scattering particles.

As follows from Fig. 6.3c, the experimental arrangement actually demonstrated antiphase change of signals from the photomultiplier (the first section of oscillograms) related to the decrease in size of cavitation nuclei, which was

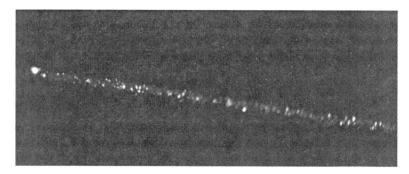

Fig. 6.5. Tracks of diffraction spots crossing the laser beam due to natural convection of water in the cuvette

6.1 Dynamics of Liquid State in Pulsed Rarefaction Waves

Table 6.1. Experimental results on acoustic diagnostics of free gas microbubbles obtained by various researchers

Author	R_0, μm	N_0, cm^{-3}	k_0
	6	≤ 1	$6 \cdot 10^{-10}$
Strasberg (1956) [8]	22	$\ll 1$	$2 \cdot 10^{-10}$
	13	$\ll 1$	$3 \cdot 10^{-10}$
Gavrilov (1970) [23]*	50–0.5	≤ 1	10^{-8}–10^{-12}
	6	$\simeq 1$	
Hammitt et al. (1974) [7]	3	$\simeq 100$	
Besov at al. (1984) [9]	1.5	10^3–10^4	$< 10^{-6}$
Kedrinskii (1989) [14]	range (Fig. 2,b)	10^5–10^6	range

* The size range observed during settling of a sample of fresh tap water over several hours

a direct prove of the existence of free gas microbubbles among the cavitation nuclei. The third section of oscillograms accounted for the synchronous growth of microbubbles in the rarefaction waves (after reflection of the shock wave from the free surface) to the initial sizes up to 4–5 μm. In the case of growth to these sizes, the upper beam registered two explicit bursts, while the lower beam registered three intensity bursts against the background of growth. Then the bubbles collapsed, which was proved by the intensity decay.

Two problems associated with the state of a real liquid are the stabilization of nuclei and their density N in a unit volume, which is directly associated with the mechanism of bubble cluster formation in rarefaction waves. Several models have been proposed to solve the first problem:

- Fluctuating cavities: $R_* = \sqrt{kT/\sigma}$ (Frenkel, 1945 [15])
- Hydrophobic particles with nuclei in the interstices: $R_* = 2\sigma/p_0$ (Harvey, 1944 [16])
- Surface organic films (Herzfeld and Fox, 1954 [17])
- Ionic mechanism (Blake, 1949 [18], and Akulichev, 1966 [19])
- Solid-particle nuclei (Plesset, 1969 [20])
- Microbubbles when the influence of heat fluxes, Stokes forces, and buoyancy forces is balanced: $R_* = (\nu^2 kT/\rho_0 g^2)^{1/7}$ (Kedrinskii, 1985 [21])
- Combination structures ("microparticles-microbubbles", Fig. 6.2; Besov, Kedrinskii and Pal'chikov, 1991 [22])

6.1.2 Formation Mechanism of Bubble Clusters

All of the above types of microinhomogeneities with their stabilization properties can exist in a real liquid and hence there is a wide spectrum of size distribution from nanometers to tens of microns (Fig. 6.2). Another problem is determining the number of nuclei N_0 in a unit of volume of cavitation zone and their volume concentration k_0.

Results on acoustic diagnostics of free gas microbubbles obtained in particular by Strasberg [8] (experimental data are given in Table 6.1) indicated very low value of N_0 and contradict to experimental data observed in developed cavitation clusters. This fact became a base to suggest an avalanche-like population of the zone of developing bubble cavitation by nuclei. For example, this model was considered in [24] to explain the development of cavitation in the focal zone of an ultrasonic concentrator ($f = 550$ kHz) registered using high-speed photography.

Figure 6.6 presents several frames of high-speed recording. One can see in the frames that immediately after application of the field only a single bubble appeared in the frame. Strasberg's results seemed to be confirmed, but after 10 periods, cavitation bubbles formed a dense cloud near the focus. It was assumed that there was an avalanche-like multiplication of nuclei caused by the instability of the shape and fragmentation of the bubbles under intense collapse. The fragments acted as new cavitation nuclei. Then the process repeated over and over again: number of nuclei in ultrasonic field was increasing continuously. Thus, one can conclude that *process of cavitation development consists in a peculiar ultrasonic pumping of the liquid by nuclei*. This mechanism is quite probable and can be physically justified for cavitation in ultrasonic fields. However, two facts fail to fit the scheme (Fig. 6.6):

1) The velocity of fragments must be high enough so that they could quickly and uniformly distribute in space (usually a dense cluster of fragments instead of bubbles appears in the shot);
2) A dense cluster of bubbles is also observed in the field of the *single rarefaction pulse* (for example, in the case of a shock wave produced by underwater explosion reflected from the free surface) when the effect of the pumping by nuclei is unable.

It should be noted that the first four results for N_0 in Table 6.1 were associated with gas nuclei (Fig. 6.2, 1), the latter accounted for microinhomogeneities of any nature, including solid nuclei (and their combinations with gas nuclei 2) at which gas–vapor bubbles could develop under the effect of tensile stresses.

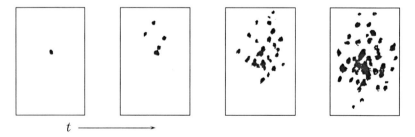

Fig. 6.6. "Reproduction" of cavitation nuclei in the focal zone of the concentrator [24] (the interval between shots is three periods, vertical size is 6 mm)

6.1 Dynamics of Liquid State in Pulsed Rarefaction Waves

A fundamentally new mechanism for the generation of cavitation zones was proposed by the author in [25]. The essence of the mechanism was as follows:

a) It was assumed that a real liquid contained a spectrum of nuclei of radii 10^{-7}–10^{-3} cm and constant density of 10^{5}–10^{6} cm^{-3};
b) The concept of visible (detectable) size of cavitation nuclei was introduced;
c) The apparent multiplication of cavitation nuclei in relatively weak ultrasonic fields was explained by the successive saturation of the zone by bubbles that needed different times for attaining visible size, depending on their initial size (Fig. 6.2b);
d) For strong rarefaction phase the entire spectrum of nuclei could reach visible size simultaneously;
e) Behind the front of strong rarefaction wave the density of saturation of the cavitation zone by nuclei reached maximum, the spectrum transformed into a monodisperse structure.

The last three assertions were based on theoretical and numerical analysis of the effects of initial size on the time required for a nucleus to grow to a given size [18]. Let us consider the simplest case of constant tensile stresses. The dynamical behavior of a bubble of volume V is described by the first integral of the Rayleigh equation

$$\dot{V}^2/V^{1/3} = 6\eta F(V, V_0, p), \tag{6.2}$$

where at $\gamma \neq 1$,

$$F = \frac{1 + \text{We}\, V_0^{-1/3}}{\gamma - 1} V_0^\gamma (V_0^{1-\gamma} - V^{1-\gamma}) - p(V - V_0) - \frac{3}{2}\text{We}\,(V^{2/3} - V_0^{2/3})$$

and dimensionless variables and parameters were introduced as follows: $V = (R/R_v)^3$, R_v is the visible bubble radius, R_0 is the radius of the nucleus, $V_0 = (R_0/R_v)^3$, $p = p_\infty/p_0$, p_∞ is pressure at infinity, $\text{We} = 2\sigma/p_0 R_v$, $\eta = p_0/\rho_0 c_0^2$, and the dot denotes a variable with respect to $\tau = t'c_0/R_v$. The function F in the right-hand part of (6.2) describes a family of curves dependent on V_0 and p. A solution exists only for those parts of the curves where $F \geq 0$.

To enable qualitative analysis of potential solutions we write the derivative as

$$F_V = (1 + \text{We}\,V_0^{-1/3})\left(\frac{V_0}{V}\right)^\gamma - \text{We}\,V^{-1/3} - p.$$

The functions F and F_V possess the following properties:

$$F(0, V_0, p) = -\infty, \quad F(V_0, V_0, p) = 0, \quad F_V(0, V_0, p) = \infty,$$

$$F_V(V_0, V_0, p) = 1 - p, \quad \text{and} \quad F_V(\infty, V_0, p) = -p.$$

232   6 Problems of Cavitative Destruction

We split the range of p into two parts: $0 \leq p < 1$ and $p < 0$. If the first inequality is valid the function F becomes curve 3 in Fig. 6.7a. Here $p = 0$, $V_0 = 0.01$ and is practically zero in the scale of the figure. The region of bubble oscillations is restricted on the right by the second point of intersection of the function F with the axis V. If $V = 1$ in this point (it is assumed that the particles of $R_v = 0.01$ cm can be seen with the naked eye), then from $F(1, V_0, p) = 0$ one can easily get the expression

$$p = \frac{1 + \text{We } V_0^{-1/3}}{\gamma - 1} V_0^\gamma \frac{V_0^{1-\gamma} - 1}{1 - V_0} - \frac{3}{2} \text{We } \frac{1 - V_0^{2/3}}{1 - V_0} = f(\text{We}, V_0),$$

which defines the threshold at which bubbles (in the oscillation mode) become visible: visible size is attained at $p \leq f$ and not attained at $p > f$.

Two types of solutions can exist in the second case for different values of $p < 0$ (Fig. 6.7). At $p = -0.01$ for any V the function $F > 0$ (Fig. 6.7a, curve 1), which implies unbound growth of the bubble. The asymptotic limit of reaching the visible size at infinite time is determined by the conditions that both the function F and its derivative F_V vanish to zero. For the above values of the parameters, this corresponds to $p = -0.0081$ and $V = 6$ (Fig. 6.7, curve 2).

Integration of (6.2) from V_0 to 1 allows us to determine the time τ_v required for the bubbles to reach visible size:

$$\tau_v = \int_{V_0}^1 \frac{V^{-1/6} dv}{\sqrt{6\eta |p|}} \quad \text{or} \quad \tau_v \simeq \frac{3}{\sqrt{6\eta |p|}} \quad \text{at} \quad R_0 \ll R_v.$$

Figure 6.7b shows these results as a function of R_0/R_v for $p = -1$ and $p = -10$ (points 1 and 2, respectively) beginning at about 10 μm (the visible size is assumed to be 100 μm). For smaller sizes the quantity τ_v at given p and $R_0 \ll R_v$ is practically independent of the initial size of cavitation radii and is defined by the above analytical dependence.

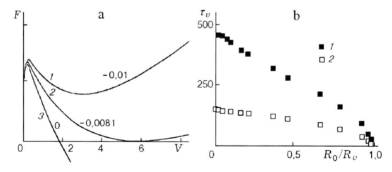

Fig. 6.7a,b. Qualitative analysis of the dynamics of cavitation nuclei in the field of tensile stresses (a) and of the time of reaching the detectable size R_v (b): $R_v = 0.01$ cm, We $= 0.015$, $\gamma = 1.4$, and $V_0 = 0.01$

For the set of points 1, we see that nuclei within $10 < R_0 < 100$ μm reach visible size gradually and approximately 12 μs after application of the stress the whole range of initial sizes within this range becomes visible. When the amplitude of tensile stresses increases by an order of magnitude, τ_v for the entire spectrum changes insignificantly (points 2, Fig. 6.7b), i.e., the visible size is achieved almost simultaneously.

At low-amplitude tensile stresses, certain effects lead to experimentally-established low saturation of the cavitation zone by detectable bubbles. For example, at $p = -0.1$, the nuclei with initial size $5 \cdot 10^{-3}$ cm become visible in 12.5 μs, the nuclei $5 \cdot 10^{-4}$ cm in size become visible after over 47.5 μs, while the nuclei about $4.3 \cdot 10^{-4}$ cm in size are a kind of threshold: the time of their growth to visible size is infinitely long. Naturally, below this threshold the nuclei are not detected, which explains the low number of bubbles in the visible cavitation cluster.

It was shown in [18] that the above reasoning and conclusions have some relation to ultrasonic cavitation. Let us consider some features of the dynamics of a single bubble in the ultrasonic wave field $p_\infty = p_0 + Ap_0 \cos(\omega t)$, where $A > 0$ and interaction starts at the compression stage (quarter-period). At $A = 1.5$ ($p = -1.5$ at the maximum of the negative phase) and $\omega = 116$ kHz the time it takes for the nuclei of the spectrum $8.6 \cdot 10^{-3}$–$8.6 \cdot 10^{-4}$ cm to reach visible size varies within 20–42 μs. They become visible during the first unloading phase (the wave period is 54 μs). As A increases, even in the case of simultaneous growth of the field frequency, a tendency to simultaneous achievement of visible size by a rather wide spectrum of cavitation nuclei is observed. Thus, for $A = 5$ and $\omega = 450$ kHz, as the nucleus radii change by a factor of 30 ($R_0 = 7 \cdot 10^{-3}$–$2.2 \cdot 10^{-4}$ cm), the time of achievement of visible size varies only within the range 7.4–12.3 μs.

6.1.3 Mathematical Model of Cavitating Liquid

The above data support the unambiguous interpretation of a real liquid as a two-phase medium, in spite of the negligibly small initial gas content (volume concentration of $k_0 \simeq 10^{-8}$–10^{-12}). In this case, it is logical to assume that the transformation of rarefaction waves (phases) in a real liquid with regard to inversion is similar to the propagation of shock waves in bubbly media. It also make sense to use the mathematical model of such a medium for describing cavitating liquid [6, 27]. This model is a combination of conservation laws for the average density ρ, pressure p, and mass velocity v, and a subsystem describing the dynamics of the bubbly medium state. The pk-model as system of equations for two basic characteristics, pressure p and volume concentration k of the gas–vapor phases, is the most convenient form to describe processes in cavitating media:

$$\Delta p - c_0^{-2} \frac{\partial^2 p}{\partial t^2} = -\rho_0 k_0 \frac{\partial^2 k}{\partial t^2}, \qquad (6.3)$$

and
$$\frac{\partial^2 k}{\partial t^2} = \frac{3k^{1/3}}{\rho_0 R_0^2}(p_0 k^{-\gamma} - p) + \frac{1}{6k}\left(\frac{\partial k}{\partial t}\right)^2. \tag{6.4}$$

The advantage of the system (6.3)–(6.4) is that it enables analytical solutions, if the appropriate physical conditions are formulated in the problem. In particular, one can assume for a real liquid that at negligible initial gas content and micron size of nuclei the bubbly system immediately reaches equilibrium behind the shock-wave front and is completely "ignored" by the shock wave.

The changing state of the medium with microinhomogeneities behind the rarefaction-wave front (or in the external ultrasonic field [28]) makes it possible to consider the assumption on the insignificance of gas in expanding bubbles of the cavitation cloud and the inertia term $k_t^2/6k$ in (6.4) as the above physical conditions. Then the dynamics of volume concentration in a bubble cluster can be described by the equation

$$\frac{\partial^2 k}{\partial t^2} \simeq -\frac{3k^{1/3}}{\rho_0 R_0^2} p.$$

Changing the right-hand part of (6.3), neglecting the compressibility of the liquid component of the medium (carrier phase), and introducing the spatial variable $\eta = \alpha r k^{1/6}$ under the assumption on validity of $|p_{\eta\eta}\eta_r^2| \gg |p_\eta \eta_{rr}|$ and $k \gg |rk_r/6|$ (where $\alpha = \sqrt{3k_0}/R_0$), we get the equation of the Helmholtz type for pressure in a cavitating liquid

$$\Delta p \simeq p. \tag{6.5}$$

Such a "heuristic" approach justified its validity in mathematical modeling of the transformation of shock waves in bubbly systems [29].

The simultaneous solution of (6.4) and (6.5) determines the parameters of rarefaction waves and the dynamics of the cavitation process [4, 27, 30]. It should be noted that comparison of these approximations with the numerical solutions of the complete system of equations and with experimental data proved that this model provides reliable estimates for the main characteristics of wave processes in cavitating liquids.

6.1.4 Dynamic Strength of Liquid

It is known that the tensile stresses at the front of a rarefaction wave are continuous and reach the maximum value after a final time Δt_*, which can be defined as the slope of the front. Solution of the axisymmetrical problem for the cavitation development at underwater explosion near free surface shows that this fact is essential in determining the limiting stress the liquid can withstand [30]. During the time Δt_* the volume concentration k of the gas–vapor phase increases by several orders of magnitude with respect to k_0 and significantly changes the state of the medium and the applied stress field. As

6.1 Dynamics of Liquid State in Pulsed Rarefaction Waves

a result, the maximum negative pressure amplitudes in a cavitating liquid can decrease by several orders of magnitude [30] with respect to those in an ideal one-phase model.

Figure 6.8a shows the profiles of shock and rarefaction waves calculated for three points at the symmetry axis (r is taken with respect to the charge radius R_{ch}, $\Delta t_* = 0.1$ μs) using the two-phase model (6.4)–(5.5) for underwater explosion of a 1.2-g charge at a depth of 5 cm (which is almost $10R_{ch}$). The rarefaction wave profiles shown by the dashed line correspond to the one-phase model, those shown by the solid line were calculated using the two-phase model of a cavitating liquid. Let us compare the data for the encircled parameters (for $r = 3.5$). Here the dashed line corresponds to the profile of the rarefaction wave with the front slope $\Delta t_* = 0.1$ μs. The maximum tensile stresses decreased to $p_{min} \simeq -17$ MPa against -100 MPa in the one-phase liquid. If the front slope increased to 1 μs, the nuclei would grow considerably over time and the cavitating liquid would keep out the growth of the tensile stresses above -3 MPa (solid line).

Figure 6.8b presents a comparison of numerical estimates of the profiles of rarefaction waves obtained at different conditions and the experimental data (dots) at a point with the coordinate $r = 4.5$ cm at the symmetry axis: a 1.2-g charge was exploded at a depth of 18.5 cm. The curve 0 was for the one-phase model, the curves 1–3 were obtained for $R_0 = 5$ μm, $k_0 = 10^{-11}$, and $\Delta t_* = 0$, 1 and 5 μs, respectively, the curve 4, for $k_0 = 10^{-10}$ and $\Delta t_* = 1$ μs. The results 2 and 4 had different volume concentrations. One can see that the experimental points were near the curve 3, which justifies the feasibility of estimating the wave field parameters.

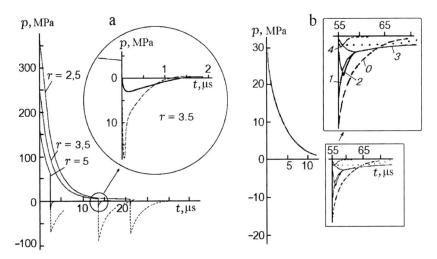

Fig. 6.8a,b. Profiles of shock and rarefaction waves at underwater explosion (a), comparison of numerical and experimental data for rarefaction wave profiles (b)

236   6 Problems of Cavitative Destruction

6.1.5 Tensile Stress Relaxation (Cavitation in a Vertically Accelerated Tube)

An interesting experimental and numerical estimate of the relaxation time of tensile stresses in a cavitating liquid can be considered by the example of cavitation produced near the bottom of a vertical tube filled with the liquid when the tube is suddenly accelerated downward by an impact (Fig. 6.9) [28].

According to the experimental data obtained by the author together with I. Hansson and K. Morch at the Danish Technical University, a zone of intense bubble cavitation developed at the bottom of the tube (Fig. 6.9). For sufficiently high acceleration amplitude, this cavitation zone could transform into a continuous gas–vapor layer, which would result in detachment of the liquid column from the bottom. The experiment was modeled numerically using the one-dimensional system (6.4)–(6.5) for the following boundary condition at the tube bottom ($z = 0$):

$$\frac{\partial p}{\partial z} = -\rho_0 a(t) \ .$$

Here $a(t)$ is the acceleration of the tube and z is the vertical coordinate. It was assumed for simplicity that the liquid occupied the entire half-space $z \geq 0$. Solving (6.5), we get an analytical dependence for $p(k)$

$$p = -\rho_0 \frac{|a(t)| \exp(-\alpha k^{1/6} z)}{\alpha k^{1/6}} \ , \qquad (6.6)$$

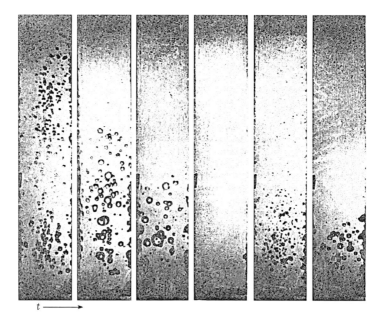

Fig. 6.9. Dynamics of the cavitation zone in a liquid that experiences downward acceleration pulse in the tube

6.1 Dynamics of Liquid State in Pulsed Rarefaction Waves

which, when substituted in (6.4), reduces the equation for volume concentration near the tube bottom to

$$\frac{d}{dt}\left(k^{-1/6}\frac{dk}{dt}\right) = \frac{3\,a(t)}{\alpha R_0^2}. \tag{6.7}$$

We want to remind readers that the gas pressure inside the bubbles has been neglected, which is justified for the purpose of estimating the relaxation time, the above equation can be solved analytically, given the explicit dependence $a(t)$.

We estimate the transformation of the pulse profile, as a model of reflection of an underwater shock wave from a free surface, assuming that the acceleration and hence the unloading pulse have exponential form. In this case, the solution of (6.7) has the form

$$k^{5/6} = 1 + \frac{5t_* a_{\max}}{2\alpha R_0^2}[t - t_*(1 - \exp(-t/t_*))]. \tag{6.8}$$

Using (6.6), we obtain main estimates for the relaxation processes in a cavitating liquid.

If the maximum tensile stress is instantaneous and the acceleration has the form $a(t) = a_{\max}e^{-t/t_*}$, one can estimate:

- The relaxation of tensile stresses (by the time $t = t_*$),

$$\frac{p}{p_{\max}} \simeq \frac{1}{e}\left[\frac{2\alpha R_0^2 e}{5t_*^2 a_{\max}}\right]^{1/5}$$

- The time of relaxation of stresses (e-fold),

$$t_{\text{cav}} \simeq 14.3(k_0 R_0^2 a_{\max}^{-2})^{1/4}$$

Simple estimates done using these equations for $k_0 = 10^{-10}$, $R_0 = 1$ μm, $a_{\max} = 5 \cdot 10^7$ cm/s² (corresponding to -30 MPa), and $t_* = 10$ μs proved that

- The amplitude of the rarefaction wave in the cavitating liquid decreased 20-fold as compared to $p/p_{\max} = 1/e$ in the one-phase liquid;
- The relaxation time of tensile stresses in the cavitating liquid was less than 0.1 μs.

If the slope of the front of tensile stresses is defined as a linear growth of the amplitude of the applied acceleration $a(t) = a_{\max}(t/t_*)$, one can easily find:

- The ultimate tensile stress,

$$p \simeq -\rho_0\left(\frac{a_{\max}^2 R_0^3}{3k_0}\right)^{2/5} t_*^{-2/5}.$$

Following this estimate, the cavitating liquid at $t = t_*$ accepts only -3 MPa instead of -30 MPa prescribed by the one-phase model. We used the following parameters: $k_0 = 10^{-11}$, $R_0 = 0.5$ μm, and $t_* = 1$ μs for the front slope.

The calculations in [30] yielded a quantity of the same order, which showed the reliability of the expressions for estimating both the relaxation time and the limiting tensile stresses.

The complete system of equations for the cavitating liquid without any assumptions (including those on the state of the gas) and with regard to heat transfer in the calculation of pressure inside a bubble was applied to the problem of a shock rarefaction tube. This is an analog of the classical shock tube with a liquid in the high-pressure chamber and a gas in the low-pressure chamber, and also that of the problem on the decay of an arbitrary discontinuity with essentially unsteady and nonlinear conditions. In (6.4), the pressure in gas p_g was determined from the known equation

$$\frac{dp_g}{dt} = 3(\gamma-1)\frac{q}{4\pi R^3} - \frac{3\gamma p_g}{R}\frac{dR}{dt},$$

using the following dependences proposed by Nigmatullin to simplify the problem on heat transfer:

$$q = 4\pi R^2 \lambda_g \, \text{Nu}\, \frac{T_0 - T}{2R}, \quad T = \frac{p_g}{(\gamma-1)c_v \rho_g} = T_0\left(\frac{R}{R_0}\right)^3 \frac{p_g}{p_{go}},$$

$$\text{Nu} = \sqrt{\text{Pe}} \quad \text{at} \quad \text{Pe} > 100, \quad \text{Nu} = 10 \quad \text{at} \quad \text{Pe} < 100,$$

and

$$\text{Pe} = 12(\gamma-1)\frac{T_0 R|S|}{\nu|T_0 - T_g|}.$$

The dynamic structure of the wave field in a cavitating liquid is shown in Fig. 6.10 [31]. It was calculated for the following conditions: $\nu = 0.001$ cm^2/s, $c_v = 0.718 \cdot 10^7$ cm^2/(s$^2 \cdot$deg), $\lambda_g = 2470$ g·cm/(s$^3 \cdot$deg), $k_0 = 10^{-4}$, and $R_0 = 50$ μm. The rarefaction wave formation at initial moments ($t = 3.3$, 10, 20, 30, and 40 μs) after the discontinuity decay is shown in Fig. 6.10a. Two wave profiles with regard to heat exchange (curve 1) and for adiabatic version (curve 2) are compared for the times 90 μs in Fig. 10b and 440 μs in Fig. 6.10c. One can see that the wave field split into two characteristic parts: a precursor formed by a centered rarefaction wave and propagating with the "frozen" speed of sound over the unperturbed liquid, and the main disturbance in the form of a wave with an oscillating front and propagating with the equilibrium speed characteristic of the two-phase bubbly medium. We note the analogy with the separation of shock waves into a precursor and a wave packet observed for bubbly media by the author in [29].

6.1.6 Transition to the Fragmentation Stage (Experimental Methods)

As noted above, for sufficiently intense rarefaction waves, bubble cavitation is characterized by unbound growth of the nuclei from the entire theoretically

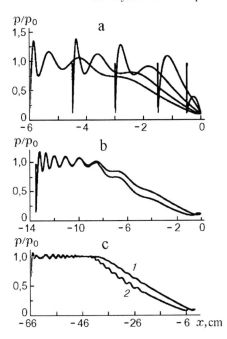

Fig. 6.10a–c. Calculation of the wave pattern in the high-pressure chamber of the hydrodynamic rarefaction tube: (a) initial moments after the discontinuity decay, (b) 90 µs later, (c) 440 µs later; curve 1 with regard to heat exchange and curve 2 for adiabatic process

possible size spectrum. Short relaxation times of tensile stresses in the cavitation zone, as compared with the typical times required for the zone to reach the volume concentration of several dozen percent, brought researchers to the idea of a two-phase model with instantaneous relaxation [32]. In this model, tensile stresses disappear instantaneously behind the rarefaction wave front, and the process of development of the cavitation zone is determined by the inertial properties of the system. This approach removes the restrictions on the volume concentration and the zone can build up until a foamy structure is formed.

Unfortunately the essentially nonlinear processes of unbound growth of bubbles in cavitating clusters, their hydrodynamics interactions in dense-packed structures with volume concentration 0.5–0.75, and the transition through the foamy structure to the separation of spalls and the formation of the droplet phase have not been completely understood yet. Therefore, more experimental information is needed and new methods should be developed to determine the features of the process. Two types of failure under pulse loading are usually considered in solid-state mechanics: plastic and brittle failure. The latter case is characterized by the discontinuities (spalls) in the form of surfaces on which critical stresses arise.

Studies of the structure of a cavitating liquid flow at underwater explosion near the free surface showed that the liquid in the intense unloading waves failed by both mechanisms and cavitating spalls were observed (Fig. 6.11) [33]. Figure 6.11a illustrates one of the initial moments of development of the cavitation zone at underwater explosion: dark bow-shaped zone below the horizontal free surface and cavity with detonation products adjacent to zone. Later (Fig. 6.11b) the cavitation zone disintegrated into several cavitating layers.

A cavitating liquid therefore displays brittle properties, which is not characteristic of a liquid. The nature of this object is still to be explained. There is still no answer to the question of where and why the spalls are formed in a large volume of the cavitating liquid; the prerequisite for their arising is possibly formed by a wave field from the very beginning.

Let us next discuss a number of methods which enable estimates of basic elements of the disintegration process. The experiments of [32] with an exploding wire showed that under axial loading of a cylindrical liquid sample, the flow split up into two zones: cavitating external and inner zones as a continuous liquid ring enclosing the cavity containing explosion products. As the process developed, the external cavitating layer should breakup into fragments (the medium is no longer continuous) and therefore a detector measuring dynamical pressure of flux at a certain distance (from the initial position of the free surface) should record a smooth transition from one state to another.

Figure 6.12b shows the "two-pulse" structure of the brake pressure in the form of three-dimensional diagrams of pressure as a function of time and the distance δ to the free surface. One can see that the pulse 1 corresponding to the flow of the cavitating liquid decreased sharply with increasing distance from the free surface. It vanished after 150–200 μs, which could be considered as a result of the formation of a foamy structure, an intermediate step in the fracture process. The pulse 3 corresponded to the layer of continuous liquid, which broke up much later due to the developing instability of the continuous liquid layer that thinned as the explosion cavity expanded. The experiment

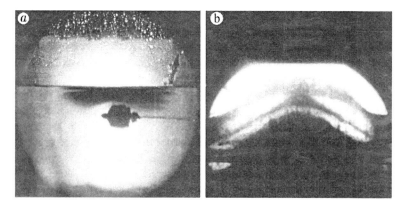

Fig. 6.11a,b. The cavitation zone under the free surface (**a**) and two stratified cavitating spalls (**b**) at shallow underwater explosion (early and late stages of the cavitation process, respectively)

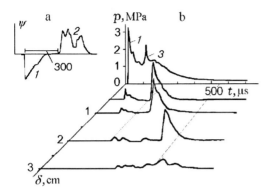

Fig. 6.12a,b. Dynamics of electric potential (**a**) and two-pulse structure of the brake pressure in the cavitating liquid shell (**b**): (1) cavitating liquid, (2) "gas-droplets" system, and (3) one-phase liquid

showed that cavitation in viscous media (for example, water solution of glycerin) could be reversible, bubbles collapsed in the cavitation zone, no fracture occurred, and two pressure pulses merged somewhat later.

The inversion can be detected more exactly by measuring the electric potential ψ in the medium (see Fig. 6.12a). It is well known that a potential difference occurs in a dispersed system when there is relative motion between the phases. The potential difference is in the direction of the relative velocity. The cause is the presence of a double electric layer at the interface and the flow-induced separation of ions adsorbed on the surfaces of bubbles or drops. Naturally, the signs of the ions will differ for each two-phase structure: the cavitating liquid (1) and the gas-droplet system (2). Thus, by measuring the instant when the potential changes sign, one can determine when inversion of the two-phase state has occurred. The measurements were carried out in [34] using a setup similar to that used in [32]. They showed that after approximately 500–600 μs the inversion process was completed.

The opacity of the intensely developing cavitation zone and the screening of its inner structure by a bubble layer on the shock tube wall made the use of standard high-speed optical recording less efficient than the use of flash X-rays [35].

The later optically-opaque stage of the cavitation process in intense rarefaction waves was studied using three X-ray devices and a two-diaphragm hydrodynamic shock tube (Fig. 6.13a). The tube consisted of three sections: a high-pressure chamber (6), an evacuated acceleration channel (with the driving piston p and two separating diaphragms d at the ends), and a working section (4) made of duralumin and containing the liquid under study. The lower diaphragm was ruptured by an electromagnetic system with a needle n. The velocity of the piston before the rupture of the upper diaphragm separating the channel from the liquid was measured by fiber-optic sensors (5). To arrange rapid-series frame-by-frame radiography, the system was synchronized with the three X-ray devices started with different preset time lags.

The dynamics of the inner structure of the cavitation zone (cluster) was studied using pulsed X-ray devices PIR-100/240 designed at the Labora-

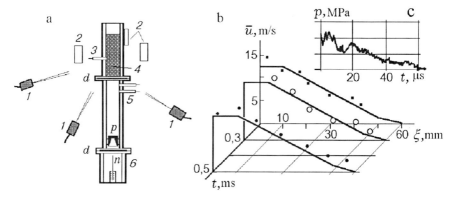

Fig. 6.13a–c. Shock tube experiments using the X-ray pulse technique (**a**): (1) X-ray apparatuses, (2) photo film, (3) pressure gauge, (4) working channel with the liquid, (5) optical sensor for measuring the piston velocity, (6) high-pressure chamber; the effect of "freezing" of the mass velocity profile in the cavitating liquid (**b**); shock wave profile (**c**)

tory of Pulse Electrophysics of the Lavrentyev Institute of Hydrodynamics (Novosibirsk). The spectrum of the γ-quantum energy of the devices was adjusted to the resolution of structural inhomogeneities in the few-centimeters-thick layers of a cavitating liquid: the average radiated energy was 70 keV, the maximum energy was up to 200 keV, and the duration of a single burst in the pulse half-height was 80 ns. The X-ray/optical schemes were as follows: a signal from the piezosensor (3) placed near the diaphragm d in the working channel 4 of the shock tube triggered both the oscillograph and the pulse generator, the latter triggered with a preset delay that the X-ray devices and the flash lamp used for illumination at the early "optical" stage.

A shock wave in the liquid sample was produced by an impact of the piston. Its positive phase lasted for several tens of microseconds and the amplitude varied within 5–30 MPa. Reflection of the shock wave from the free surface of the liquid resulted in a strong cavitation, as shown by the X-ray photographs in Fig. 6.14 for different times from zero to 1 ms. The first frame corresponds to the arrival of the shock wave front to the free surface (the diameter of the liquid column was 30 mm).

We see that after 600 μs the cavitation zone already reached dense packing with coarse gas–vapor cells. Computer processing of the density distribution of the X-ray negatives enabled analysis of the process dynamics without interfering the medium by sensors. Computer-generated versions of the experimental data could be easily handled and offered a unique opportunity to reproduce the dynamic structure of the cavitation zone (the lower limit of the resolution in the concentration k was about 2 %) in any cross section.

Figure 6.15 shows a typical computer version of one cavitation stage. The dark vertical bands at both sides of the photo are the walls of the duralumin

6.1 Dynamics of Liquid State in Pulsed Rarefaction Waves   243

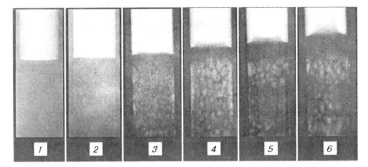

Fig. 6.14. X-ray images of the dynamics of the cavitation zone (the interval between frames is 200 μs)

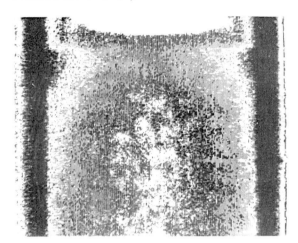

Fig. 6.15. Computer version of the X-ray imaging of the cavitation zone

tube, the zone of light spots in the center, and at the tube walls are the zones of intense bubble cavitation.

The method of construction of the computer version based on X-ray photographs shown in Fig. 6.16a was thoroughly described in [36]. It consisted in subsetting the object in a finite number of elementary cells, whose density was calculated from the digital X-ray image of the zone based on the laws of absorption and photometric relations. The path of an elementary beam of intensity I_0 (Fig. 6.16a) through the cylindrical specimen placed in the shock tube (T) included a number of elementary sections, i.e., the cross-sections of the radial line over the concentric rings that formed the sample cross-section. Radiation passed though the object onto the film, forms an image consisting of the appropriate number of pixels of size equal to the cross-section of the elementary beam. The intensity of the jth beam at the tube exit was determined by the mean density of the medium $\bar{\rho}_i$ in the ith ring and by the effective mass radiation absorption factor, which depended on the effective radiation energy and differed for the elementary beams passing through the

244   6 Problems of Cavitative Destruction

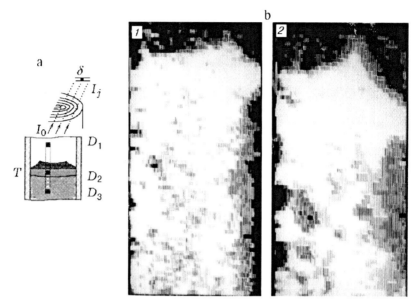

Fig. 6.16a,b. Construction scheme for a computer analog of the cavitation zone (**a**); computer version of the X-ray image of a later stage of the process (400 and 800 µs) (**b**)

walls of the duralumin tube and the cavitating sample. It was also determined by the trajectory of the jth beam in the ith ring of the liquid sample and in the shock tube wall. Naturally, all these parameters should be summed over i in the expression for I_j.

Figure 6.16b shows two frames of cavitation zones to the right of the symmetry axis after computer processing and "removing" the tube wall: the density of the dark and light zones corresponded to that of air and the light, respectively. After 800 µs there was already an intense disintegration of the sample: large voids near the axis and the sample practically separated from the tube wall, near which a dense cavitation zone was observed.

Scanned and computer-processed X-ray negatives of the cavitation zone developing behind the front of the shock wave reflected from the free surface were used. They provided information on the dynamics of the average density $\bar{\rho}$ of the cavitation zone and the time t^* required for the cavitating zone to reach the bulk density of bubbles as functions of the deformation rate of the medium $\dot{\varepsilon}$ [36] (Fig. 6.17):

$$\bar{\rho} = \rho_0(1 + \dot{\varepsilon}t)^{-1} \ . \tag{6.9}$$

Hence one could easily estimate the value of t^* as $t^* \simeq (\dot{\varepsilon})^{-1}$, which was convenient for analysis when the profile and parameters of the shock wave incident on the free surface were known.

6.1 Dynamics of Liquid State in Pulsed Rarefaction Waves   245

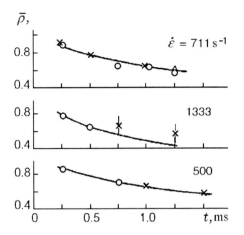

Fig. 6.17. Dynamics of the mean density of the cavitation zone based on the computer-aided analysis of X-ray photographs

Following the experimental data, the relaxation time of the medium to the foam-like state for $\dot{\varepsilon} \simeq 1330$ c^{-1} was approximately 700 μs, while for $\dot{\varepsilon} \simeq 500$ c^{-1} it was about 2000 μs, which was consistent with the above dependence.

Using this method, a tomographic image of the cavitation zone was produced, which demonstrated that the cavitation zone developed mainly in the central part of the liquid volume and was practically axisymmetrical (Fig. 6.18).

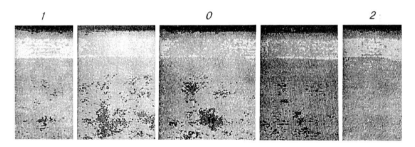

Fig. 6.18. X-ray tomogram of the cavitation zone: (0) center, (1) and (2) the left and right parts

6.1.7 "Frozen" Mass Velocities in a Cavitation Zone

Experimental studies proved that the process of cavitation damage in liquids, unlike failure in solids, may be an order of magnitude longer than the duration of the unloading wave. Nevertheless, as shown above, the spall layers can occur in cavitating liquids as well. Considering the fast relaxation of stresses behind the unloading wave produced by intense growth of cavitation nuclei, it is natural to assume that these processes are due to high gradients of mass velocities in certain parts of the cavitation zone. Distribution of these places in the unloading wave, which is apparently associated with the shock wave profile, characterizes the process of deformation of the two-phase medium.

The dynamics of distribution of mass velocities throughout the cavitation zone at initial stages of the process was studied experimentally using the successive flash X-ray recording of a sample with special lead foil traces under loading. Some traces were made of $25 \times 0.5 \times 0.04$ mm bands attached by small pieces of flimsy to lavsan bands, other 1×1 mm traces were attached by thin threads suspended independently in parallel to the sample axis [36, 37].

In all experiments, the process was recorded under identical conditions at fixed times with preliminary registration of initial position of the traces. Six photos were obtained for each experiment: three for the initial state from different observation angles and three at specific moments. The immovable trace at the external side of the shock tube fixed the initial position of the traces with respect to the tube and their shift over the given times. X-ray tubes were located at the arc of a circle with a pitch of $60°$ in the plane perpendicular to the tube axis. This enabled control over spatial arrangement of the traces and their departure from the tube walls and provided the surveillance of the zone from different sides.

The mass velocity of the medium was determined from the displacement of traces at given intervals between the X-ray frames, including the initial state. The plane of linear traces was arranged horizontally to observe the dynamics of the velocities in the cross-section of the cavitation zone. The reliability of the trace method was checked by means of an electromagnetic sensor used to measure the mass velocity profile in short shock waves [38]. The sensor was made of an aluminum-backed lavsan film whose size was close to that of linear traces. The film was placed in the field of a permanent magnet. In the experiments, we compared the data from the electromagnetic sensor and from a variable-capacitance transducer that registered the shift of the free surface. At pre-threshold loads, when no cavitation was developed, the wave profiles determined using the traces and the electromagnetic transducer coincided, while the short wave reflected from the free surface had practically the same form as the incident wave. Hence we concluded that the velocity of the traces was the same as the mass velocity of the ambient liquid.

Since we measured only three values of the average mass velocity including the static frame, the recording moments were selected so that the data obtained in different experiments partially overlapped. The thickness of the image of traces in X-ray images was determined by the out-of-focus degradation due to the finiteness of the focus of the X-ray tube (for linear traces it was 0.2 mm, for pointwise traces it depended on their size and was about 0.5 mm), which restricted the accuracy of calculations of the mean velocity. In addition, the measurement accuracy was determined by the averaging time interval. Since the velocity of traces was restricted, the time lapse between X-ray flashes was chosen so that it did not affect the resolution of displacement of the traces. Relative error related to the time lapse between flashes varied from 7 % for the interval of 0.3 ms to 20 % for the 0.1 ms interval. One of the reasons for the random inaccuracies was the formation of bubbles on the trace surface that caused their local shift.

The time dependence of the average mass velocity determined from the results of one set of experiments with identical loading conditions showed that over the first 0.6–0.8 ms, the velocity of particles was practically invariable and the difference in the trace velocities remained invariant with depth. In this run of experiments, the average amplitude of shock waves was 11.4 ± 0.6 MPa, the duration was 70 μs, and the wave profile near the front had a "step" (Fig. 6.13c).

The position of linear traces in the photos was determined by the center of lead bands. Although the radial distribution of velocity was not measured, at later times experiments revealed a bend in linear traces, which qualitatively characterized the radial structure of the flow and yielded the velocity profile typical of a viscous liquid flow.

To show the depth distribution of mass velocities, the measurement results of two identical experiments are shown by dots in the coordinates "velocity–Lagrangian coordinate–time" (Fig. 6.13b). The point of origin was related to the free surface. These results were compared with calculations following the model of instantaneous relaxation of tensile stresses behind the rarefaction wave front [32] in the following statement. Let a shock wave, whose profile is given as a certain monotone function $p_{sh} = p_{fr} f(x)$, where p_{fr} is the pressure in the shock-wave front, reaches the free surface (coordinate $x = x_0$) at the time $t = 0$. At $t > 0$ a rarefaction wave propagates with the speed of sound c_0 in the liquid with monodisperse distribution of microinhomogeneities as free gas microbubbles of 1 μm radius and nuclei concentration of $N = 10^4$ cm^{-3}. Instantaneous relaxation of tensile stresses occurs behind the wave front.

Then following the analysis of Chernobaev [32], we write the system of equations describing a flow of cavitating liquid in Lagrangian mass coordinates:

$$\frac{\partial x}{\partial t} = u, \quad \frac{\partial 1/\bar{\rho}}{\partial t} = \frac{1}{\rho_0} \frac{\partial u}{\partial \xi}, \quad \frac{\partial u}{\partial t} = 0,$$

$$p = p_0, \quad \text{and} \quad \bar{\rho}^{-1} = \rho_0^{-1} + v_{bubl},$$

where ξ is the Lagrangian coordinate, u is the mass velocity, $\bar{\rho}$ and ρ_0 are the density of the mixture and the liquid, respectively, v_{bubl} is the volume of cavitation bubbles per mixture mass unit, and p_0 is the atmospheric pressure.

If we take $\xi = x_1$, within the accepted approximation at the moment of arrival of the rarefaction wave, the pressure in the point x_1 instantaneously decreases from $p_1 = p_{fr} f(2\xi - x_0)$ to p_0, the mass velocity u_1 becomes equal to $2p_1/\rho_0 c_0$, and its distribution (by virtue of the law of conservation of momentum $\partial u/\partial t = 0$) in the cavitation zone becomes equal to $u = u(\xi)$ and is determined by the profile of the initial shock wave:

$$u(\xi) = \frac{2p_{\text{fr}}}{\rho_0 c_0} f(2\xi - x_0).$$

Now one can easily find the deformation rate of the cavitating liquid $\dot{\varepsilon} = \partial u/\partial \xi$ as

$$\dot{\varepsilon} = \frac{2p_{\text{fr}}}{\rho_0 c_0} f_\xi(2\xi - x_0)$$

and, after integrating the equation of continuity, obtain the equation for the function $\bar{\rho}(t)$:

$$\frac{1}{\bar{\rho}} = \frac{1}{\rho_0}\left[1 + \dot{\varepsilon}(\xi)\left(t - \frac{\xi}{c_0}\right)\right].$$

The subscript ξ at the function f denotes a derivative with respect to the Lagrangian coordinate. Based on the experimental pressure profile (Fig. 6.13c), we present the model velocity distributions for the appropriate times in Fig. 6.13b (solid lines). One can see that the calculated curves are adequate to the experimental data, support the tendency to the preservation of the velocity profile for different times, and suggest "freezing" of the mass velocity profile in the cavitating liquid. At later times and greater depths from the free surface one can notice the deviation of calculated curves from the experimental data due to the idealization of the pressure profile or the approximate character of the model neglecting the inhomogeneity of the macrostructure of the cavitation zone.

6.1.8 Model of "Instantaneous" Fragmentation

The experiments showed that the minimum surface density of the energy required for cavitation damage of a liquid under pulse loading was 0.15 ± 0.03 J/cm^2, which was sufficient to break a 16-mm long sample. In this case, the critical value of energy was only sufficient to bring the medium to the state of a foam structure.

Following the data of Chernobaev and Stebnovskii [32, 34], the minimum specific energy of the cavitation damage was 0.4 J/g in the case of explosive loading. Though the estimate was of general character and disregarded the inhomogeneities of the flow structure along the radius, it was valid for determining the amount of energy to be released in the center of the liquid sample to disperse it.

It follows from the experiments on explosive loading that the zone of intense cavitation can be defined by sight as the zone densely packed by bubbles. In this case, the chains of adjacent bubbles can fill the mixture volume with the shift for a bubble radius or by parallel transfer. Assuming invariable spherical shape from the initial number of nuclei of the order of 10^6 cm^{-3}, one can readily estimate that the size of liquid fractions appearing between

6.1 Dynamics of Liquid State in Pulsed Rarefaction Waves   249

touching bubbles in such structures varies within 7–25 μm, depending on the filling configuration. This estimate is done based on the central nucleus, however, the filling configuration suggests that it should have several thin trains, which in the process of dispersion and disintegration can bring micron-size droplets to the spectrum of the drop structure of the forming inversion two-phase medium.

Main features of the failure of a finite volume of liquid with the free surface (called "cavitation damage" in [4]) as a result of explosive loading can be described as follows. The reflection of a strong shock wave from the free surface leads to the formation of an unloading wave. Behind its front, a strong bubble cavitation develops on microinhomogeneities that act as cavitation nuclei.

The unbound growth of cavitation bubbles leads to the formation of the foamy structure in the "boiling" liquid. The latter is finally transformed into the gas–droplet structure in the course of inertial expansion after relaxation of tensile stresses. The duration of each phase of disintegration can differ and depend strongly on the loading dynamics. Based on the typical time scales of the process determined experimentally and numerically, one can cite the typical times obtained in laboratory conditions: the tensile stress relaxation is on the order of microseconds, dense cavitation cluster develops over tens of microseconds, the foamy structure forms over hundreds of microseconds and falls to liquid (possibly cavitating) fragments over a few milliseconds. Let us then consider the problem of simulation of the last phase of the process.

Neglecting the details of the transition from the foamy structure to the droplet state, we will assume that fragmentation occurs instantaneously as soon as the cavitation zone structure reaches dense packing. We then assume that the dense packing of bubbles is instantaneously transformed into that of non-coagulating spherical elastic liquid drops. The size spectrum of the droplets was estimated above. This model was called the sandy model, since the structure of the medium was characterized only by elastic interactions between particles [39].

Preliminary experimental comparison of the characteristic details of the process of explosive disintegration of the continuous liquid and natural sandy shells was carried out, keeping all geometrical parameters of the charge–shell system. High-speed recording of the process of failure of both types of cylindrical shells under axial explosive loading showed the identity of basic structural features of the resulting two-phase flows [6]. The high-speed video frames presented in Fig. 6.19 illustrate the characteristic dispersion pattern of the sandy shell with the flow decay into a streamer structure for a relatively small (about 3) and large (of the order of 10) calibers of the HE charge–shell assemblies.

Comparing the dispersing sandy and water shells of equal thickness, one can notice that the streamer structures of gas–droplet and sandy flows (typ-

250   6 Problems of Cavitative Destruction

Fig. 6.19a,b. Dispersion of a sandy shell under explosive loading for (**a**) relatively small (of the order of 3) calibers and (**b**) large (of the order of 10) calibers

ical of thin shells) that occurred at the same times were identical. Note also that the spherical shape was preserved at large calibers.

To study the fine structure of the flow during the formation of a sandy cloud a special device was developed for measuring the particle distribution at an arbitrary local point of the flow. The device consisted of a disk with a two-centimeter-high flat rim of about 15-cm radius mounted on a motor shaft. The disc could rotate with a linear velocity of up to 150 m/s. An adhesive tape was

6.1 Dynamics of Liquid State in Pulsed Rarefaction Waves

attached to the external perimeter of the disk with the adhesive compound outward. The entire device was mounted inside a hermetically sealed housing with a small 2×2 cm window opposite the tape and a ventilator. The trap was placed at a given distance from the charge-shell assembly and was accelerated to the required velocity in accordance with the flow duration in order to eliminate the possibility of repeated superposition of the flow. The duration of the flow was determined by analyzing the evolution of the process in the vicinity of the given point. Since the trap window was always open, it was not necessary to synchronize the startup of the trap with the process. Under explosive loading a part of the flow entered the trap, tiny particles were caught on the rotating strip. The time variation of the concentration and distribution of particles as the flow passed by the given point were thus measured.

Figure 6.20 shows two scans of the structure of the sandy flow for the caliber $6R_{\rm ch}$ recorded at distances of 0.5 and 1 m from the assembly; the time interval between frames was 1 ms. It is interesting to note that particles in the flow were stratified according to size. At 0.5 m, smaller particles were in the tail of the flow (recorded at later times), while at 1 m the trap picked up only large particles. Thus the sandy cloud to the stop instant turned out to be strongly stratified: small particles occupied the central zone and large particles were distributed at the periphery. Under careful investigation of the internal structure of the dispersing shell one can notice its turbidity in the central part due to or a return of some particles to the center from periphery or greater inertia of particles in a flow.

The observed stratification effect was studied within the frame work of above sandy model (model of instantaneous inversion of two-phase state) for a spherical HE charge with density $\rho_{\rm ch}$ and radius r_1 surrounded by two-phase shell "liquid particles–air" of external radius r_2 [39]. A volume concentration of the dispersed phase in the shell, 74 %, was assumed to be equal to the concentration of dense packing of spheres. Detonation of the charge was modelled by an instantaneous explosion at constant volume. The pressure of the detonation products was taken to be an average value and the density was $\rho_{\rm ch}$. The spherically symmetrical motion of the two-phase mixture can be described by the following system of equations of the mechanics of heterogeneous media written separately for each component:

$$\frac{\partial \rho_1}{\partial t} + r^{-2}\frac{\partial}{\partial r}(r^2 \rho_1 u_1) = 0 \,, \quad \frac{\partial \rho_2}{\partial t} + r^{-2}\frac{\partial}{\partial r}(r^2 \rho_2 u_2) = 0 \,,$$

$$\frac{\partial \rho_1}{\partial t} + r^{-2}\frac{\partial}{\partial r}(r^2 \rho_1 u_1) + \alpha_1 \frac{\partial p_2}{\partial r} + \frac{\partial}{\partial r}\alpha_1(p_1 - p_2) = -f \,,$$

$$\frac{\partial \rho_2}{\partial t} + r^{-2}\frac{\partial}{\partial r}(r^2 \rho_2 u_2) + \alpha_2 \frac{\partial p_2}{\partial r} = f \,,$$

$$\frac{\partial \rho_2 e_2}{\partial t} + r^{-2}\frac{\partial r^2 \rho_2 e_2 u_2}{\partial t} = \frac{\alpha_2 p_2}{\rho_2^0}\left(\frac{\partial \rho_2^0}{\partial t} + u_2 \frac{\partial \rho_2^0}{\partial r}\right) \,,$$

252   6 Problems of Cavitative Destruction

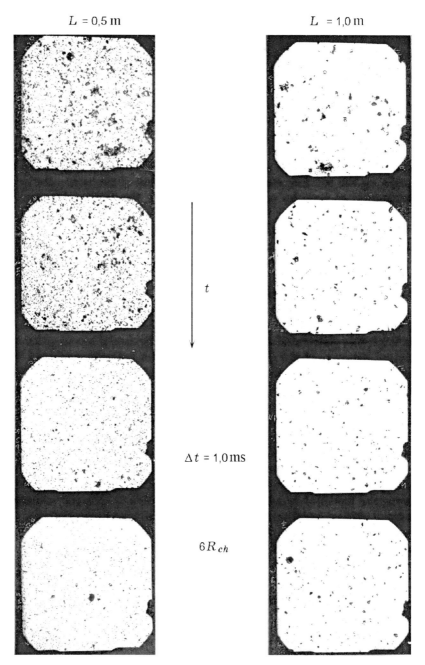

Fig. 6.20. Dispersion of a sandy shell under explosive loading: dynamics of the flow structure

6.1 Dynamics of Liquid State in Pulsed Rarefaction Waves

and

$$\frac{\partial}{\partial t}(\rho_1 E_1 + \rho_2 E_2) + r^{-2}\frac{\partial}{\partial r}[r^2(\rho_1 u_1 E_1 + \rho_2 u_2 E_2) + r^2(\alpha_1 u_1 p_1 + \alpha_2 u_2 p_2)] = 0 \; .$$

Here $\rho_i = \rho_i^0 \alpha_i$ ($i = 1, 2$), $\alpha_1 + \alpha_2 = 1$, $f = 0{,}75(\alpha_2 \rho_1^0 C_{\mathrm{d}}/d)|u_1 - u_2|(u_1 - u_2)$, where subscripts 1 and 2 refer to the gas and dispersed phases; d is the particle diameter, ρ_i, ρ_i^0, α_i, u_i, p_i, E_i, and e_i are the average and true densities, the volume concentration, the velocity, the pressure, and the total and internal energies of the ith phase, respectively. The expressions for the drag coefficients can be found in [39]. The Tait equation was used for the dispersed phase. The system of equations was closed by the assumption on simultaneous deformation of the phase: when $\alpha_2 > 0.74$ the particles deformed stacking at the vertices of regular tetrahedrons, while the surface of contact between two particles was planar, but beyond the contact points the particles preserved their spherical shape.

The particle-in-cell method was used for the numerical calculations. In view of high velocity nonequilibrium of the phases to ensure stability of calculations, the right-hand part of the momentum conservation equations was approximated as follows: one factor was chosen from the lower (in time) layer of the difference grid and another factor from the upper layer. The explosive was hexogen with density $\rho_{\mathrm{ch}} = 1.65$ g/cm^3 and caloricity of 1.32 kcal/g. The other initial parameters of the problem were $r_1 = 0.3$ cm, $r_2 = 1.5$ cm, $e = 5526$ J/g, $\rho_1 = 1.65$ g/cm^3, $\alpha_2 = 0.74$, $\rho_2^0 = 1$ g/cm^3, $u_2 = 0$, and $e_2 = 0$. The ambient medium of the droplet shell was air with initial parameters $\rho_1 = 0.001$ g/cm^3, $u_1 = 0$, and $e_1 = 250$ J/g.

The calculations were performed for three types of liquid particles: $d = 1$, 6, and 60 µm. Three stages of the process could be identified from the numerical results. The first stage was associated with the discontinuity decay on the inner boundary of the two-phase region (Fig. 6.21a), which led to

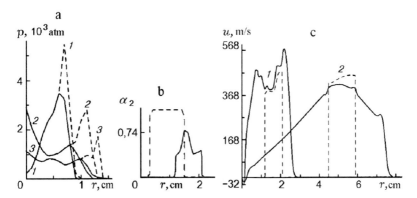

Fig. 6.21a–c. Distribution of pressure (**a**), concentration (**b**), and mass velocity (**c**) under explosive launching of a shell composed of liquid droplets

254   6 Problems of Cavitative Destruction

a rarefaction wave in the detonation products and a shock wave in the gas and dispersed phases (broken lines). In this case, the shock wave in the gas (solid line) was somewhat delayed and its amplitude was much smaller than in liquid particles of diameter 1 µm (lines 1–3 correspond to times 3, 6, and 9 µs).

The shock wave compressed the dispersed phase, which acquired a velocity that was higher than the gas velocity: the boundary of the detonation products did not keep up with the inner edge of the dispersed layer. After reaching the external boundary a diverging shock wave was produced in the air, while a rarefaction wave propagated through the particles. Together with divergent effects, the rarefaction wave led to a rapid drop in tensile stress in the particles, whose density in the shell dropped below the bulk value (the particles divided and the shell became permeable for gas phase): Fig. 6.21b shows the distribution of the concentration of the dispersed phase in the layer for the times 0 and 20 µs (solid line).

At the second stage, the kinetic energy was transferred to the dispersed phase. This stage ended after 60–70 µs, when the pressure in the detonation products went down to atmospheric pressure and the particles began to decelerate: Fig. 6.21c shows the distribution of mass velocities for $t = 20$ and 100 µs (curves 1 and 2, respectively); the diameter of particles was 6 µm.

The third stage was characterized by the cumulation of the rarefaction wave toward the center and a strong return flow of the gas (Fig. 6.22, curves 1 and 2) that decelerated and then entrained small particles toward the center (Fig. 6.22, curves 3 and 4, the velocity of liquid particles was negative), thus producing the size stratification observed experimentally. The pressure in the inner cavity increased again, the particles stopped and were then carried off by the diverging gas flow. As a result, attenuating oscillations of the two-phase flow appeared with light particles.

The dynamics of the flow external and inner boundaries for $d = 1$ µm (broken curves 1) and $d = 6$ µm (dash-dot curves 6) is shown in Fig. 6.23.

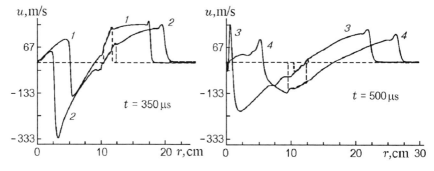

Fig. 6.22. Distribution of mass velocity of the gas and the dispersed phase (broken curves) under explosive launching of the shell of liquid droplets (later times, the third stage of the process, particles 1 µm in size): (1) $t = 300$ µs, (2) $t = 350$ µs, (3) $t = 400$ µs, (4) $t = 500$ µs

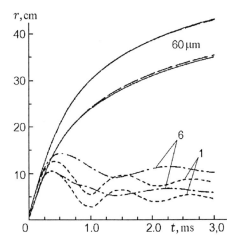

Fig. 6.23. Expansion and structure of a sandy shell under explosive loading

The amplitude and frequency of oscillations of the dispersed zone in the first case ($d = 1$ μm) was greater than in the second ($d = 6$ μm) in view of lower permeability. Thus, the wave processes in the gas phase influenced the dynamics of the inner layer boundary, which consisted of small particles and oscillated with the periodicity of reflection and focusing of waves typical for the diverging two-phase shell.

In the case of large particles where $d = 60$ μm, the resistance of the cloud was sufficiently low and though it led to rarefaction in the inner region (of the order of 50 kPa), the return flow of the gas did not effect its motion. Therefore the trajectory of heavy particles determined the external boundary of the drop cloud. This effect also produced a sharp (by about a factor of 40) increase in the thickness of the two-phase layer under dispersion, during which particles of the size range 1–60 μm by the time $t = 3$ ms occupied the space between $r_{in} = 5$ cm to $r_{ex} \simeq 45$ cm (Fig. 6.23).

The sandy model of one stage of disintegration of a liquid under explosive loading may find an interesting extension within the experimental studies of the flow structure of a dusty layer in a rarefaction wave [32]. The following experiment was performed at the Karman Laboratory of Fluid Mechanics and Jet Propulsion (California Institute of Technology) by Sturtevant: a 60-cm layer of glass balls of diameter 125 μm was packed on the bottom of a cylindrical chamber filled with air at atmospheric pressure and separated from a low-pressure chamber by a diaphragm. After rupture of the diaphragm a rarefaction wave propagated in the air gap, reached the layer boundary, reflected from it, and was partially refracted inside the layer. The experiments showed that the refracted wave induced a rapid extension of gas into the pores between the particles. Horizontal spall discontinuities of the type seen in liquids (Fig. 6.24) were observed, which then transformed into a system of cavities forming a honeycomb structure (see comparison of the X-ray shots of the cavitation zone) [40].

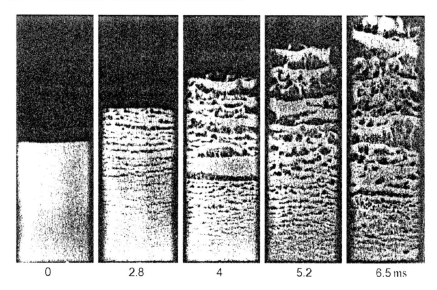

Fig. 6.24. Structure of a 125-μm layer of glass balls in the rarefaction wave

According to the experimental data of [40], the thickness of the spalls in the dusty layer was several times the particles diameter. In the upper part of the layer, the particles obtained an average acceleration of about 275 g in the first 5 ms and reached velocities of about 15 m/s. We note that the density in the dusty layer was about $5 \cdot 10^5$ cm^{-3}, which corresponded to the density of cavitation nuclei, while the layer itself simulated the state of a cavitating liquid when the bubble zone reached the bulk density.

The above analysis of the essentially nonlinear effects determining the behavior of real liquids under explosive loads showed that, in spite of the complexity of the problem, adequate physical-mathematical models describing wave processes in cavitating and disintegrating liquids could be constructed. We can conclude that some progress shown in the problem of inversion of the two-phase state was noteworthy. The above experimental methods and mathematical models provided answers to a number of fundamental questions related to the disintegration mechanics of a liquid. Unsolved problems include the mechanism of "brittle" failure of the foamy structure and the foam-droplet transition, the development of methods suitable for the entire spectrum of nuclei, the stability of combination structures of the "gas nucleus-solid particles" type, the metastable state of a liquid in the "deep" negative phase, and the kinetics of the formation of vapor centers in the front of an intense rarefaction wave.

However, it is noteworthy that lately some progress has been shown in the problem of inversion of the two-phase state concerned with the peculiarities of the flow structure of a cavitating medium at late stages of disintegration.

6.2 Disintegration of a Liquid, Spalls: Experiments and Models

6.2.1 Cavitative Destruction of a Liquid Drop

After digitizing at a laser scanner (FEAG), the X-ray images of the late stage of disintegration were processed to analyze the dynamics of density distribution. Analysis revealed fluctuations of local density, proving a strong inhomogeneity of the disintegration zone. The results were in line with the accepted model of inversion of the flow structure through an intermediary foamy state, but provided no information on the character of its structure and physical mechanism of transition to the droplet state. The following experimental setup made it possible to considerably restrict the disintegration zone and study the main elements of the flow structure. The approach was based on the disintegration of a small water drop under a microsecond-long shock. The short duration was of principal importance, since the shock wave generated in the drop should be ultrashort to result in cavitation damage.

The experimental setup (Fig. 6.25) consisted of a load-bearing element of an electromagnetic shock tube. A liquid drop whose radius varied from several millimeters to one centimeter was placed on the membrane. A pulsed magnetic field was generated between a plane spiral coil and a conducting membrane when a capacitor bank discharged on the plane spiral coil placed between the membrane and the disk D. Oscillogram of the current in an electrical circuit and membrane velocity had weakly oscillating character. The shock wave of amplitude 5–6 MPa and duration of about 3–5 μs was generated in a drop by a pulse membrane motion. The wave parameters were estimated by the dynamics of the free surface of the thin liquid layer on the membrane recorded by a variable-capacitance transducer.

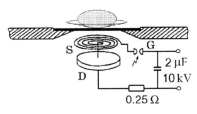

Fig. 6.25. Experimental setup for studying cavitation damage of a liquid drop

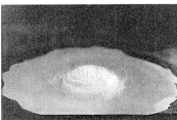 

Fig. 6.26. Disintegration of a liquid drop under shock-wave loading (general view)

258   6 Problems of Cavitative Destruction

Figure 6.26 shows two successive frames of the disintegration. The light oval with uneven edges is the membrane with a drop (in the center) at one of the intermediate disintegration stages.

The first photo shows the drop with disturbed surface in the center (approximately 100 µs after generation of the shock wave). The second shows one of the states of the disintegrating volume with spherical contours of the voids, the dispersed lower part of the drop and growing bubble clusters on the surface. Special recording with nanosecond-long flashes made it possible to study the dynamics of the fine structure of the disintegrating drop (Figs. 6.27 and 6.28) [41].

According to the experiments, the disintegration process consisted of two stages. A dense cavitation zone formed at the initial stage (Fig. 6.27). The first frame shows the initial state: the drop is transparent and through it one can clearly see the surface structure of the duralumin membrane. The impact of the diaphragm on the drop (diameters of drop was about 2 cm, of membrane was 8 cm) produced a circular cumulative jet as a thin veil and the first microclusters of cavitation bubbles of millimeter size (Fig. 6.27, frame 2). The structure of the cavitation zone was determined by the increasing number of macroclusters apparently formed during the growth of microinhomogeneities and their fusion.

To 70 µs the drop "boils up" (Fig. 6.27, frame 4) acquiring a distinct honeycomb structure. Then the cavitation zone is developing by inertia and

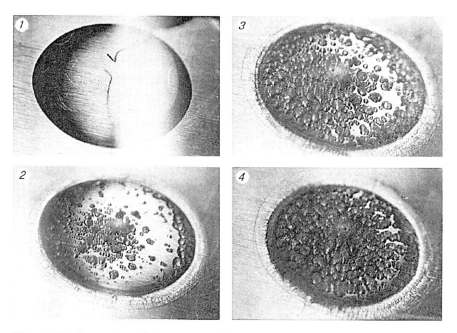

Fig. 6.27. Formation of the system of bubble clusters in a drop loaded by an ultrashort shock wave

6.2 Disintegration of a Liquid, Spalls: Experiments and Models   259

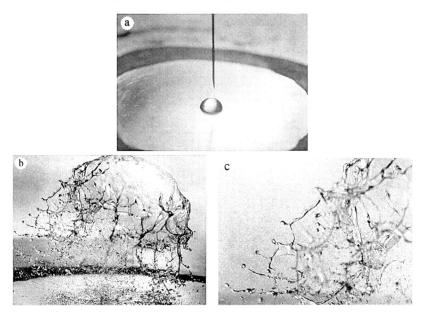

Fig. 6.28a–c. Transformation of the cavitating drop into the honeycomb structure as a grid of liquid bunches (**a**) and disintegration into liquid bunches and then into microdrops (**b**)

the drop is transforming into a spatial grid with clear zones as cells of liquid bunches (Fig. 6.28b), wherein the main mass of the initial drop was probably concentrated. The elements of the liquid grid under further stretching were gradually separated into individual fragments, small jets, which due to instability disintegrated into individual drops (Fig. 6.28c is the magnified left part of Fig. 6.28b). It is noteworthy that even in the droplet state the flow retained the cellular pattern of the structure over a long time.

Thus, the cavitation zone generated by strong rarefaction wave was a system of bubble microclusters formed at the initial stage of expansion and integration of cavitation nuclei. The inertial development of these microclusters led to the formation of a dome as a honeycomb structure of the type of a liquid grid with a thin film. The disintegration of the grid cells into individual jets and then into droplets (Fig. 6.28c) formed the basis of the inversion mechanism of the two-phase state (foam-droplet transition) under dynamic loading of a liquid [41].

6.2.2 Cavitative "Explosion" of a Liquid Drop

The dynamics of the fracture of a semispherical drop placed on the membrane of an electromagnetic shock tube was numerically studied by combining the IKW-model and the model of "frozen" profile of mass velocities [24]. The latter are realized at fast relaxation of tensile stresses in the cavitation zone.

6 Problems of Cavitative Destruction

The initial stage of cavitation and the wave field in the cavitating liquid are calculated using the IKW model which includes:

- The system of gas dynamics equations for average p, ρ, u, and v:

$$r\frac{\partial \rho}{\partial t} + \frac{\partial r\rho u}{\partial r} + r\frac{\partial \rho v}{\partial z} = 0,$$

$$\frac{\partial u}{\partial t} + u\frac{\partial u}{\partial r} + v\frac{\partial u}{\partial z} + \frac{1}{\rho}\frac{\partial p}{\partial r} = 0,$$

$$\frac{\partial v}{\partial t} + u\frac{\partial v}{\partial r} + v\frac{\partial v}{\partial z} + \frac{1}{\rho}\frac{\partial p}{\partial z} = 0$$

- The equations of state of the liquid phase and bubbly liquid:

$$p = p_0 + \frac{\rho_0 c_0^2}{n}\left[\left(\frac{\rho}{\rho_0(1-k)}\right)^n - 1\right],$$

$$\rho = \rho_l(1-k), \quad k = \frac{k_o\rho}{\rho_l(1-k_0)}\left(\frac{R}{R_0}\right)^3, \quad p_{gas} = p_0\left(\frac{R}{R_0}\right)^{3\gamma},$$

- The equation of bubble dynamics:

$$R\frac{\partial^2 R}{\partial t^2} + \frac{3}{2}\left(\frac{\partial R}{\partial t}\right)^2 = \frac{p_{gas} - p}{\rho_{liq}}.$$

This model can be considered as a classical model proven by many problems of the dynamics of the state of bubbly media. It is based on the assumption on the validity of the two-sided inequality $L \gg l \gg R$ (where L and l are the characteristic wavelength and the distance between the bubbles), though as shown in Chap. 5, it can be used with a slight violation of the condition $l \gg R$.

The process is axisymmetrical, therefore a section of the circle (half-section of the hemisphere, see Fig. 6.29) can be used as a calculation domain. In Fig. 6.29, the x-axis lies on the membrane, the z-axis is the symmetry axis, and the appropriate boundary conditions are specified on the axes. A circular arc is the external boundary of the drop at which the condition on the free boundary is specified.

The shock wave is generated in the drop by the membrane moving in the positive direction of the z-axis. In the case shown in Fig. 6.29, the membrane velocity grew linearly over 1 μs from 0 to 10 m/s, then went down to 0 over 2 μs (the shock wave amplitude is 15 MPa). The radius of the drop was 0.5 cm, the initial radius of cavitation nuclei was $R_0 = 1.5$ μm, and the volume concentration of the gas–vapor phase was $k_0 = 10^{-6}$. The data were taken from the above experiments.

The calculations show that after 2.5 μs (Fig. 6.29a) the shock wave reflects from the free surface of the droplet, a rarefaction wave propagats near the membrane from the free surface inside the bubble, and behind the rarefaction wave front an abrupt growth of bubbles and a jump of the mass velocity are observed. By $t = 14$ μs, a cloud of cavitation bubbles reached

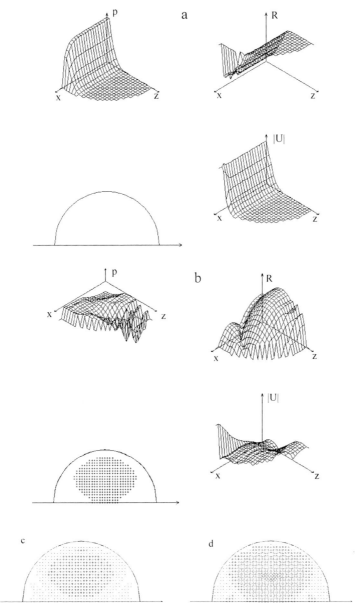

Fig. 6.29a–d. Dynamics of the distribution of the bubble radius, mass velocity, and pressure in the cavitation zone: (a) $t = 2.5$ μs; (b) $t = 14$ μs; (c) the general picture of bubble distribution in the cluster, $t = 20$ (c) and 40.5 μs (d)

the visible size $R \geq 0.1$ mm. By this moment, the tensile stresses in the cavitation zone were almost completely relaxed, the pressure gradient was taken equal to zero and thus the mass velocity field became "frozen". Under these conditions the model of frozen mass velocity can be applied and the cavitation process acquirs inertial character. The form of the cavitation zone for the two subsequent time moments ($t = 20$ and 40.5 μs, Fig.6.29c,d) shows that a zone of strong bubble cavitation was formed in the center of the drop. According to the calculation results, 60 μs after the cavitation cluster with bulk-density bubbles was formed in the very center of drop. Thus, with regard to the further development of the cavitation zone, one can assume that the dome as a grid of liquid bunches is formed as a result of a kind of "cavitative explosion" inside the drop.

6.2.3 Spall Formation in a Liquid Layer

As shown above, a liquid breakdown under explosive loading can be accompanied by spall formation. This is in essence a cavitation process, since its development, including the formation and dynamics of the wave field and the flow structure, is totally governed by the cavitation effects developing behind the front of strong rarefaction waves. It was well known that similar effects (spallation) occur in brittle solids under dynamic loading. The discovery of these effects in cavitating liquids came as a surprise. Moreover, in liquid media, the spalls have a foamy structure that later transforms into a gas-droplet system, thus completing the inversion of the two-phase state of the liquid under pulse loading.

Formation of spalls involves a series of essentially nonlinear phenomena, with fast relaxation of tensile stresses behind the unloading wave as one of principal stages, and the resulting stabilization ("freezing") of the mass velocity field in the cavitation zone.

Let's consider a shock wave reflection from free surface of thin (1 *cm* length) liquid layer [35]. Shock wave is generated by membrane motion and propagates from the left to the right. Numerical simulation of this process is based on combinatorial method, which as in the previous problem, consists in the determination of the structure and parameters of the mass velocity field in the cavitation zone using the IKvanW model and assuming that after field stabilization the cavitation process is developing by inertia to dense packing of bubbles (to the so-called bulk density when the volume concentration of vapor-gaseous phase reaches 75–80%).

Figure 6.30 shows the shock wave profile (1) before reflection from the free surface, the rarefaction wave profile (2), and the structure of the wave field in the layer to 14.7 μs (3), when the tensile stresses in a right-half of the layer (from 0.5 to 1 cm) were practically relaxed. The left-hand side of the diagram is the wave packet (rarefaction wave with an oscillating profile) reflected from the membrane, whose front reached the coordinate $x = 0.4$ cm. One could see that ahead of the wave packet, up to the free surface the pressure field was practically uniform. Hence, after 14.7 μs one could introduce the condition of stabilization of the mass velocity field. The calculations were performed for the initial concentration $k_0 = 10^{-7}$ and the initial bubble

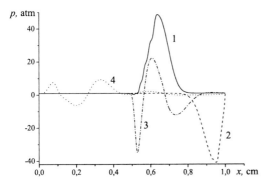

Fig. 6.30. Pressure profiles for: (1) 4.6 μs, (2) 7.4 μs, (3) 9.8 μs, and (4) 14.7 μs.

radius $R_0 = 1.5 \cdot 10^{-4}$ cm. The amplitude of the shock wave was 50 atm, the pulse duration was about 1 μs.

The results of the compatibility control of the two models above are presented in Fig. 6.31 as the values of volume concentration in an arbitrary zone calculated using the IKvanW model and the model of the "frozen" field of mass velocity. For the above conditions it was assumed that the relaxation process (pressure amplitude decreased from the initial value to 5%) was over by the time $t = 14.7$ μs, beginning from this moment the development of the cavitation zone was calculated simultaneously using both models.

We see that the calculation results almost coincided until $t = 45$ μs, when the volume concentration was somewhat higher than 20%. The lack of hydrodynamic interaction of bubbles was one of the reasons why the IKW model could not be applied to calculations of the late cavitation stages in the disintegration problem.

The characteristic velocity profiles are shown in Fig. 6.32: there was a velocity jump at the free boundary; the solid and broken lines present the velocity profile before and after the relaxation of tensile stresses, respectively.

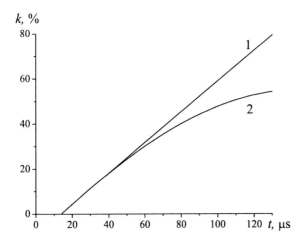

Fig. 6.31. Calculations of concentration using the frozen field model (1) and the IKvanW model (2)

264   6 Problems of Cavitative Destruction

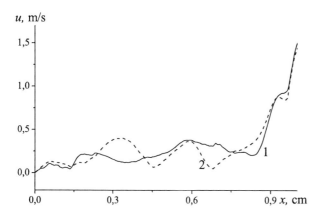

Fig. 6.32. Velocity profiles

The diagram in Fig. 6.33 shows the dynamics of distribution of concentration for three moments: 45.7, 255, and 380 μs, when the concentration corresponded to the bulk density of bubbles.

One can see two distinct cavitation zones that developed into spalls. These results allow one to make conclusion that: 1. there is the tendency to spall formation of foamy structure, 2. the calculation gives enough well qualitative description of experimental data (Fig. 6.1), and 3. the mechanism and nature of formation of spall series are determined by the oscillating structure of rarefaction wave (Fig. 6.30, curve 3, $t = 9.8$ μs) [35].

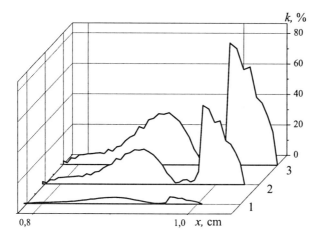

Fig. 6.33. The regions of volume concentration growth: for 45.7 μs (1), 255 μs (2), and 380 μs (3)

6.2.4 Initial Stages of Disintegration: Solids and Liquids

As we emphasized already, cavitation is a discontinuity of liquid (the initial stage of transformation of its state) in the tensile stress field accompanied by the growth of gas–vapor bubbles on cavitation nuclei. From the viewpoint of physical acoustics, one of the parameters that characterizes the cavitation strength of water is the cavitation threshold: if the pressure exceeds negative pressure, an intense growth of cavitation nuclei starts. As a result, the dynamics of the free surface of the liquid changes abruptly [71]. Depending on the measuring technique and the purification efficiency of water, the cavitation threshold varies from one to several hundreds of atmospheres [71–75], while the statistical interval of the measured quantities registered using standard methods varies within 50–100 % [74, 75], depending on the size of cavitation nuclei, their distribution fluctuations, the nonlinear dynamics of microbubbles, and the measuring technique. The pressure profile of the incident shock wave is reconstructed from the dynamics of the free surface of water for pre-cavitation reflection period.

The statistical scattering of the experimental cavitation threshold values can be considerably decreased using the capacitance technique for recording the threshold from the dynamics of the free surface of the liquid under reflection of a shock wave from it [71, 76]. As a result of reflection of the wave, a pulsed rarefaction wave is formed near the surface. The wave propagating downward initiates the growth of cavitation nuclei and the formation of a cavitation cluster (Fig. 2.5). The dynamics of the cluster, as found in [76], determines the inadequate (in terms of a uniform medium) behavior of the free surface of the liquid, which depends essentially on the amplitude of the incident shock wave and is characterized by unjustifiably (at first sight) prolonged reaction of the surface to the loading. As follows from experimental data [76], the shift of the surface is registered at a longer interval than the shock wave duration.

From the experimental data for distilled water [76], three effects can be distinguished: pre-threshold (p_{sh} = 2.9 MPa) when no cavitation was observed, threshold (p_{sh} = 3.4 MPa) when the first reaction to the shock wave loading was manifested, and cavitation (p_{sh} = 5.9 MPa), which resulted in significant changes of the dynamics of the free surface. In that case the shift of the free surface due to the growth of the cavitation cluster was comparable with its shift at the moment of reflection of the shock wave. The velocity of the free surface U corresponding to the completed reflection of the shock wave from the free boundary could be associated with the specific volume ϕ (gas content) of the formed bubbles by the simple relation $\phi = U/c$, where c is the speed of sound in water [72].

Practical observations showed that for threshold loading the value of ϕ was within 10^{-4}–10^{-3}. Table 6.2 presents the data on the dependence of the amplitude of the cavitation threshold displayed by the abrupt increase of the slope angle of the curves of the free surface shift on the duration of the SW positive phase.

These data suggest an increase in the liquid strength with decreasing time of exposure to the pulse loading. A similar effect is observed under disinte-

Table 6.2. Cavitation thresholds versus SW duration

SW duration, μs	Threshold, MPa
2.9	6.5
3.2	5.25
4	3.4
4.5	3.1
4.8	2.7
6	1.8

gration of solids. The closest phenomenon to cavitation in liquids is spall fracture in solids, when the material strength increases as the loading time decreases. For the case of threshold pulses, a weak dependence of the failure time on the amplitude of the initial loading pulse is established. The effect is called a dynamic branch. In [77,78], a structure-time failure criterion was proposed that allows one to calculate the strength increase observed in the experiments and explain the phenomena in brittle materials observed in experiments on spallation [77]. A similar criterion is used for materials displaying plastic properties, which enables effective prediction of the behavior of the dynamic yield stress [79]:

$$\frac{1}{\tau}\int_{t-\tau}^{t}\left(\frac{\sigma(t)}{\sigma_c}\right)^{\alpha}dt < 1 \qquad (6.10)$$

where σ_c is the static yield stress, τ is the incubation time related to the dynamics of the dislocation process. The parameter α characterizes the sensitivity to the stresses that lead to irreversible deformation. It is greater than or equal to one in solids and up to several tens in some materials, for example, $\alpha = 10$–30 for steels and alloys. It should be noted that the condition of irreversible growth of cavitation bubbles under critical loading that causes disintegration of the sample is applied to liquids as well [48].

To analyze the initial stage of cavitation damage we use the criterion (6.10) considering that in solids mechanics tensile stresses are positive, while in fluid mechanics tensile stresses are negative. Furthermore, one should take into account the contribution of compression. Therefore, the relation for liquids become

$$\frac{1}{\tau}\int_{t-\tau}^{t}\text{sign}(\sigma(t'))\cdot\left(\text{abs}\left(\frac{\sigma(t')}{\sigma_c}\right)\right)^{\alpha}dt' < 1 \qquad (6.11)$$

If we assume σ_c is equal to 0.1 MPa for distilled water, since the intense growth of the cavitation nuclei under the effect of tensile stresses starts with neutralizing the forces of surface tension $\sigma_c = 2\sigma_w/r$, where σ_w is the surface tension in water and $r = 1.5$ μm is the radius of cavitation nuclei [43]. We select τ and α using experimental points. The load applied in the experiments we approximate using the formula

$$\sigma(t) = -P_A \cdot \sin\left(\frac{\pi T}{T}\right) \cdot e^{-t/T_1} \quad (6.12)$$

where P_A is the pulse amplitude, T is the pulse duration, and T_1 characterizes the subsidence ratio. The maximum absolute value of the amplitude in the pulse P_m is achieved at $t = (T/\pi)\mathrm{arctg}(\pi T_1/T)$:

$$P_m = \frac{\pi T_1}{\sqrt{T^2 + \pi^2 T_1^2}} \cdot \exp\left(\frac{T}{\pi T_1}\mathrm{arctg}\left(\frac{\pi T_1}{T}\right)\right)$$

The applied load $\sigma(t)$ induces a pressure wave moving toward the surface. If the surface coordinate is zero and the time is reckoned from the moment when the wave arrives at the surface, the wave can be written as $\sigma(t + x/c)H(t + x/c)$. Upon reflection from the surface the wave acquires the form $-\sigma(t - x/c)H(t - x/c)$ and pressure in the liquid will be determined by the sum of two waves. Here $H(t)$ is the Heaviside step function, c is the velocity of propagation of the wave in the liquid ($c = 1500$ m/s). The calculations show that the experimental loading curves are well described by the formula (6.12) at $T_1 = 2.85 \cdot 10^{-6}$ s.

To calculate the critical strength we substitute normalized values of pressure for all times for which pressure is nonzero in the integral of (6.11). Then we find the time and the coordinate at which the maximum value of the integral is achieved. The sought value of P_m is the value at which in the given time instant the criterion (6.11) becomes equality in the given coordinate. The calculations show that the best agreement with the experiment is observed for α within the range from 0.4 to 0.5. In particular case at $\alpha = 0.5$ and $\tau = 19$ μs the experimental data turned out to practically coincide with the theoretical values of strength as a function of the pulse duration $P_m(T)$ calculated from (6.11).

Thus, it was established experimentally that the cavitation strength of water grows as the duration of the loading pulse decreases; the appropriate dependence is nonlinear. Application of the structure-time criterion makes it possible to calculate the experimentally observed increase in the cavitation threshold with decreasing pulse duration. The data evidence the fundamental character of the structure-time approach that provides adequate reflection of the dynamics of failure of solids and the initial disintegration stage of liquids [45].

6.3 Cluster, Cumulative Jets, and Cavitative Erosion

The experiments on laboratory modeling of ultrasonic cavitation erosion described in [28] showed that a bubble cluster formed in the liquid interlayer between the horn and the sample under the effect of tensile stresses in the rarefaction zones. Figure 6.34 illustrates a typical picture of the process recorded by high-speed camera. The frequency of horn oscillations was 20 kHz, the amplitude was 3.9 µm, and the volume of the cavitation zone was about 0.22 cm^3. One could see that bubbles throughout the zone oscillated synchronously, thus determining the dynamics of the entire cavitative cluster: its origination, attaining maximum density of the gas–vapor phase ($t = 32$ µs, Fig. 6.34), and collapse ($t = 100$ µs). However, it appeared that the frequency of cluster oscillations did not coincide with that of the external field produced by the horn.

As noted above, the gas–vapor bubbles were formed on the so-called cavitation nuclei, i.e., heterogeneous inclusions which were almost always presented in real liquids. Thus, for example, the Harvey model [46] is often used, which assumes the existence of cavitation nuclei as solid hydrophobic microparticles with slots that can contain gas or vapor nuclei. Another type of structures, which are called combinative structures, was registered experimentally in [47]. The authors showed that microbubbles could "attach" to the highly uneven surface of solid nuclei, thus ensuring their steady suspension in the liquid. This structure explained the effect of clarification of a liquid after passage of a shock wave as a result of failure of the combinative structure and settling of solid nuclei free from gas bubbles.

Numerous studies of cavitation erosion, for example, [48–53], have confirmed the importance of analysis of the interaction of a single bubble with a solid surface, whose damage was usually associated with the effect of shock waves and cumulative microjets resulting from the bubble collapse. Of particular interest were the experimental data of [50–52] on the correlation of the fine structure of the local damage zone with the hydrodynamic parameters of oscillation of a single bubble in the cavitating liquid (Fig. 6.35).

These data evidence in particular the existence of a threshold energy barrier [51] and a monotone dependence of the mass loss on the maximum bubble diameter $D_{\max}$, i.e., on the initial potential energy of the system, $U_{\max}$ [52]. The energy was converted into the energy of a conversion wave and the energy of the cumulative jet, which arises during collapse of the bubble. It is noteworthy that, following [54], the amplitude of the shock wave generated by the collapse of a single cavitation bubble was so small at a distance on the order of the initial bubble radius that the shock wave could not cause disintegration of the sample.

The data of [52] were generalized in [55] as two relations for the mass loss per single loading pulse:

$$\Delta g_1 \simeq 0.14 U_{\max} \quad \text{and} \quad \Delta g_2 \simeq 3 \cdot 10^8 U_{\max}^3 , \tag{6.13}$$

where Δg is measured in milligrams and $U_{\max}$ is measured in Joules. The first relation is used for $U_{\max} \geq U_*$ and the second for $U_{\max} \leq U_*$, where $U_* \simeq 2.1610^{-5}$ J, which, according to the authors of [48], determine the threshold of brittle fracture. The numerical coefficients and the value of $U_{\max}$ were obtained for aluminum.

As stated above, the state of the medium and its parameters changed essentially during the development of bubble cavitation. This led to the

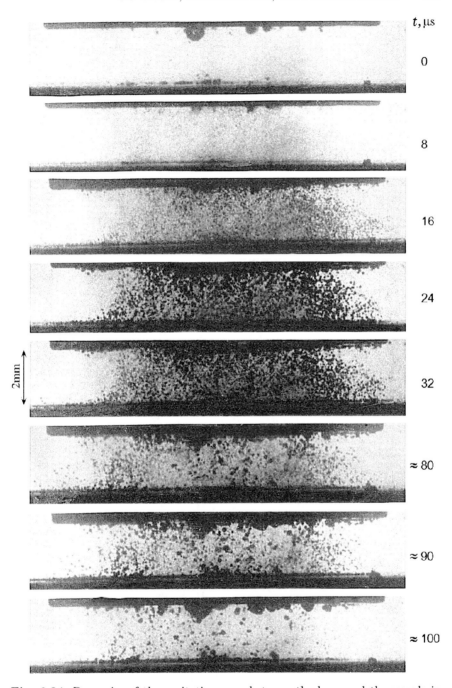

Fig. 6.34. Dynamics of the cavitation zone between the horn and the sample in the erosion tests

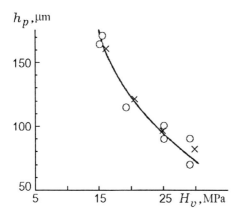

Fig. 6.35. Calculated (*crosses*) and experimental (*circles*) parameters of the damage zone under the impact of a cumulative microjet on the sample

idea of using the two-phase model of a real liquid to estimate the erosion test results. This approached is of importance, because, in spite of the local character of erosion effects, frequency of pulse loadings and their intensity should depend on the hydrodynamic characteristics of the bubble cluster and the structure of the wave field in it. The inhomogeneity of the medium in erosion tests was first considered in [13, 28]. In the following section using the two-phase model of a cavitating liquid and generalized experimental and numerical data on local fracture of the sample by cumulative microjets, we considered different approaches to the estimate of erosion effects.

6.3.1 Single Cavity and Cumulative Jets: Experiment and Models

Above we mentioned the interesting experimental results for an "incident shock wave–bubble–specimen" system obtained in [47], where the dependence of the pit depth of the cavity h_p formed by an impact of a cumulative microjet on the shock wave amplitude p_{sh} compressing the bubble near the specimen and on the hardness H_v of the specimen material were studied for aluminum (Fig. 6.35). We use these results to determine the relation between the penetration depth of the jet and the mass loss under a single load on the specimen. Obviously the frequency of the effect and its intensity can be determined only within the two-phase model of the dynamics of a cavitating liquid.

Over a fairly wide range of parameters the pit depth is $h_p \propto p_{sh}/H_v$. We should note that in the classical problems on cumulative-jet penetration into targets the microhardness H_v is involved in the condition of equality of pressures at the jet–target interface and is considered as a parameter responsible for dissipative processes:

$$\rho_j(V_j - V_p)^2/2 = \rho_m V_p^2/2 + H_v .$$

Here V_p and V_j are the penetration and the jet velocities, respectively. This condition makes it possible to determine the minimum velocity of the cumula-

6.3 Cluster, Cumulative Jets, and Cavitative Erosion

tive jet at which the jet does not penetrate into the target: $V_{j,min} = \sqrt{2H_v/\rho_j}$. Thus, for values $H_v = 14$–70 used in [41] the minimum velocity of the jet acting on the specimen varies within 170–375 m/s.

The experimental data of [41] showed that when the shock wave amplitude was $p_{sh} \leq 35$ MPa, an aluminum specimen remained intact for any H_v within the above range: $h_p = 0$. Obviously for these values of p_{sh} and bubble radius $R_0 = 0.85$ mm, the velocity of the cumulative microjet did not exceed the strength threshold of the target. These data can be interpreted as a threshold value of the initial potential energy of the system, which, in this case, was determined by $U_* \simeq 0.1$ J.

In the experiments of [41], the initial bubble volume V_0 was fixed, therefore practically $h_p \propto p_{sh} V_0$ or $h_p \propto U_{max}$, which was in agreement with the data of [41]. Taking into account the character of flow transformation under cumulation, it is reasonable to consider the dependence of h_p on the relation U_{max}/S_j, where $S_j = \pi d_j^2/4$ is the cross-sectional area of the jet. Thus, a new parameter appears that can be estimated as follows. Using the data of [59], we determine the pit diameter d_p as a function of the penetration depth h_p:

$$d_p \simeq \frac{2h_p}{0.06 + 5.6 \cdot 10^{-3} h_p} . \tag{6.14}$$

Here the dimensions are given in micrometers. The relationship between d_j and d_p is estimated within the framework of the classical cumulation theory for the problem of an infinite incompressible flow over a plate [52]

$$\frac{d_j}{d_p} = 1 - \frac{2\mu(1 + \operatorname{tg}\mu)}{\pi},$$

where in terms of the cumulation parameters $\operatorname{tg}\mu = V_p/(V_j - V_p) = \lambda$, V_p is the penetration velocity of the jets, $\lambda = \sqrt{\rho_j/\rho_m}$, ρ_m is the density of the specimen material (for equal densities, $V_p = V_j/2$). Substitution of the data on aluminum yields $d_j \simeq d_p/3$.

Finally, it is logical to assume that the jet diameter is proportional to the maximum bubble size R_* (or $y_* = R_*/R_0$). Then the semi-empirical dependence of the penetration depth on the basic parameters of the problem and on the integral of the potential energy of the system is defined by

$$h_p \simeq 11.6 R_0 \int_{y_{min}}^{y_*} \frac{p y^2 dy}{H_v} \tag{6.15}$$

(h_p and R_0, μm; p and H_v, MPa). The coefficient 11.6 is determined from the experimental data of [41].

The numerical data obtained using (6.15) are shown in Fig. 6.35 by crosses. The agreement is quite satisfactory. Following [52], when the target is made of soft materials, about 20 % of the specimen mass is ejected from the pit produced by the cumulative jet. If the pit is a cone, the ejected volume can be calculated using (6.14),

$$V_{er} \simeq \frac{h_p^3}{(0.134 + 0.0125 h_p)^2} ,\qquad (6.16)$$

where h_p is determined from (6.15). Under the assumption that precisely this mass $(\rho_m V_{er})$ determines the erosion effect (mass loss by the specimen) for plastic materials, Eqs. (6.13) allow one to estimate numerically the damage dynamics, if the density of bubbles per unit area and their oscillation frequency are known. These parameters are found by analysis of the initial state of microinhomogeneities in the liquid and by solution of the problem of a cavitation cluster.

It should be noted that the above-mentioned threshold is not the only restriction in the estimate of the erosion effect. The second basic constraint is the length of a cumulative jet. Following [53], the penetration depth L_p of the cumulative jet, the jet length L_j and density ρ_j, and the density of the target material ρ_m are related by

$$L_p = \lambda L_j .$$

Thus, a cumulative jet penetrates into a liquid only to the depth equal to its length ($\lambda = 1$). The presence of an interlayer between the cavitation bubble and the specimen wall reduces considerably the effectiveness of its action. As is clear from Kurbatskii's calculation of the axisymmetrical problem [46] of the collapse of an initially spherical hollow cavity near a solid wall, a microjet forms near the initial position of the center of the bubble even when the interlayer thickness is $L = 0.5 R_{max}$. The jet length is on the order of its half-radius ($L_j = L$), while the distance to the specimen is about three lengths of the jet. Although the velocity is fairly high (greater than 180 m/s), such a jet cannot produce any effect on the specimen.

When a bubble comes in contact with the surface, a jet with length $L_j \simeq R_{max}$ and diameter $d_j \simeq 0.2 R_{max}$ and with an almost uniform velocity field is generated. Figure 6.36 shows the dynamics of the bubble profile and the particle trajectories for various times. Profile b in Fig. 6.36a corresponds to a dimensionless time 0.812 5, profile h is for 1.090 6 and the time of collapse of a hollow cavity in an unbounded liquid is 1.092 9 (0 is the initial profile). The velocity of the jet tip at the moment of contact with the bottom boundary can be estimated in m/s as $V_j \simeq 1.3 \cdot 10^4 \sqrt{p/\rho_j}$, if pressure and density are measured in MPa and kg/m^3, respectively. Thus, the above strength threshold for the contact of a bubble with a surface can be overcome for the softest material only when the external pressure p is no less than 0.2 MPa. The kinetic energy of the jets is only 0.4 % of U_{max}, which exceeds 0.01 % markedly [48].

As to the external field, the above reasoning dealt with hydrostatics and arbitrary bubble shape before the collapse. However, in reality, a cavitation

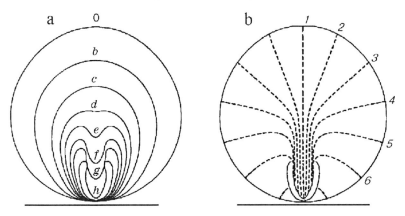

Fig. 6.36a,b. Formation of a cumulative microjet (calculations for axisymmetrical statement) (a) and particle trajectories on the cavity surface (b)

bubble grows from a nucleus near the wall in the rarefaction phase of the ultrasonic field. At the moment of maximum expansion the bubble takes the shape of an ellipsoid, whose lower part is distorted versus the initial position of the nucleus. In this case, the bubble can move some distance away from the wall. Another feature of the real process is that the horn-induced external pressure field is considerably distorted during the development of a cavitation zone because of energy consumption.

6.3.2 Bubble Cluster Effect

As was mentioned above, erosion results from the collective action of a cavitation cluster, whose dynamics determines the characteristic collapse times, the dynamics of the pressure field in the cavitation zone, and the flow structure in the vicinity of an individual bubble near the wall. To use the above estimates, one should calculate the mentioned characteristics within a two-phase mathematical model.

6.3.2.1 Settling Time and State of a Liquid

Calculations of cavitation should be based on reliable data on the initial parameters of gas content: the volume concentration k_0 and the nucleus size R_0. To study the distribution of microinhomogeneities, we performed experiments with distilled, fresh, and settled tap water using the Malvern Instruments M 6.10 equipment with a magnetic mixer [51]. The measurements were based on the method of light diffraction by microinhomogeneities in liquids. The results are summarized in Tables 6.3 and 6.4, where t^* is the settling time in hours, k^* is the volume concentration of nuclei, D is the size range in μm, R^* is the size of a nucleus in μm, and β is the percent of bubbles of the given size.

274 6 Problems of Cavitative Destruction

Table 6.3. Dynamics of gas content in a settling liquid

Medium	t^*, hour	k^*	D, µm
Tap water	0	$2.8 \cdot 10^{-4}$	40–120
	2	$8 \cdot 10^{-6}$	1.2–15
	17	$1 \cdot 10^{-6}$	1.2–10
	21	$< 1 \cdot 10^{-6}$	1.2–11
Distilled water	0	$< 1 \cdot 10^{-6}$	1.2–13
	24	$< 1 \cdot 10^{-6}$	1.2–3.4

Table 6.4. Size distribution of cavitation nuclei for settled samples

R^*, µm	β, %	
	Tap water	Distilled water
11.1	0.1	–
8.3	8.3	–
6.2	18.8	–
4.6	13.5	–
3.0	3.5	0.2
2.6	2.3	1.7
2.2	1.3	15.0
1.9	0.8	33.9
1.6	0.6	30.0
1.4	0.3	13.8

Table 6.3 presents data on the dynamics of gas content under settling of a liquid. The size distribution of cavitation nuclei for settled samples in percent is given in Table 6.4. One can see that the upper bound for tap water was much higher, and the spectra overlapped. In distilled water, the particles of sizes of 1.4–2.2 µm accounted for more than 90% of the volume content of microinhomogeneities allowed by the method. We note that the equipment had the resolution limit of the volume concentration 10^{-6}, which did not permit one to obtain qualitative data for the number of particles in the distribution because it was difficult to select a standard test sample. An estimate of the number of particles was obtained by analysis of the tracks of diffraction spots of microinhomogeneities, which proved that the total number of microinhomogeneities of any character in a sample of distilled water was up to 10^5–10^6 cm^{-3}.

6.3.2.2 Two-Phase Model. Statement of the Problem

The development of cavitation in thin liquid layers will be studied within the framework of a simple scheme that enables one to consider several variants of axisymmetrical and plane statements: an immovable rigid sphere with radius a is placed within a hollow sphere (horn of radius a_{ex}) filled with liquid and oscillating with frequency f. The size of the gap δ between them is controlled by shifting their centers L.

As mentioned above, the cavitating flow is described by the conservation laws for average characteristics and is closed by the relations for concentration k in the monodisperse mixture of bubbles. As a governing system of equations we use the system (6.4)–(6.5) [28, 29]

$$\Delta p \simeq p \qquad (6.17)$$

and

$$\frac{\partial^2 k}{\partial t^2} = \frac{3k^{1/3}}{\rho_0 R_0^2}(p_0 k^{-\gamma} - p) + \frac{1}{6k}\left(\frac{\partial k}{\partial t}\right)^2. \qquad (6.18)$$

with the spatial variable $\eta = \alpha r k^{1/6}$ [28, 29].

For the axisymmetrical problem of the development of cavitation in a narrow clearance between two spherical surfaces, the solution of (6.17) is presented as a combination of the Bessel functions $K_{n+1/2}(\zeta r)$ and the Legendre polynomials $P_n(\cos\Theta)$ and becomes

$$p = \sum_n B_n r^{-1/2} K_{n+1/2}(\zeta r) P_n(\cos\Theta).$$

Here $\zeta = \alpha k^{1/6}$ and $\alpha = \sqrt{3k_0/R_0^2}$. Below we shall restrict ourselves to two terms in the series. Then, the approximate solution of (6.17) determining the analytical dependence $p(k)$ is found as

$$p \simeq \sqrt{\pi/2\zeta}\exp(-\zeta r) r^{-1}[B_0 + B_1(1 + 1/\zeta r)\cos\Theta].$$

The coefficients in the expansion are found from the boundary conditions

$$\frac{\partial p}{\partial r} = 0 \quad \text{at} \quad r = a \quad \text{and} \quad \mathbf{n}\nabla p = -\rho_0 \beta(t) \quad \text{at} \quad r = r_*,$$

where $\mathbf{n}$ is the unit normal to the surface a_{ex}, $\beta(t) = -\mathbf{n}b\omega^2\sin(\omega t)$ is the acceleration of the surface, and b is the amplitude of the acceleration.

The pressure in the cavitation zone near the horn surface ($r = r_*$) is finally determined from the relation

$$p = p_0 + \rho_0 b r_* \omega^2 \sin(\omega t)[1 - (2+\zeta a)r_*/(1+\zeta a)a]/N, \qquad (6.19)$$

where $N = \cos\gamma\{r_*(1+\zeta r_*)[1+(1+\zeta a)^2]/a(1+\zeta a) - [1+(1+\zeta r_*)^2]\} + \sin\gamma\,\text{tg}\,\Theta(1+\zeta r_*)$, $\cos\gamma = (a_{ex}^2 - L^2 + r_*^2)/2r_*a_{ex}$, $r_* = \sqrt{a_{ex}^2 - L^2\sin^2\Theta} - L\cos\Theta$.

The substitution of (6.19) into (6.18) reduces the problem to the solution of an ordinary second-order differential equation, in which the spatial coordinate r acts as a parameter. The angle Θ is reckoned from the centerline, the coordinate origin is placed at the sphere center a, and r_* is the coordinate of the point at the radius center a_{ex}.

6.3.2.3 Analysis of the Calculation Results

According to the above data, the volume concentration k_0 was considered within the range 10^{-6}–10^{-12}. All calculations were done for $R_0 = 1$ μm. Generally speaking, with regard to the above experimental data for the nuclei range, the system of equations (6.17)–(6.18) should be complicated: a special equation of the type of (6.18) should be written for each part of the spectrum. However, as shown in [13], an original polydisperse distribution structure became monodisperse in intense ultrasonic fields (Fig. 6.37).

The calculations results for the expansion and collapse phases for bubbles (a) with various initial size (from 6 for R_1 to 2.88 μm for R_{10}) obtained within the one-dimensional two-phase model, and the pressure profile in the cavitation zone (b) are given in Fig. 6.37b. The synchronous collapse showed that bubbles became equal in size by the moment the maximum size was reached.

Using the above estimates of the erosion damage, we try to simulate the experimental result of [46]. In this case the scheme of two spheres should have the following geometrical parameters: $a_{\mathrm{ex}} = 20$ cm, $a = 1$ cm, displacement of the centers $L = 18.95$ cm, and the gap between the "horn" surface and the sample $\delta = 0.5$ mm. The equipment frequency is 14.5 kHz.

A study of fine structure of the pressure profile as a function of k_0 shows that an increase in its values from 0 (or 10^{-12}), at which the field amplitude decreases appreciably although the wave profile remains unchanged) to 10^{-6} results in the gradual formation of the loading pattern "allowed" by the cavitating liquid [51]. In cavitating liquids, the prevailing part of the horn-induced waves is adsorbed, and the loading rate of the sample increases noticeably. This gives rise to irregular peak loads with a fairly high amplitude in the

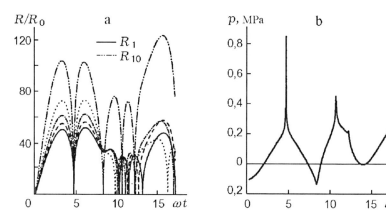

Fig. 6.37a,b. Numerical results for the bubble spectrum in the cavitation zone (a) and for pressure in the center of the cavitation zone (b): $f = 20$ kHz, $b = 3$ μm, $R_1 = 6$ μm, and $R_{10} = 2.8$ μm.

gap. The behavior and parameters of the cavitation bubble change abruptly as compared with the dynamics of a single bubble [51].

We use the numerical interpretation of experimental data of [48] in (6.13) from the estimate of the sample mass loss per single loading pulse and try to relate the erosion rate to the potential energy of the bubble $U_{\max}$. It should be noted that the values are due to the fatigue effects emerging under long cyclical loading of the sample, therefore the estimates for a single load may appear strongly averaged.

The potential energy $U_{\max}$ is determined by the integral $\int p\,dv$ and summation of it over all oscillations gives the dynamics of the mass loss rate W_t^*. The calculation results for an oscillation amplitude $b = 25$ μm and $k_0 = 10^{-8}$ averaged over the entire current time interval show that with time the value of W_t^* tends to stabilize at the level of about 0.1 mg/s. Taking into account the closeness of the microhardness of the samples $H_v = 140$ [48] and $H_v = 133$, these calculations can be extrapolated to the tests of [45] with soft cermet materials. The dynamics of the mass loss $W(t)$, which, following [45], is close to the linear function $W \simeq 2(t-5)/3$ (t, min; W, mm^3) within the range 10–30 min can be estimated from the function W_t for the time 10–30 min.

Figure 6.38 shows the data of [49] on the surface profile subject to erosion for the times 2, 6, and 18 min with different scales of vertical erosion depth. The data on the dynamics of the weight loss $W(t)$ and the loss rate W_t (mm^3/min) for this experiment are shown in Fig 6.39. The numerical estimates for $W^*(t)$ (dashed lines and crosses) and the value of W_t^* (encircled asterisk) for $t = 40$ min obtained under the above assumptions are also presented here. Evidently, the orders of magnitude are almost the same.

The analysis shows that the two-phase model combined with approaches of the classical cumulation theory allows one to get estimates of erosion effects on the extension of test results with accuracy to an order of magnitude,

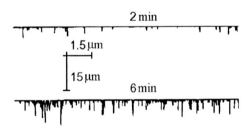

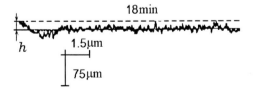

Fig. 6.38. Experimental data on surface erosion: characteristic dynamics of the damage structure [43]

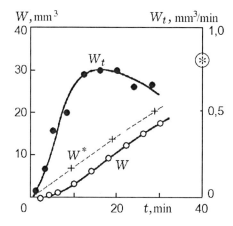

Fig. 6.39. Experimental (*circles*) [43] and numerical (*dashed line* and *crosses*) data on the velocity W_t^* and the volume of the erosion zone (*encircled cross*)

without considering the complete problem of the cavitation damage of the samples. The calculation of the development of cavitation bubbles at microinhomogeneities [58] with allowance for the transformation of the pressure field in the cavitating liquid shows that there is a certain position of the nuclei that determines an optimal (in terms of cavitation erosion) relation between the velocity of the cumulative jet, its length, and the distance to the sample surface.

6.4 Cavitative Clusters and Kidney Stone Disintegration Problem

Numerous experimental investigations confirm that the physics of cavitation is directly related to the wide spectrum of phenomena studied within the frame work of medical applications of shock-wave and ultrasonic diagnostics for special aims. Micro bubbles and cavitation clusters, their interaction with shock waves (SW) and SW generation, micro bubble pulsations and cumulative micro jets play both the positive (a kidney stone disintegration by ESWL SW) and undesirable (destruction of tissues and cells) roles in the processes of damage, destroying and treatment.

The effect of rarefaction phases of ESWL shock waves as well as its focusing on the target as one of elements of the mechanism of kidney stone destruction will be discussed. And as the preface the short survey of some biomedical problems will be presented to show the important role of SW and bubble clusters in their solution.

6.4.1 Shock Waves, Bubbles and Biomedical Problems

It's well known that the tensile stresses arising in a human body under the application of shock-wave and ultrasonic methods of treatment or diagnostics

6.4 Cavitative Clusters and Kidney Stone Disintegration Problem

can tend to cavitative phenomena as well as in a case of normal liquids. Often for the study of these phenomena it's required to determine the special (including 3D) mathematical models. With this point of view the results published in [59–61] become interesting.

Taking into account the translational motion and the deformation of each bubble induced by mutual interactions among the bubbles the governing equations of three-dimensional motions for an arbitrary configuration of N bubbles were derived by Takahira et al. [59]. The authors have investigated the dynamics of a bubble cluster in a liquid by means of a series expansion of spherical harmonics and have shown, in particular, that the lowest natural frequency of the bubbles is much lower than the frequency of an isolated bubble. It was shown that the heat transfer inside the bubble is important for investigating the nonlinear bubble oscillations and the subharmonic oscillation of interacting bubbles occurs more easily than that of an isolated bubble due to bubble–bubble interactions [60].

Chapter 5 [1,3] an effect of collective behavior of gas bubbles in two-phase layer under a shock wave was discussed. One can note that the authors [61] studying numerically nonlinear oscillations of spherical bubbles in a sound field have detected the analogous results. It was shown that when bubbles with different initial radii oscillate in a sound field, the independent oscillation of each bubble is suppressed by bubble interactions and the bubbles take on a collective behavior.

Interesting experiments were performed by Ceccio et al. [62] to examine differences in bubble cavitation inception, form, and acoustic emission in fresh and salt water. Differences detected are attributed to variations in the freestream nuclei population, and it is hypothesized that the solution of salt reduced the number and size of freestream nuclei. Differences in the overall acoustic emission were attributed to variation in the bubble event rate and average maximum bubble size between the fresh and salt water cavitation.

An overview of the basic physical principles of shock waves, and the history and basic research behind shock-wave use in medicine for disintegration of kidney and ureteral stones, in orthopaedics and traumatology was presented in [63]. One noted that the theory of shock-wave therapy for orthopaedic diseases involves the stimulation of healing processes in tendons, surrounding tissue and bones. This is a completely different approach from that of urology, where shock waves are used for disintegration. Thiel considers [64] that the concept of orthopaedic disorders is that shock waves stimulate or reactivate healing processes in tendons, surrounding tissue and bones, probably through microdisruption of avascular or minimally vascular tissues to encourage revascularization, release of local growth factors, and the recruitment of appropriate stem cells conducive to more normal tissue healing.

It's well known that lithotripsy is a common effective treatment for kidney stones. However, focal volumes are often larger than stones, and surrounding tissue is often injured. Sokolov, Bailey and Crum [65] have suggested new dual-pulse lithotripter consisting of two opposing, confocal and simultaneously triggered electrohydraulic sources to accelerate stone fragmentation and to reduce cell lysis in vitro. Model gypsum stones and human erythrocytes were exposed to dual pulses or single pulses. The results of tests in vitro have shown that at the focus, model stones treated with 100 dual pulses at a charging voltage of 15 kV broke into eight times the number of fragments as stones treated with 200 single pulses at 18 kV.

Using an experimental system that mimics stone fragmentation in the renal pelvis, the role of stress waves and cavitation in stone comminution in shock-wave lithotripsy (SWL) was investigated in [66]. Spherical plaster-of-Paris stone phantoms (D = 10 mm) were exposed to 25, 50, 100, 200, 300 and 500 shocks at the beam focus of a Dornier HM-3 lithotripter operated at 20 kV and a pulse repetition rate of I Hz. This apparent size limitation of the stone fragments produced primarily by stress waves (in castor oil) is likely caused by the destructive superposition of the stress waves reverberating inside the fragments, when their sizes are less than half of the compressive wavelength in the stone material. On the other hand, if a stone is only exposed to cavitation bubbles induced in SWL, the resultant fragmentation is much less effective than that produced by the combination of stress waves and cavitation. It is concluded that, although stress-wave-induced fracture is important for the initial disintegration of kidney stones, cavitation is necessary to produce fine passable fragments, which are most critical for the success of clinical SWL. Stress waves and cavitation work synergistically, rather than independently, to produce effective and successful disintegration of renal calculi in SWL. One can mention that an analogous effect was predicted in the results of the numerical studies in [67].

Several mechanisms of kidney stone fragmentation in extracorporal shock-wave lithotripsy (ESWL) were also under discussion in [68]. As a new mechanism, the circumferential quasistatic compression or "squeezing" by evanescent waves in the stone has been introduced. The fragmentation was studied in the experiments with self-focusing electromagnetic shock-wave generators with focal diameters comparable to or larger than the stone diameter. A quantitative model of binary fragmentation by "quasistatic squeezing" was developed. This model predicts the ratio of the number of pulses for fragmentation to 2-mm size and of the number of pulses required for the first cleavage into two parts. This "fragmentation ratio" depends linearly on the stone radius and on the final size of the fragments.

Stone comminution and tissue damage in lithotripsy are sensitive to the acoustic field within the kidney, but yet knowledge of shock waves in vivo is limited. In this connection Cleveland et al. [69] have made measurements of lithotripsy shock waves inside pigs with small hydrophones. A hydrophone was positioned around the pig kidney following a flank incision. It appeared that a combination of nonlinear effects and inhomogeneities in the tissue broadened the focus of the lithotripter. The shock rise time was on the order

6.4 Cavitative Clusters and Kidney Stone Disintegration Problem 281

of 100 ns, substantially more than the rise time measured in water, and was attributed to higher absorption in tissue.

A novel, less invasive, shock wave source that can be introduced into an arbitrary position in a human body percutaneously has been developed [70]. The shock wave source consists of an explosive, an optical fiber, a balloon catheter, and a Nd:YAG laser, which generates a spherical explosive shock wave. The destructive potential of the present source for injuring tissue was confirmed. The subsequent cell elongation and split in the direction of the shock wave has been observed.

According to the prior study [70], a rapid recanalization therapy of cerebral embolism, using liquid jet impacts generated by the interaction of gas bubbles with shock waves, can potentially penetrate through thrombi with very efficient ablation. The studies in [71] were undertaken to examine the liquid jet impact effect on fibrinolysis in a tube model of an internal carotid artery. The shock wave was generated by detonating a silver azide pellet weighing about a few μg located in a balloon catheter. Thrombi were formed using fresh human blood and gelatin. The fibrinolysis induced by the liquid jet impact with urokinase was explored as the percentage of the weight loss of the thrombus. The results suggest that liquid jet impact thrombolysis has the potential to be a rapid and effective therapeutic modality in recanalization therapy for patients with cerebral embolism and other clinical conditions of intra-arterial thrombosis.

The interaction of air bubbles attached to gelatin surfaces, extirpated livers or abdominal aortas of rats with underwater shock waves was investigated in [72] to help clarify the tissue-damage mechanism associated with cavitation bubbles induced during extracorporeal shock-wave lithotripsy. The bubble attached to gelatin or a rat's liver surface migrates away from the surface with an oscillatory growth/collapse behavior after the shock-wave interaction. The penetration depth of the liquid jet into the gelatin and the radius of the subsequent damage pit on the surface depend on the initial bubble radius.

The effect of extracorporeal shock waves on hemoglobin released from red blood cells was recently found to be minimized under minute static excess pressure [73]. The experiments carried out have shown that a dominant mechanism of shock wave action is a shock wave–gas bubble interaction.

A method for real-time in vitro observation of cavitation on a prosthetic heart valve has been developed. Cavitation of four blood analog fluids (distilled water, aqueous glycerin, aqueous polyacrylamide, and aqueous xanthan gum) has been documented for a Medtronic/Hall prosthetic heart valve [74]. The observations were made on a valve that was located in the mitral position, with the cavitation occurring on the inlet side after valve closure on every cycle. Stroboscopic videography was used to document the cavity life cycle. Bubble cavitation was observed on the valve occluder face. For each fluid, cavity growth and collapse occurred in less than 1 ms, which provides strong evidence that the cavitation is vaporous rather than gaseous. The cavity duration time was found to decrease with increasing atrial pressure at constant aortic pressure and beat rate. The area of cavitation was found to decrease with increasing delay time at a constant aortic pressure, atrial pressure, and bear rate. Cavitation was found to occur in each of the fluids, with the most cavitation seen in the Newtonian fluids (distilled water and aqueous glycerin).

As a rule, in studies of cells or stones in vitro, the material to be exposed to shock waves (SWs) is commonly contained in plastic vials. It is difficult to remove air bubbles from such vials. The attempt to determine whether the inclusion of small, visible bubbles in the specimen vial has an effect on SW-induced cell lysis was made [75]. It was found that even small bubbles led to increased lysis of red blood cells and that the degree of lysis increased with bubble size. Thus, bubble effects in vials could involve the proliferation of cavitation nuclei from existing bubbles. Whereas injury to red blood cells was greatly increased by the presence of bubbles in vials, lytic injury to cultured epithelial cells was not increased by the presence of small air bubbles. This suggests different susceptibility to SW damage for different types of cells. Thus, the presence of even a small air bubble can increase SW-induced cell damage, perhaps by increasing the number of cavitation nuclei throughout the vial, but this effect varies with cell type.

Impulsive stress in repeated shock waves administered during extracorporeal shock-wave lithotripsy (ESWL) causes injury to kidney tissue. In a study of the mechanical input of ESWL, the effects of focused shock waves on thin planar polymeric membranes immersed in a variety of tissue-mimicking fluids have been examined [76]. A direct mechanism of failure by shock compression and an indirect mechanism by bubble collapse have been observed. Thin membranes are easily damaged by bubble collapse. After propagating through cavitation-free acoustically heterogeneous media (liquids mixed with hollow glass spheres and tissue), shock waves cause membranes to fail in fatigue by a shearing mechanism. As is characteristic of dynamic fatigue, the failure stress increases with strain rate:, determined by the amplitude and rise time of the attenuated shock wave. Shocks with large amplitude rind short rise time (i.e., in uniform media) cause no damage. Thus the inhomogeneity of tissue is likely to contribute to injury in ESWL. A definition of dose is proposed which yields a criterion for damage based on measurable shock wave properties.

The role of shock waves in the treatment of soft tissue pain is at present unknown. There is a potential for further therapeutic applications of shock waves since shock waves exert a strong biological effect on tissue, which is mediated by cavitation. Experiments using shock waves for tumor therapy have shown some promising results [77], yet devices that generate waveforms other than lithotripters are probably better suited. Shock waves cause a transient increase in the permeability of the cell membrane, and this might lead to further applications of shock waves.

6.4 Cavitative Clusters and Kidney Stone Disintegration Problem   283

Characteristics of the underwater shock waves and of ultrasound focusing were studied by Takayama [78] by means of holographic interferometric flow visualization and polyvinyliden-difluoride (PVDF) pressure transducers. These focused pressures, when applied to clinical treatments, could effectively and noninvasively disintegrate urinary tract stones or gallbladder stones. However, as it was noted, despite clincal success, tissue damage occurs during ESWL treatments, and the possible mechanism of tissue damage is briefly described.

Miller and Song studied lithotripter shock waves with cavitation nucleation agents and shown that they can produce tumor growth reduction and gene transfer in vivo [79]. Cavitation nucleation agents (CNA) can greatly enhance DNA transfer and cell killing for therapeutically useful applications of nonthermal bioeffects of ultrasound (US) and shock waves (SW). Either saline, Optison US contrast agent, a vaporizing perfluoropentane droplet suspension (SDS) or air bubble were injected intratumorally at 10% of tumor volume as a CNA. In some tests, droplets or contrast agent were injected IV. Shock waves (SW) were generated from a spark-gap lithotripter at 7.4 MPa peak negative pressure amplitude. IV injection of Optison or droplet nucleation agents before SW treatment reduced tumor growth to factors of 1.0 and 0.7, but did not increase transfection. These results demonstrate the efficacy of CNA in vivo and should lead to improved strategies for simultaneous SW tumor ablation and cancer gene therapy.

6.4.2 Some Results on Modelling of ESWL Applications

In his survey [55] Groening gave a brief review of extracorporeal shock wave lithotripsy (ESWL) and distinguished the works of Jutkin, Goldberg (1959), Forssman (1975), Chaussy et al., Reichenberger and Naser, Ziegler and Wurster (1986), Muller (1987), Delius and Brendel, Takayama et al., and Coleman and Saunders (1988), in which the structure and parameters of the wave field in the focus zone were studied in detail. But the cavitation effects in them were simulated primarily as a result of the interaction of shock waves with artificially produced gas bubbles and had no connection to cavitation, as a new medium state, which considerably varies the wave-field parameters [6] and the conditions at which bubble collapse with cumulative microjet formation is possible.

Physical aspects of wave focusing in the context of ESWL were analyzed by Sturtevant [56], who raised the question on the mechanism of disintegration of kidney stones and on the role of the rarefaction wave in it. According to [56], strong rarefaction waves following the shock wave front lead to reduced duration of its positive phase, the sharply peaked waveform, and to limitation of the amplitude. When such a wave penetrates into the target, it induces shear stresses leading to spalls.

Kuwahara [57] cited experimental data for the disintegration process recorded by the high-speed video camera. Following his data, when a shock wave falls on the target, first an intense cavitation zone develops around the stone; thereafter, the stone deforms and disintegrates. In Kuwahara's opinion, the disintegration mechanism is governed by the cavitation bubbles.

Similar effects were observed by Kitayama et al. [58], who noted that after the wave focusing, pressure around the target reduces sharply, which leads to the formation of a cavitation zone. It is assumed that over this time cracks grow around the stone, then they are widened to a certain maximum size, and finally a collapse occurs. Then the second stage of expansion starts when the cracks grow further and the stone disintegrates. This is only a suggestion, since there is no observational evidence for this process, because the object is shielded by the ambient cavitation cloud.

Grunevald et al. [59] proposed an acoustic scheme for calculating the pressure field in the shock wave focus that takes into account the shift of the source. Delius [60] considered three possible mechanisms of tissue damage by shock waves: thermal effect, direct mechanical effect, and the indirect effect called cavitation. His estimates showed that for typical wave forms generated in the systems one can expect a temperature decrease by about 2 °C for a pulse frequency of 100 Hz. Delius related the damage to the formation of cumulative microjets produced either by asymmetrical collapse of the bubble near the target solid wall or by the interaction of the cavitation bubble with the shock wave [61]. It should be noted that this is one of the typical mechanisms of the cavitation erosion in ultrasonic fields, but under ultrasonic erosion there arise cumulative microjets whose frequency is on the order of

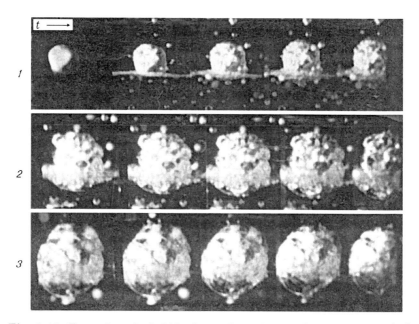

Fig. 6.40. Formation of a bubble cluster formation on the stone surface [60]

6.4 Cavitative Clusters and Kidney Stone Disintegration Problem 285

tens of kilohertz, which does not happen in lithotriptor systems due to the low-frequency generation of the sequence of shock waves and the complex structure of the cavitation zone near the target that has the form of a dense cloud of bubbles [58].

Figure 6.40 presents three intervals (1–3) of the formation of a bubble cluster on the stone surface under shock-wave focusing (Delius' experiments, 1990). The time between frames in each interval was 40 μs and the time lag between successive intervals was 200 μs. The shock wave was generated by Dornier Lithotriptor with the amplitude of 65 MPa in the front and a rarefaction phase of maximum amplitude −6 MPa. A stone was positioned onto a metal foil in the geometrical focus of the lithotriptor.

Typical shock wave profiles with the rarefaction phases (Fig. 6.41) and their interaction with microbubbles of micron sizes are presented in [62, 63]. Prat [64] concluded that a shock-induced cavitation has a potential for selective and noninvasive destruction of deep stomach tumors. For this purpose, an outgoing pulse is a sequence of rarefaction and compression phases.

Experimental investigation of the "pressure–density" diagram in human tissues and tissue-simulating gelatin and the numerical simulation aimed at predicting the level of high pressure generated in tissues and its effect on the tissues were reported in [65]. In particular, the paper described hysteresis effects found under blood examination when the density of the medium did not come back to the initial values after removal of the load. It should be noted that a similar effect for distilled water was first described in [42]. Its nature appeared to be determined by the two-phase state of the medium containing microinhomogeneities that played the role of cavitation nuclei and changed their structure under the effect of the wave field. Probably such a mechanism

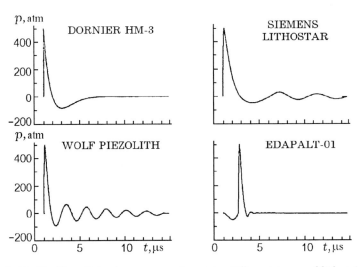

Fig. 6.41. Profiles of shock waves produced by different types of lithotriptors

determines the nature of hysteresis found in blood [65]. On the other hand, some researchers, for example, [66], doubt the key role of cavitation in the disintegration of kidney stones.

Sato et al. [67] proposed using soft-flash radiography to visualize shock-induced cavitation zones in living tissues. Earlier, the method of flash X-rays was used for the same purposes in studying the dynamics of irreversible cavitation processes in distilled water with high-density gas–vapor phase [35, 68].

Isuzugawa et al. [69] studied experimentally different types of reflectors in order to generate spherical shock waves in water and focus them in certain types of oils.

The list of references could be continued, but the explanations of the reasons for disintegration would remain the same as mentioned above: cumulative jets arising under collapse of cavitation bubbles and/or shear stresses produced by ultrashort strong shock waves inside the target and resulting in development of microcracks in it.

On the basis of the above statements and known experimental facts, we shall analyze other possible mechanisms of target disintegration.

6.4.3 Hydrodynamic Model of the Disintegration in the Cavitation Zone

Figure 6.40 shows that focusing of the shock wave with the rarefaction phase results in total locking of the target by a dense bubble layer already at the initial stage of evolution of the cavitation cluster. Future behavior of the cluster and the effect of its dynamics on the target cannot be analyzed and predicted directly.

However, some principal points of the probable effect of the cluster on the target can be simulated by the example of the development of cavitation effects on the bottom of a vertical tube containing liquid and driven by a downward acceleration from an impact [28]. Obviously, in this case, due to inertial properties the liquid will tend to budge, while the motion of the tube bottom will simulate a piston removed from the liquid: a rarefaction wave will propagate in the liquid. Upon sufficiently strong impact (and accordingly at a certain critical amplitude of the rarefaction wave) the bottom can separate from the liquid. This effect can be considered as an unbound development of the cavitation cluster near the tube bottom up to the formation of a foam-like structure or practically continuous gas–vapor layer. Thus, we return to the inversion of the two-phase state of the liquid. However, now we are interested in other consequences of the effect.

Figure 6.42 demonstrates the same sequence of the events near the bottom of the tube that received the acceleration a_* of the type of a symmetrical half-wave with an amplitude of about $2 \cdot 10^4$ m/s^2 and the pulse width of about 300 μs. Acceleration and pressure gauges were built in the tube bottom. Their oscillograms are presented in Figs. 6.43 and 6.44, respectively. The experiments were run by the author at the Danish Technical University within

6.4 Cavitative Clusters and Kidney Stone Disintegration Problem 287

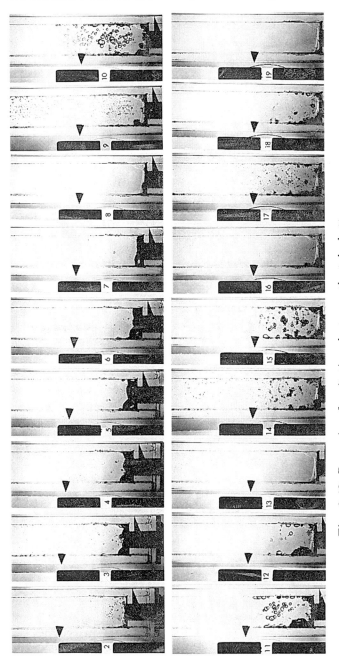

Fig. 6.42. Dynamics of cavitation cluster on the tube bottom

288   6 Problems of Cavitative Destruction

a joint research project on cavitation erosion executed in collaboration with
I. Hansson and K. Morch.

The interval between frames in Fig. 6.42 was 1 ms, the frame number
minus one was the time (in ms). One can see that under the effect of tensile
stresses a dense cavitation zone was formed near the tube bottom, which by
4 ms transformed into a gas–vapor layer and practically separated the liquid
from the tube bottom. Then, under the effect of the pressure difference the
layer started collapsing and disappeared after 7–8 ms. At this moment the
gauges recorded an abrupt jump of acceleration with an amplitude exceeding a_* (Fig. 6.43, $t = 7.735$ ms), and pressure pulse with an amplitude of
about 1.5 MPa and duration of 1.5 ms (Fig. 6.44b). In Figure 6.44 part (a) is
the oscillogram with a very short sweep time of about 2 ms (the recorded profile of tensile stresses on the tube bottom was transformed by the cavitation
process as compared to the above profile); part (b) is the oscillation series
against the background of the rarefaction phase (full scan was 10.235 ms), of
which the strongest oscillation coincided in time with the second acceleration
jump (see Fig. 6.43); and part (c) are two strong successive oscillations.

It is obvious that the second acceleration jump and pressure pulse were
caused by the hydraulic impact of the liquid on the tube bottom under the
collapse of the gas–vapor layer. It should be noted that the trace (Fig. 6.42,
black triangle on the left) fixed to the liquid moving down shifted only downward in all the frames. Therefore, an assumption on tube oscillations due to
elasticity of the spring supporting it (see the photos) was excluded.

As a result of the hydraulic shock, the tube received a second acceleration
(Fig. 6.43) and a new cavitation cluster started developing near the tube
bottom (Fig. 6.42, frames 9–12). This process was repeated several times:
the acceleration (Fig. 6.43) and pressure (Fig. 6.44) oscillograms recorded
recurring hydraulic shocks.

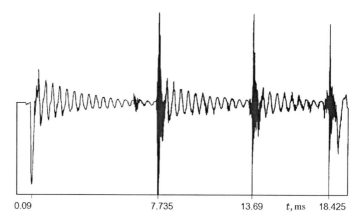

Fig. 6.43. Oscillogram of the acceleration of the tube bottom recording the recurring hydraulic shocks

6.4 Cavitative Clusters and Kidney Stone Disintegration Problem 289

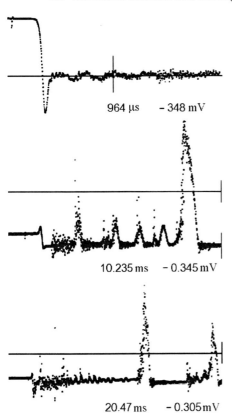

Fig. 6.44. Pressure oscillogram recorded at the tube bottom (the peaks correspond to repeated hydraulic shocks)

Thus, one can expect that focusing of only one pulse with the rarefaction phase in lithotriptor systems can induce a series of hydraulic impacts on the target, which can be considered as one of real disintegration mechanisms.

Following Fig. 6.44, a series of intermediate weak shocks can occur in the interval between main shocks due to the oscillation of separate large gas–vapor bubbles.

We note that the numbers presented in Fig. 6.44 correspond to the $p(t)$ curve parameters (time in milliseconds and amplitude in millivolts) at the place of location of the vertical trace in the oscillogram. The minus sign stands for negative pressure at the tube bottom. One can see that the first curve (a) provides information on the structure of tensile stresses. The second curve (b) shows the first strong positive pressure pulse as a result of the hydraulic shock. The third oscillogram (c) confirms the availability of a series of successive shocks.

It should be noted that the rough surface of the target in which protrusions may alternate with depressions favors the disintegration, following the mechanism proposed by Field et al. [70] to explain the surface erosion under the impact of liquid droplets. Indeed, according to the model, the impact and flowing of the droplet may result in formation of convergent shock waves and cumulative jets (Fig. 6.45).

With this connection the Sreiber's results [96] can be interesting. They concern the model for the formation of a vapor bubble cavitation zone within a water layer (the state is close to the saturation line) due to a rarefied wave propagating in the layer. Models are presented for vapor cavitation zone development, pressure field formation and the phenomenon of coalescence. The specific feature of the cavitation problem is the multiply reflected rarefied wave at the bottom and the formation of the most intensive cavitation zone near the bottom.

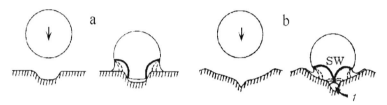

Fig. 6.45a,b. Unevenness of the target surface favors its disintegration [70] under the impact by a drop: central hollow (1) and focused shock wave (SW)

6.4.4 Rarefaction Phase Focusing and Cluster Formation

Let us consider the problem on focusing of a shock wave with a rarefaction phase. There is no need to consider the complete statement beginning with the generation of a shock wave in one focus (or on the surface) of an elliptic reflector, its propagation, reflection, and diffraction. One can use the abundant experimental data, for example, those of [55, 56, 62], on the structure and parameters of "standard" profiles of shock waves (Fig. 6.41) generated in ESWL and consider only the region of the flow near the focus. In this case, the problem is reduced to the statement, within which a one-dimensional cylindrical wave is focused on a solid nucleus a. It is important to choose the appropriate law of pressure fluctuation at the external boundary of the region of the one-dimensional flow, which would enable modeling of a real shock wave structure in the vicinity of the target. The cylindrical statement was selected because the real focus in the ellipsoid (when recording the wave focusing without a target) differed considerably from the point focus, not to mention that the real wave was focused on the object of finite dimensions.

Another feature of the statement is of principal importance: let us assume that the liquid filling the space around the target is a real liquid, i.e., a two-phase medium containing microinhomogeneities as free gas microbubbles that act as cavitation nuclei. The cavitating flow, as shown above, is described by the system of conservation laws for average characteristics of the medium,

6.4 Cavitative Clusters and Kidney Stone Disintegration Problem

while its state is described by a dynamic subsystem that includes the relation between the density ρ, the volume concentration of the gas phase k, and the average pressure p in the medium. In the Lagrangian coordinates, the system becomes

$$\frac{\partial u}{\partial t} = -\frac{1}{\rho_0}\left(\frac{x(r,t)}{r}\right)^2 \frac{\partial p}{\partial r}, \quad \frac{\partial x}{\partial t} = u, \quad \frac{1}{\rho} = \frac{1}{\rho_0}\left(\frac{x(r,t)}{r}\right)^2 \frac{\partial x}{\partial r},$$

$$p = 1 + \frac{\rho_0 c_0^2}{n p_0}\left[\left(\frac{\rho}{1-k}\right)^n - 1\right], \quad k = \frac{k_0}{1-k_0}\rho\beta^3,$$

$$\beta\frac{\partial S}{\partial t} + \frac{3}{2}S^2 = C_1\frac{T}{\beta^3} - C_2\frac{S}{\beta} - p, \quad \frac{\partial \beta}{\partial t} = S,$$

$$C_1 = \frac{\rho_{g0}T_0 B}{p_0 M}, \quad \text{and} \quad C_2 = \frac{4\mu}{R_0\sqrt{p_0\rho_0}}.$$

Here T is the temperature of gas in the bubble, $\beta = R/R_0$, B is gas constant, μ is the viscosity factor, and M is the molar mass of the gas.

Figure 6.46 shows the dynamics of spatial distribution of bubble radii and pressure during focusing of the shock wave from the distance of 10 cm for the times 50, 60, and 170 µs from the onset of the process ($k_0 = 10^{-4}$, $R_0 = 1$ mm). One can see that already after 60 µs the shock wave reached the target surface (dotted line, the target radius was 1 cm). Its parameters were close to those mentioned in [56], including the rather long rarefaction phase with an amplitude of several MPa. The distribution of bubble radii by that time showed no peculiarities: they collapsed intensely under the effect of the shock wave near the target and their radius was somewhat greater at the periphery (the boundary of "launching" of the shock wave toward the cylindrical target).

After 170 µs the situation changed considerably: as a result of reflection a dense bubble cluster (about 1–2 cm thick) was formed on the target. By this

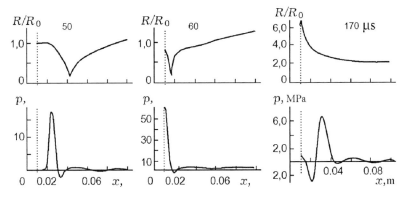

Fig. 6.46. Development of the cavitation zone near the focus (relative change of the bubble radius R/R_0 in the zone) and the dynamics of the shock wave profile

time, the volume concentration k in the cluster increases by about a factor of 300 and pressure in the bubbles was almost zero. One could expect that according to the above experimental data the pressure difference at the external boundary of the bubble cluster and the mean pressure inside it should lead to the cluster collapse and the mentioned hydraulic effect.

References

1. D.H. Trevena: *Cavitation and Tension in Liquids* (Hilger, Bristol Philadelphia 1987)
2. D.A. Wilson, J.W. Hoyt, J.W. McKune: *Measurement of Tensile Strength of Liquid by Explosion Technique*, Nature **253**, 5494 (1975)
3. G.A. Carlson, K.W. Henry: *Technique for Studying Tension Failure in Application to Glycerol*, J. Appl. Phys. **42**, 5 (1973)
4. V.K. Kedrinskii: *Surface Effects at Underwater Explosion (Review)*, Zh. Prikl. Mekh. i Tekh. Fiz. **19**, 4, pp. 66–87 (1978)
5. R. Cole: *Underwater Explosions* (Dover, New York, 1965)
6. V.K. Kedrinskii: *Nonlinear Problems of Cavitative Disintegration of Liquid at Explosive Loading*, Zh. Prikl. Mekh. i Tekh. Fiz.**34**, 3, pp. 74–91 (1993)
7. F.G. Hammitt, A. Koller, O. Ahmed, J. Pjun, E. Yilmaz: Cavitation threshold and superheat in various fluids. In: Proc. of Conf. on Cavitation (Mech. Eng. Publ. Ltd, London, 1976), pp. 341–354
8. M. Strasberg: Undissolved air cavities as cavitation nuclei. In: *Cavitation in Hydrodynamics* (National Phys. Lab., London, 1956)
9. A.S. Besov, V.K. Kedrinskii, E. I Pal'chikov: *Studying of Initial Stage of Cavitation Using Diffraction-Optic Method*, Pis'ma Zh. Exp. Teor. Fiz. **10**, 4, pp. 240–244 (1984)
10. K.S. Shifrin: *Light Diffusion in Turbid Media* (Gostekhizdat, Moscow-Leningrad, 1951)
11. H.C. van de Hulst: *Light Scattering by Small Particles* (John Wiley, New York, 1957)
12. L.D. Volovets, N.A. Zlatin, G.S. Pugachev G.S.: *Arising and Development of Micro-Cracs*, Pis'ma Zh. Exp. Teor. Fiz., Arising and Development of Micro-Crack in Plexiglas at Dynamic Tensile (Spell), 4, pp. 1079–1084 (1978)
13. V.K. Kedrinskii: Peculiarities of bubble spectrum behavior in the cavitation zone and its effect on wave field parameters. In: Proc. Conf. Ultrasonics Intern. 85 (Gilford, London, 1985), pp. 225–230 (1985)
14. V.K. Kedrinskii: On relaxation of tensile stresses in cavitating liquid. In: Proc. 13th Intern. Congress on Acoustics, vol 1 (Dragan Srnic Press, Sabac, 1989) pp. 327–330
15. R.L. Gavrilov: *Content of Free Gas in Liquids and Methods of its Measurement*, In: L.D. Rozenberg (ed.) *Physical Base of Ultrasonic Technology* (Nauka, Moscow, 1970), pp. 395–426
16. M.G. Sirotyuk: *Experimental Studying of Ultrasonic Cavitation*, In: L.D. Rozenberg (ed.) *Strong Ultrasonic Fields*, part 4 (Nauka, Moscow, 1968) pp. 75–81
17. V.K. Kedrinskii: On multiplication mechanism of cavitation nuclei. In: E. Shaw (ed.) Proc. of 12th Int. Congress on Acoustics (Toronto, 1986), pp.14–18 (1986)

18. V.K. Kedrinskii, V.V. Kovalev, S.I. Plaksin: *On Model of Bubbly Cavitation in a Real Liquid*, Zh. Prikl. Mekh. i Tekh. Fiz. **27**, 5, pp. 81–85 (1986)
19. V.K. Kedrinskii: *Dynamics of Cavitation Zone at Underwater Explosion Near Free Surface*, Zh. Prikl. Mekh. i Tekh. Fiz. **16**, 5, pp. 68–78 (1975)
20. I. Hansson, V. Kedrinskii, K. Morch: *On the Dynamics of Cavity Cluster*, J. Phys. D Appl. Phys. 15, pp. 1725–1734 (1982)
21. V.K. Kedrinskii: *Perturbation Propagation in Liquid with Gas Bubbles*, Zh. Prikl. Mekh. i Tekh. Fiz. **9**, 4, pp. 29–34 (1968)
22. V.K. Kedrinskii: *Negative Pressure Profile in Cavitation Zone at Underwater Explosion Near Free Surface*, Acta Astron. **3**, 7–8, pp. 623–632 (1976)
23. V.K. Kedrinskii, S. Plaksin: Rarefaction wave structure in a cavitating liquid. In: V.K. Kedrinskii (ed.) Problems of Nonlinear Acoustics: Proc. of IUPAP-IUTAM Symposium on Nonlinear Acoustics, Part 1 (Novosibirsk, 1987), pp. 51–55
24. N.N. Chernobaev: Modeling of shock-wave loading of liquid volumes. In: S. Morioka, L. van Wijngaarden (eds.) Proc. IUTAM Symposium on Adiabatic Waves in Liquid–Vapor Systems, (Springer, Berlin Heidelberg New York, 1989), pp. 361–370
25. V.K. Kedrinskii: *The Experimental Research and Hydrodynamic Models of a "Sultan"*, Arch. Mech. **26**, 3, pp. 535–540 (1974)
26. S.V. Stebnovskii: *On Mechanism of Pulse Fracture of Liquid Volume*, Zh. Prikl. Mekh. i Tekh. Fiz. **20**, 2, pp. 126–132 (1989)
27. A. Berngardt, E. Bichenkov, V. Kedrinskii, E. Pal'chikov: Optic and X-ray investigation of water fracture in rarefaction wave at later stages. In: M. Pichal (ed.) Proc. IUTAM Symp. on Optical Methods in the Dynamics of Fluids and Solids, (Prague, 1984, Springer, Berlin, Heidelberg, New York, 1985), pp. 137–142
28. A.R. Berngardt, V.K. Kedrinskii, E.I. Pal'chikov: *Evolution of Internal Structure of Zone of Liquid Fracture at Pulse Loading*, Zh. Prikl. Mekh. i Tekh. Fiz. **36**, 2, pp. 99–105 (1995)
29. Berngardt A.R.: Dynamics of the cavitation zone under impulsive loading of a Liquid. PhD Thesis, Novosibirsk (1995)
30. A.S. Besov, V.K. Kedrinskii, E.I. Pal'chikov: On threshold cavitation effects in pulse rarefaction waves. In: P. Pravica (ed.) Proc. of 13th Int. Congress on Acoustics vol. 1 (Dragan Srnic Press, Sabac, 1989), pp. 355–358
31. I.G. Getz, V.K. Kedrinskii: *Dynamics of Explosive Loading of Two-Phase Volume*, Zh. Prikl. Mekh. i Tekh. Fiz.**30**, 2, pp. 120–125 (1989)
32. A.V. Anilkumar: Experimental studies of high-speed dense dusty gases: Thesis, Pasadena (1989)
33. V.K. Kedrinskii, A.S. Besov, I.E. Gutnik: *Inversion of Two-Phase State of Liquid at Pulse Loading*, Dokl. RAN **352**, 4, pp. 477–479 (1997)
34. M.N. Davydov: *Development of Cavitation in a Drop at Shock-Wave Loading*, Dynamics of Continue Medium (Lavrentyev Institute of Hydrodynamics, Novosibirsk, 2001) **117**, pp. 17–20
35. M.N. Davydov, V.K. Kedrinskii: *Two-Phase Models of Cavitative Spell Formation in Liquid*, J. Appl. Mech. Techn. Phys. **44**, 5 (2003) pp. 72–79
36. A.S. Besov, V.K. Kedrinskii, E.I. Palchikov: *On Threshold Effects in Pulse Rarefaction Wave*, Pis'ma Zh. Exp. Teor. Fiz. **15**, 16, pp. 23–27 (1989)
37. M. Cornfeld: *Elasticity and Strength of Liquids* (Inostrannaya Literatura, Moscow, 1951), p. 46

294   6 Problems of Cavitative Destruction

38. L.Y. Briggs: Appl. Phys., 26 (1955); **21** (1950)
39. R. Knepp, J. Daily, F. Hammit: *Cavitation* (Mir, Moscow, 1974)
40. A.D. Pernik: *Cavitation Problems* (Sudostroenie, Leningrad, 1966)
41. A. Besov, V. Kedrinskii: Dynamics of bubbly clusters and free surface at shock wave reflection. In: J. Blake, J. Boulton-Stone, N. Thomas (eds.) Proc. Intern. Symp. on Bubble Dynamics and Interface Phenomena (Birmingham, 6–9 Sept. 1993, Kluwer Academic Publisher, 1994), pp. 93–103
42. N.F. Morozov, Yu.V. Petrov, A.A. Utkin: Dokl. Akad. Nauk **313**, 2 (1990)
43. N.F. Morozov, Yu.V. Petrov: *Problems of Dynamics of Failure of Solids* (St. Petersburg University Publishers, St. Petersburg, 1997) p. 132
44. A.A. Gruzdkov, Yu.V. Petrov: Dokl. Akad. Nauk **364**, 6 (1999)
45. A.S. Besov, V.K. Kedrinskii, N.F. Morozov, Yu.V. Petrov, A.A. Utkin: *On Analogy of Initial Stage of Fracture of Solids and Liquids at Pulse Loading*, Dokl. Akad. Nauk **378**, 3, pp. 333–335 (2001)
46. E.N. Harvey, A.H. Whiteley, W.D. McElroy et al.: *Bubble Formation in Animals. II. Gas Nuclei and Their Distribution in Blood and Tissues*, J. Cell. Compar. Physiol. **24**, 1 (1944)
47. A.S. Besov, V.K. Kedrinskii, Y. Matsumoto et al.: Microinhomogeneity structures and hysteresis effects in cavitating liquids. In: Proc. 14th Int. Congress on Acoustics (Beijing, 1992), pp. 1–3
48. T. Okada, Y. Iwai, A. Yamamoto: *A Study of Cavitation Erosion of Cast Iron*, J. Wear, 84 (1983)
49. T. Okada, Y. Iwai, Y. Hosokawa: *Comparison of Surface Damage Caused by Sliding Wear and Cavitation Erosion on Mechanical Face Seal*, J. Tribology, 42 (1984)
50. Y. Tomita, A. Shima, K. Takayama: Formation and limitation of damage pits caused by bubble-shock wave interaction. In: K. Takayama (ed.) Proc. National Symp. Shock Wave Phenomena, (Tohoku 1989)
51. N. Sanada, A. Asano, J. Ikeuchi et al.: Interaction of a gas bubble with an underwater shock wave, pit formation on the metal surface. In: Proc. 16th Int. Symp. Shock Tubes and Waves (VCH Publ., Weinheim, 1988)
52. V. Makarov, A.A. Kortnev, S.G. Suprun, G.I. Okolelov: Cavitation erosion and spectrum analysis of pressure pulse heights produced by cavitation bubbles. In: Proc. 6th Int. Symp. Nonlinear Acoustics, vol. 2 (Moscow State Univ., Moscow, 1976)
53. S. Fujikawa, T. Akamatsu: *Experimental Investigations of Cavitation Bubble Collapse by a Water Shock Tube*, Bull. ASME **21**, 152 (1978)
54. R. Ivany, F. Hammitt: *Cavitation Bubble Collapse in Viscous Compressible Liquids Numerical Analysis*, Trans. ASME. Ser. D. 4 (1965)
55. V.K. Kedrinskii, V.A. Stepanov: Cavitation effects in thin films In: M. Hamilton, D. Blackstock (eds.) Proc. 12th ISNA, Frontiers of Nonlinear Acoustics (Elsevier Applied Sci., London New York, 1990), pp. 470–475 (1990)
56. V.P. Alekseevskii: *On the Theory of Armor Perforation by Cumulative Jets* (UkrSSR Academy of Sciences Publishers, Kiev, 1953)
57. M.A. Lavrentiev: *Cumulative Charge and Principle of its Action*, Uspekhi Mat. Nauk **12**, 4, pp. 41–52 (1957)
58. K.A. Kurbatskii, V.K. Kedrinskii: Collapse of a bubble in the cavitation zone near a rigid boundary. In: Abstr. 124th Meeting of ASA (New Orleans, 1992)
59. H. Takahira, T. Akamatsu, S. Fujikawa: JSME Intern J, Series B-Fluids and Therm. Eng. **37**: 2 (1994) pp. 297–305

60. H. Takahira: JSME Intern J, Series B-Fluids and Therm. Eng. **40**: 2 (1997) pp. 230–239
61. H. Takahira, S. Yamane, T. Akamatsu: JSME Intern J, Series B-Fluids and Therm. Eng. **38**: 3 (1995) pp. 432–439
62. S. Ceccio, S. Gowing, Y.T. Shen: Journal of Fluids Engineering-Transactions of the ASME **119**, 1 (1997) pp. 155–163
63. M. Thiel, M. Nieswand, M. Dorffel: Minimally Invasive Therapy and Allied Technologies **9**, 3–4 (2000) pp. 247–253
64. M. Thiel: Clinical Orthopaedics and Related Research **387**, (2001) pp. 18–21
65. D.L. Sokolov, M.R. Bailey, L.A. Crum: Ultrasound in Medicine and Biology, **29**, 7 (2003) pp. 1045–1052
66. S.L. Zhu SL, F.H. Cocks, G.M. Preminger, P. Zhong: Ultrasound in Medicine and Biology, **28**, 5 (2002) pp. 661–671
67. V.K. Kedrinskii: On a mechanism of target disintegration at shock wave focusing in ESWL.: In: P.K. Kuhl, L. Crum (eds.) Proc.16th Intern. Congress on Acoustics, Seattle, USA, vol. 4 (University of Washington, Washington, 1998) pp. 2803–2804
68. W. Eisenmenger: Ultrasound in Medicine and Biology **27**, 5 (2001) pp. 683–693
69. R.O. Cleveland, D.A. Lifshitz, B.A. Connors et al.: Ultrasound in Medicine and Biology **24**, 2 (1998) pp. 293–306
70. T. Kodama, H. Uenohara, K. Takayama: Ultrasound in Medicine and Biology **24**, 9 (1998) pp. 1459–1466
71. T. Kodama, M. Tatsuno, S. Sugimoto et al.: Ultrasound in Medicine and Biology, **25**, 6 (1999) pp. 977–983
72. T. Kodama, K. Takayama: Ultrasound in Medicine and Biology **24**, 5 (1998) pp. 723–738
73. M. Delius, F. Ueberle, W. Eisenmenger: Ultrasound in Medicine and Biology **24**, 7 (1998) pp. 1055–1059
74. C.M. Zapanta, E.G. Liszka, T.C. Lamson et al.: J. Biomech. Engin.-Trans. Asme **116**, 4 (1994) pp. 460–468
75. J.C. Williams, M.A. Stonehill, K. Colmenares et al.: Ultrasound in Medicine and Biology **25**, 3 (1999) pp. 473–479
76. D. Howard, B. Sturtevant: Ultrasound in Medicine and Biology **23**, 7 (1997) pp. 1107–1122
77. M. Delius: Zentralblatt für Chirurgie, **120**, 4 (1995) pp. 259–273
78. K. Takayama: Japan. J. Appl. Phys., Part 1, Regular Papers Short Notes and Review Papers, **32**, 5B (1993) pp. 2192–2198
79. D.L. Miller, J.M. Song: Ultrasound in Medicine and Biology, **28**, 10 (2002) pp. 1343–1348
80. H. Grönig: Past, present and future of shock focusing research. In: Proc. Intern. Workshop on Shock Wave Focusing (Sendai, 1989) pp. 1–38
81. B. Sturtevant: The physics of shock focusing in the context of ESWL. In: Proc. Intern. Workshop on Shock Wave Focusing (Sendai, 1989) pp. 39–64
82. M. Kuwahara: Extracorporeal shock wave lithotripsy. In: Proc. Intern. Workshop on Shock Wave Focusing (Sendai, 1989) pp. 65–89
83. O. Kitayama, H. Ise, T. Sato, K. Takayama: Non-invasive gallstone disintegration by underwater shock focusing. In: H. Grönig (ed.) Proc. 16th Intern. Symp. on Shock Tubes and Waves (VCH, Aachen, 1987) pp. 897–904

84. M. Grunevald, H. Koch, H. Hermeking: Modeling of shock wave propagation and tissue interaction during ESWL. In: H. Grönig (ed.) Proc. 16th Intern. Symp. on Shock Tubes and Waves (VCH, Aachen, 1987) pp. 889–895
85. M. Delius: Effect of lithotriptor shock waves on tissues and materials. In: M. Hamilton, D. Blackstock (eds.) Proc. 12th ISNA, Frontiers of Nonlinear Acoustics, (ESP Ltd, London, 1990) pp. 31–46
86. V.K. Kedrinskii, R.I. Soloukhin: *Collapse of a Spherical Gas Bubble in Water by a Shock Wave*, J. Applied Mechanics and Technical Physics **2**, 1, pp. 27–29 (1961)
87. C. Church, L. Crum: A theoretical study of cavitation generated by four commercially available ESWL. In: M. Hamilton, D. Blackstock (eds.) Proc. 12th ISNA, Frontiers of Nonlinear Acoustics, (ESP Ltd, London, 1990) pp. 433–438
88. C. Church: *A Theoretical Study of Cavitation Generated by an Extracorporeal Shock Wave Lithotripter*, J. Acoust. Soc. Am. **86**, 1, pp. 215–227 (1989)
89. F. Prat: The cytotoxicity of shock waves: cavitation and its potential application to the extra-corporeal therapy of digestive tumors. In: Brun, Dumitrescu (eds.) Proc. 19th Int. Symp. on Shock Waves, (Marseille, 1993)
90. H. Nagoya, T. Obara, K. Takayama: Underwater shock wave propagation and focusing in inhomogeneous media. In: Brun, Dumitrescu (eds.) Proc. 19th Int. Symp. on Shock Waves, vol. 3 (Marseille, 1993) pp. 439–444
91. C. Stuka, P. Sunka, J. Benes: Nonlinear transmission of the focused shock waves in nondegassed water. In: Brun, Dumitrescu (eds.) Proc. 19th Int. Symp. on Shock Waves, vol. 3 (Marseille, 1993) pp. 445–448
92. E. Sato et al.: Soft flash X-ray system for shock wave research. In: Brun, Dumitrescu (eds.) Proc. 19th Int. Symp. on Shock Waves, vol. 3 (Marseille, 1993) pp. 449–454
93. I. Bayikov, A. Berngardt, V. Kedrinskii, E. Pal'chikov: *Experimental Methods of Study of Cavitative Cluster Dynamics*, Zh. Prikl. Mekh. i Tekh. Fiz. **25**, 5, pp. 30–34 (1984)
94. K. Isuzugawa, M. Fujii, Y. Matsubara et al.: Shock focusing across a layer between two kinds of liquid. In: Brun, Dumitrescu (eds.) Proc. 19th Int. Symp. on Shock Waves (Marseille, 1993)
95. J.E. Field, M.B. Lesser, J.P. Dear: Proc. Roy. Soc. London **A**, 401 (1985)
96. I.R. Shreiber: Acustica **83**, 6 (1997) pp. 987–991

7 Jet Flows at Shallow Underwater Explosions

7.1 State-of-the-Art

Apparently the problems of liquid flows produced by shallow underwater explosions were first considered in a relatively full scope by Cole [1]. The research results obtained in this field over the next quarter of the century were surveyed in [2–4]. It has been shown that the changes in the medium structure observed near the free surface and the formation of directional high-velocity flows are associated with the reflection of shock waves and the dynamics of a cavity with detonation products. The directional throwing out on the free surface at shallow underwater explosions is called "sultan" in Russian publications [3, 4]. Sometimes sultans are refered to as "splashes" in English. Sultans are defined as a throwing out of water (vertical water column).

In the 1960s, two models were proposed to explain the formation mechanism of sultans. According to the first physical model formulated by M.A. Lavrentyev in [5], a spherical shock wave reflecting from the free surface initiates the development of intense bubble cavitation, which separates a part of the liquid beneath the charge. As a result, a cumulative semi-spherical pit is formed on the free surface, while the cavity with detonation products confining about a half of the energy of detonated HE produces a velocity field orthogonal to the pit surface. In the result the flow similar to spherical cumulation must be arisen. Thus, following Lavrentyev's model, only the joint effect of the shock wave and the cavity with detonation products can induce the formation of a sultan.

Another model was proposed by L.V. Ovsyannikov [6] based on analysis of the problem of a floating bubble in an exact mathematical statement. On the basis of approximate representation of the analytical solution (for initial moments of time), he offered a hypothesis suggesting that given the proper depth and charge weight, an upward cumulative jet is formed in the bottom part of the explosive cavity during its floating and deformation. The jet shaped by the moment of arrival of the cavity at the free surface "perforates" the explosive cavity and governs the sultan flow.

It should be noted that the term "sultan" (throwing out or fountain) is often used to denote different aspects of the phenomenon. In [5, 6], it was used to denote a cumulative jet, i.e., a continuous jet that is often observed above the separated dome under the explosion of light charges at certain depths. Kolsky et al. [7] called it "central spout" and note that it appears long before the moment when the cavity with detonation products reaches its maximum size. In [8,9], the term was used to denote a hollow liquid column at the grounds of the cloud of detonation products and water splashes thrown out to the atmosphere. In [10], the phenomenon on the whole was called sultan: a jet beneath the dome with the hollow liquid column at the grounds (experiments were performed for the charge weights $W = 0.075$–136 kg). There was some uncertainty in understanding of the nature of the vertical jet by the authors of [10]: on the one hand they used the term for a narrow jet of detonation products burst into the atmosphere, then while studying the jet height they determined the amount of liquid in the jet as $150\,W$, which was practically unfeasible due to the splashes surrounding the gas jet only.

In terms of the above considerations on the formation of the directional throwing out on the free surface it is logical to start the prehistory of the studies from cavitation problems.

7.1.1 Irregular Reflection and Bubbly Cavitation

The generation of cavitation and the features of the structure of the wave field near the free surface were considered in [2, 11–16]. In [11], the shock wave parameters were studied experimentally and the development of the cavitation zone was analyzed for 1-g and 100-kg charges exploded at depths $(1$–$8) \cdot R_{ch}$. Measurements were performed for the depths $(1$–$16) \cdot R_{ch}$ for the distances to the explosion site up to $120 \cdot R_{ch}$ (R_{ch} is the charge radius). The zone of nonlinear interaction of the shock wave produced by the underwater explosion with the free surface was determined in [12–14]. It was found that the zone is bounded from above by the trajectory of the triple point, where the front of the relaxed shock wave, the unperturbed front of the incident shock wave, and the rarefaction wave converged. The "smooth" fall of pressure behind the front of the disturbed wave characteristic of the nonlinear zone suggested that no visible cavitation ruptures occurred in this zone [11].

Thus, the "triple point" trajectory can be considered as a theoretically possible upper boundary of the cavitation zone, which, as we show below, is consistent with experimental data. Therefore, the development of the visible cavitation zone should be observed in the region of regular reflection, for which, according to [2, 12], the acoustic method for determining pressure is valid (the method applies the principle of superposition of the pressure field produced by the explosion of a virtual charge). As a rule, the studies based on the acoustic method are reduced either to the calculation of the zone of negative pressures [15], which at best yields only the initial condition for the development of cavitation, or to the estimates of the size and number of spall zones [2, 16] using the results for strength properties of the liquid [17]. It is noteworthy that the calculations of the zone of tensile stresses made in [15] for the depths of 1–12 m and charge weights of 50, 100, and 5000 g prove that the increase in the explosion depth has practically no effect on the position of the lower boundary of the zone, and consequently, on cavitation. Thus, for a 5-kg charge with varying explosion depth from 1.5 to 12 m, the zone radius changes from 60 to 200 m, while the lower boundary remains at the level of 4 m.

However, in spite of simplicity of the acoustic method, the origin and development of the cavitation zone have long been estimated from the character of behavior of a single cavitation bubble at fixed parameters of the applied wave field. This made the estimates of the tensile stresses in the region of regular reflection (in the cavitation zone) more complicated: the acoustic method within the framework of a one-phase liquid overestimates, sometimes by orders of magnitude, the absolute negative pressures as against real values [2].

7.1.2 Directional Throwing Out on the Free Surface (Sultans)

The first studies of surface effects dating back to early 20th century were mentioned in a review to the thesis [13]. The publications [1, 3, 7] contain qualitative description of the character of throwing out on the free surface and the results of experimental investigation of their parameters. In particular, in [7] an empirical dependence of the initial velocity of the dome rise V_0 (m/s) on the charge weight (W, kg) and the depth (H, m) was revealed,

$$V_0 \simeq 50.7 \left(\frac{W^{1/3}}{H} - 0.13 \right),$$

which is valid for the range $0.4 < W^{1/3}/H < 5.2$.

In [7], an attempt was made to estimate the structure of directional throwing out using the experimental data for its height for 0.45- and 4.5-kg charges and for different explosion depths. For example, if a 4.5-kg charge is exploded at a depth of 75 cm, the height of the sultan rise reaches 100 m and is optimal. An increase or decrease of the depth leads to the reduction of the maximum height of the throwing out.

Cole [1] was the first to mention an abnormal increase of the amplitude of the first oscillation of the cavity with detonation products for explosions at certain depths. He made a disputable assumption on the possible mechanism of the effect: at the stage of maximum expansion atmospheric air comes into the cavity, and due to successive reactions extra energy is released in the resulting mixture, which results in increased pressure in detonation products (and consequently in the liquid) under the collapse of the cavity.

Few experimental works dealt with estimates of the parameters of sultan on the free surface, among them [8–10]. In [8], both the development of the gas cavity and the sultan together with the neck and the basis radius were studied. A principal feasibility of modeling the surface phenomena under laboratory conditions up to the onset of the disintegration of a sultan due to the gravity force was shown by the example of charges of weighing of 0.256 g and 100 kg. Here the thrown out liquid mass was estimated approximately from the "upper limit" and the dependence

$$m \simeq 540 \cdot W \cdot \frac{H}{R_{\text{ch}}},$$

and the empirical dependence of the height of rise of the dome center were presented as

$$h \simeq A \cdot \ln\left(1 + \frac{V_0 t}{A}\right),$$

where $A = \Delta(\rho_{\text{liq}}/\rho_{\text{air}})$ (Δ is the thickness of the emitted liquid layer). In [7], a more precise expression that better fits the experimental data was derived to estimate the size and shape of the dome at any time from the coordinate R of the point of the dome surface with respect to the symmetry axis:

$$h \simeq B \cdot \left[\frac{H^2}{D^2 + R^2} - C\right].$$

Here H is the charge depth, B, C, and D are the constants depending on the charge nature.

In [9], the research results for the parameters of a sultan and surface waves under explosion of 100-kg charges in shallow reservoirs (up to $12 \cdot R_{\text{ch}}$) were described. In [10], an empirical expression was found for the optimal explosion depth $H_{\text{opt}} \simeq W^{1/3}$ (W in kg; H in m) corresponding to the maximum water throwing out in the sultan, and the total mass of water was estimated as approximately $150W$. In [18], the results of experimental studies and calculations for liquid flows were presented for surface explosion and the explosion of a completely submerged charge.

The above brief survey makes evident that the lack of clear understanding of the structure of throwing out on the free surface leads to uncertain estimates, in particular, for parameters such as the height and velocity of rise of the sultan and the mass of thrown out liquid. Existence of vertical and lateral jets at shallow underwater explosions was established experimentally. However, the mechanism of their formation, including the abnormal effect of the first oscillation, remained unknown for long even though the problem is of principal importance for understanding the unique hydrodynamic phenomenon.

7.2 Tensile Stress, Structure of Cavitation Region, Spalls

A thorough experimental study of the change of the shock wave profile is presented in [11]. Here we shall focus on the rarefaction waves (phases), the structure and distribution of tensile stresses, and the shapes of spalls for the depth of the charge explosions of $(0-30) \cdot R_{\text{ch}}$.

7.2.1 Development of Cavitation Zone, Spalls (Experimental Studies)

The development of cavitation, the spallation, and the structure of the directional throwing out were studied experimentally in axisymmetrical and plane statements under the explosion of an HE charge (or a wire) in tanks of size $2 \times 2 \times 2$ m^3 (Fig. 7.1, left) and $0.01 \times 0.5 \times 0.5$ m^3 (Fig. 7.1, right) [19].

A capacitor bank C was charged by a high-voltage power source 2 through a charging resistance to $U \simeq 20$ kV. An HE charge (or a wire) was blasted in a result of ionizing the gap e by a high-voltage pulse of 50 kV from the control panel of the high-speed photo-recorder 6. The signal from the Rogovsky coil occurring under the discharge of the capacitor C due to the closing of the gap e was an input for the delay unit 4 and the pulse transformer 5 triggering the lighting system by a high-voltage pulse a. Recording of different stages of the explosion in liquids was arranged with the required time delay. The parameters of the lighting system could be easily adjusted to the experiment requirements by selecting the appropriate flash lamps (Fig. 7.1, IFP-2000) and capacitor parameters for "charging."

Figure 7.2 shows the frames of high-speed recording of the explosion of a wire on the surface of a liquid. The time between frames was 4 μs. One can see the shock wave (Fig. 7.2, dark semicircle in frames 5–7) propagating from the free surface deep into the liquid. Then we can observe the shock wave generated by the explosion products. This wave propagates in air and reflects from the free surface of water (frames 7 to 15). It is evident that as the incidence angle of the air shock wave increased, beginning from frame 10, the

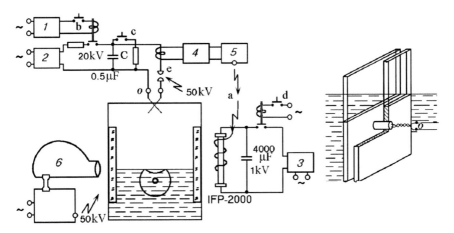

Fig. 7.1. Experimental setup for studying underwater explosions Left: (1) electromagnetic switch, (2) and (3) high-voltage power sources, (4) delay unit, (5) pulse transformer, (6) photorecorder; switches (b), (c), and (d) are the units of the interlocking and charging systems of the high-voltage schemes, (a) high-voltage pulse triggering the lighting, (e) discharge gap Right: "slot cuvette" with the charge

302   7 Jet Flows at Shallow Underwater Explosions

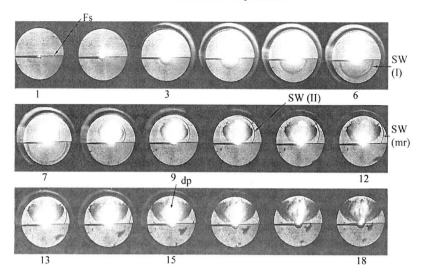

Fig. 7.2. Explosion on the surface of a liquid: irregular reflection of the air shock wave from the surface

Mach reflection (mr) was formed. Thus, the explosion on the surface of a liquid can be used as a method of studying the transition to irregular reflection of powerful impulsive shock waves from the surface in the atmosphere.

Figure 7.3 shows the explosion of a 1.2-g charge (I) at a depth of 5 cm from the free surface (Fs). The time between frames (from left to right) was 8 μs. One can see that underwater explosions near the free surface had an

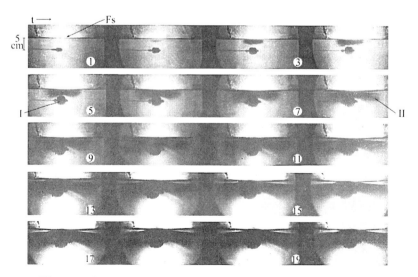

Fig. 7.3. Initial stage of the development of the cavitation zone

elongated cavitation zone (II) with dropped ends, the upper boundary of the zone moved away from the free surface, separating the zones of regular and irregular reflection. Later the cavitation will occupy all visible region: its lower boundary propagated inside the liquid.

The experimental data suggest that cavitative bubbles appeared behind the rarefaction wave in the zone of visible cavitation almost instantaneously. The rarefaction wave weakened in the course of propagation and a transparent zone arose behind the front. However, this does not mean that there was no cavitation: it is quite probable that cavitation bubbles in the zone failed to reach visible size. Nevertheless, the effect of such "invisible" cavitation on the parameters and the structure of the field of tensile stresses could still be appreciable.

7.2.2 Two-Phase Model of Cavitation Region

As was noted in the previous chapter, a real liquid always contains free gas as cavitation bubbles [20,21], and, following the contemporary concepts [22], the radii of cavitation nuclei vary over the interval $5 \cdot 10^{-7}$–$5 \cdot 10^{-3}$ cm depending on the state of the liquid, while the volume concentration of the gas is within 10^{-8}–10^{-12} cm^{-3}. The gas content parameters are very low. However, the growth of the nuclei in the rarefaction wave results in the appearance of the cavitation zone with high concentration of the gas–vapor phase, and the change of both the state of the medium and the structure of the wave field. Therefore, it is natural to believe that, as was mentioned above, a real liquid containing microinhomogeneities is a two-phase medium [23].

Let us consider the two-phase model of the development of the cavitation zone at underwater explosion [23] when a shock wave reflects from the free surface, obeying the principle of superposition of the pressure field produced by the explosion of a virtual charge. In this case, the liquid component of the two-phase liquid can be deemed incompressible, considering that the compressibility of the two-phase medium is primarily determined by the compressibility of the free gas in it, and the nonlinearity of the process depends on the dynamics of cavitation nuclei. The main problem is defining an approximate analytical dependence of the mean pressure on the volume concentration $p(k)$ of gas in the medium. As we noted, the first attempts to elaborate a mathematical model of a liquid with gas bubbles, whose dynamics is described by the Rayleigh equation, were made in [24–26]. The model was adapted to the problems of bubble cavitation in [23].

7.2.2.1 Definition of the Function $p(k)$

Let us consider the system of equations describing an axisymmetrical two-phase flow

$$\Delta \xi \simeq \xi, \quad \frac{\partial^2 k}{\partial \tau^2} = -3k^{1/3}\xi + \frac{1}{6k}\left(\frac{\partial k}{\partial \tau}\right)^2, \tag{7.1}$$

7 Jet Flows at Shallow Underwater Explosions

where $k = (R/R_0)^3$ is the relative volume concentration of the gas, $\tau = t\sqrt{p_0/\rho_0 R_0^2}$, $\xi = p - k^{-\gamma}$, p is the pressure in the medium taken with respect to the initial unperturbed p_0, and γ is the adiabatic index of the gas in the cavitation nuclei. For simplicity we neglect the nonspherical shape of the bubbles, the mass of gas in them and their motion with respect to the liquid. We shall assume that the cavitation nuclei of equal size are uniformly distributed in the liquid and their number per unit volume is constant.

During transformations we introduce the additional assumptions on the small magnitudes of terms of the $k_{rr}/k^{\gamma+1}$ type with respect to p_{rr} and the terms of the rk_r type as compared to $6k$ (the subscript denotes the appropriate derivative with respect to r). The first assumption is obvious: the multiplier $1/k^{\gamma+1}$ reduces sharply during the development of cavitation. The validity of the second assumption can be estimated using the growth of a single nucleus under the effect of negative pressure. This estimate (the degree inaccuracy is about 20–30%) is allowable for the initial stage of expansion of cavitation nuclei in the zone, when their interaction is not intense, and is apparently an upper estimate for the following stages when the interaction of bubbles with the field of tensile stresses practically levels off their concentration in the cluster and k_r decreases considerably. These conditions for the introduction of the variable $\bar{r}$ (see below) make it necessary to use the sign $\simeq$ in (7.1) to emphasize a somewhat heuristic character of the equation, in which the spatial coordinate "traces" the change of the microscale of the flow, the size of cavitation bubbles.

We want to remind the readers here that the system (7.1) is distinguished by the complex form of the equation of state of the medium, which includes the second-order equation for k (an analog of the Rayleigh equation). In the right-hand part of the equation, the pressure at infinity is presented by the averaged pressure in the two-phase medium.

In the case of explosion of a spherical charge near the free surface, it is convenient to consider the problem in an axisymmetrical statement. As spatial variables we introduce θ and $\bar{r} = r\alpha k^{1/6}$, where $\alpha = \sqrt{3k_0}/R_0$. Then for the first equation of the system (7.1) gives us

$$\frac{1}{\bar{r}^2}\frac{\partial}{\partial \bar{r}}\left(\bar{r}^2\frac{\partial \xi}{\partial \bar{r}}\right) + \frac{1}{\bar{r}^2 \sin\theta}\frac{\partial}{\partial \theta}\left(\sin\theta\frac{\partial \xi}{\partial \theta}\right) = \xi. \tag{7.2}$$

The equations permits separation of variables and the solution is sought as $\xi = \Psi(\bar{r})\Theta(\theta)$. Substituting it in (7.2) and designating the separation constant by $\nu(\nu+1)$, we get the equations

$$\frac{d}{d\bar{r}}\left(\bar{r}^2\frac{d\Psi}{d\bar{r}}\right) - [\bar{r}^2 + \nu(\nu+1)]\Psi = 0 \tag{7.3}$$

and

$$\frac{1}{\sin\theta}\frac{d}{d\theta}\left(\sin\theta\frac{d\Theta}{d\theta}\right) + \nu(\nu+1)\Theta = 0.$$

7.2 Tensile Stress, Structure of Cavitation Region, Spalls

The solution of the second equation can be expressed in terms of the Legendre spherical functions $P_\nu(\cos\theta)$ and $Q_\nu(\cos\theta)$. Substitution of the function $\Psi = \bar{r}^{-1/2}v$ into the first equation of the system (7.3) shows that its solution for the function v is the modified Bessel function. Finally, taking into account a binding solution in the region under consideration that is defined by the variation intervals $0 \leq \theta \leq \pi$ and $\bar{r} > 0$, for $\nu = n$ (the integral positive number $n = 0, 1, 2, \ldots$), the solution of (7.2) is written as

$$\xi = \bar{r}^{-1/2} \sum_{n=0}^{\infty} A_n K_{n+1/2}(\bar{r}) P_n(\cos\theta) \ . \tag{7.4}$$

The obtained expression defines the sought relation $p(k)$ within the above assumptions.

7.2.2.2 Statement of the Problem

To simulate the problem on the development of a cavitation region [23, 27], we assume that a liquid containing cavitation nuclei of radius R_0 with volume concentration of the gas k_0 consists of two identical cavities with detonation products. The cavities of radius a are placed at points O and O_1 at a distance h from each other. Both cavities can expand following the adiabatic law and the initial pressure p_α is known. Let us assume that the dynamics of a real explosive cavity $a_1(t)$ and its virtual analog $a(t)$, as well as the adiabatic index of the detonation products γ, are known. We assume that for a cavity placed at the point O (the point of location of the virtual charge) $p_\alpha < 0$, and for that at the point O_1 $p_{\alpha,1} > 0$, and make a phase displacement (by the multiplier $\sigma_0(t - (r - r_1)/c_0)$) of the effect of the virtual source at the point O with respect to the source at the point O_1, simulating the delay of the arrival of the rarefaction wave at this point:

$$\sigma_0 = \begin{cases} 0 \ , \text{ if } t < (r - r_1)/c_0 \ , \\ 1 \ , \text{ if } t \geq (r - r_1)/c_0 \ . \end{cases}$$

Here r and r_1 are physical coordinates of the point in the systems with the centers O and O_1, respectively, and c_0 is the speed of sound in unperturbed liquid. Then the pressure at any point of the system is determined by the superposition of solutions of the type (7.4):

$$\xi = \bar{r}^{-1/2} \sigma_0 \sum_{n=0}^{\infty} A_n K_{n+1/2}(\bar{r}) P_n(\cos\theta) +$$

$$+ \bar{r}_1^{-1/2} \sum_{n=0}^{\infty} B_n K_{n+1/2}(\bar{r}_1) P_n(\cos\theta_1) \ . \tag{7.5}$$

We note that the angles θ and θ_1 are reckoned from the direction $O - O_1$, each in its system. The coefficients in (7.5) are found from the boundary conditions at the surfaces of the explosive cavities:

$$\xi = p_\alpha < 0 \quad \text{at} \quad \bar{r} = a(t)\alpha k^{1/6}$$

and

$$\xi = p_{\alpha_1} > 0 \quad \text{at} \quad \bar{r}_1 = a_1(t)\alpha k^{1/6} .$$

Here we take into account the experimentally established absence of cavitation in the vicinity of the boundary of the explosive cavity, whence it follows that pressure in detonation products at the interface with the liquid and the mean pressure in the two-phase medium are equal. We note that (7.5) is also valid for the HE charge at shallow depths of $h/2$ for reasonable time intervals, when the free surface of the liquid can be considered as plane.

It is not difficult to show that in expanding (7.5) one can confine oneself to the case $n = 0$, if the ratio of radii of explosive cavities to the distance h between them is assumed to be sufficiently small. Then we have

$$\xi = \bar{r}^{-1/2} \sigma_0 A_0 K_{1/2}(\bar{r}) + \bar{r}_1^{-1/2} B_0 K_{1/2}(\bar{r}_1) . \tag{7.6}$$

The coefficients A_0 and B_0 are determined from the boundary conditions: From the first boundary condition,

$$p_\alpha (a\alpha k^{1/6})^{1/2} = A_0 K_{1/2}(a\alpha k^{1/6}) + \left(\frac{a}{r_1^*}\right)^{1/2} B_0 K_{1/2}(r_1^* \alpha k^{1/6}) ;$$

From the second boundary condition,

$$p_{\alpha_1}(a_1 \alpha k^{1/6})^{1/2} = \left(\frac{a_1}{r^*}\right)^{1/2} A_0 K_{1/2}(r^* \alpha k^{1/6}) + B_0 K_{1/2}(a_1 \alpha k^{1/6}) ,$$

where $r_1^* = \sqrt{h^2 + a_1^2 - 2a_1 h \cos\theta_1}$ and $r^* = \sqrt{h^2 + a^2 - 2ah \cos\theta}$.

Assuming $r^* \simeq r_1^* \simeq h$ and $p_\alpha = -p_{\alpha_1}$, we obtain $A_0 = -B_0$. Presenting $K_{1/2}$ as $K_{1/2}(z) = \sqrt{\pi/2z} \exp(-z)$ [28] and taking into account the above assumptions with respect to a/h for B_0 we obtain the simple expression

$$B_0 = \sqrt{\frac{2}{\pi} p_{\alpha_1} a_1 \alpha k^{1/6}} \exp(a_1 \alpha k^{1/6}) .$$

Finally, the solution (7.6) acquires the form

$$\xi = p_{\alpha_1} \left(\frac{a_1}{r_1}\right) \left[e^{-(r_1 - a_1)\alpha k^{1/6}} - \left(\frac{r_1}{r}\right) \sigma_0 e^{-(r - a_1)\alpha k^{1/6}} \right],$$

where $r = \sqrt{r_1^2 + h^2 - 2hr_1 \cos\theta_1}$, the pressure in the detonation products is determined from the adiabat $p_{a_1} = p(0)(\bar{a}_1)^{-3\gamma}$, and the dynamics of the explosive cavity is described by the empirical dependences

$$\bar{a}_1 = 1 + 0.02210^6(t/a_0) \quad \text{at} \quad t < 10^{-4}$$

and

$$\bar{a}_1 = 158.5(t/a_0)^{0.4} \quad \text{at} \quad t \geq 10^{-4},$$

where the initial radius a_0 of the HE charge is in centimeters and t is in seconds.

Thus, the initial system of equations (7.1) reduces to the form

$$\xi = p_{a_1}\left(\frac{a_1}{r_1}\right)\left[e^{-(r_1-a_1)\alpha k^{1/6}} - \left(\frac{r_1}{r}\right)\sigma_0 e^{-(r-a_1)\alpha k^{1/6}}\right] \quad (7.7)$$

and

$$\frac{\partial^2 k}{\partial \tau^2} = -3k^{1/3}\xi + \frac{1}{6k}\left(\frac{\partial k}{\partial \tau}\right)^2$$

which, if the initial data are know, allows one to solve the problem on the development of the cavitation zone and the profile of the rarefaction wave in it. Obviously, in this statement (system (7.7)) the coordinates r and r_1 act as parameters, which, in principle, reduces the problem to the solution of an ordinary differential equation for k at given r and r_1.

7.2.2.3 Calculation of the Cavitation Zone Dynamics

Before proceeding to the solution of the problem, one should know two basic physical parameters: the initial radius R_0 of the cavitation nucleus and the volume concentration k_0 of the free gas in water. For this purpose, we use the experimental data of [21, 29, 30] summed up in [22]: $k_0 \simeq 10^{-12}$ for distilled water, $k_0 \simeq 10^{-12}$–10^{-10} for tap water after 7–12 hours of settling, and $k_0 \simeq 10^{-9}$–10^{-8} for tap water after an hour of settling.

The upper limit of the size of cavitation nuclei is $R_0 \simeq 2 \cdot 10^{-2}$ cm, $R_0 \simeq 5 \cdot 10^{-3}$ cm in fresh tap water, $R_0 \simeq 5 \cdot 10^{-4}$ cm in water settled for several hours, $R_0 \simeq 5 \cdot 10^{-5}$ cm in water with stabilized content of free gas (after 7–12 hours of settling).

In the calculations we accept $k_0 = 10^{-11}$ and $R_0 = 5 \cdot 10^{-5}$ cm. Since cavitation bubble becomes visible when its radius is $R_* \simeq 10^{-2}$ cm [31], the cavitation zone is registered only if the radius of the cavitation in the rarefaction wave increases by three orders of magnitude. The calculations show that the nuclei reach this size over a short time interval.

Let us consider the explosion of a 1.2-g charge of radius $R_{\text{ch}} \simeq 0.53$ cm at a depth of $h/2 = 5.3$ cm [23, 27]. We accept the initial pressure $p(0)$ in the products of instantaneous detonation to be equal to $4 \cdot 10^4$ atm and $\gamma = 3$. We note that the dynamics of nuclei under propagation of the positive phase of a shock wave is neglected in the calculations. It is assumed that the nuclei behind the wave front instantaneously achieve uniform size distribution with respect to pressure, the gas content is extremely low and does not influence the structure of the shock wave. The system (7.7) can be solved only beginning from the moment of the onset of negative pressure in the medium.

Figure 7.4 presents the calculation results for the dynamics of the volume concentration $k(t)$ at different points near the free surface (the scheme of location of the points is shown at the top of the figure). For all constructed curves the time $t = 0$ corresponded to the moment of arrival of the rarefaction wave front at the given point. The horizontal dotted line marks the boundary of the visible size of cavitation bubbles. One can see that over the interval $2 \cdot 10^{-6} < t < 10^{-5}$ s the visible zone appeared only near the symmetry axis close to the free surface (points 2 and 3). At $t = 10^{-4}$ s the cavitation bubbles reached the visible size at all points shown in the scheme except for 1. However, the intensity of cavitation at any given moment in time in them differed essentially: for example, the value of k at points 3 and 6 differs by two orders of magnitude. The time interval between the moment when the dotted line crosses the curve $k(t)$ determined the "lifetime" of the visible bubble at the point: approximately after 400 μs the visible cavitation disappeared on the axis and at the point 4 on the ray (bubbles collapse), which was in good agreement with the experimental data.

The calculation results for the dynamics of the visible cavitation zone for the times $t = 16, 32, 48,$ and 64 μs are shown in Fig. 7.5a. The region in which bubbles at a given moment exceeded the minimum visible size is shown darkened. Figure 7.5b presents the frames of high-speed photography for the same times, while the cavitation zone developing under explosion of a 1.2-g charge at a depth of 5.3 cm.

Comparison of the numerical and experimental data shows that the two-phase model provided a rather satisfactory description of the dynamics of the cavitation zone.

Figure 7.6 presents the calculation of visible cavitation zone for the explosion at a depth of $h/2 = 3$ cm for the times 20 and 100 μs. Naturally

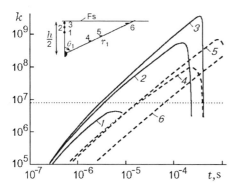

Fig. 7.4. Dynamics of $k(t)$ in the cavitation zone near the free surface

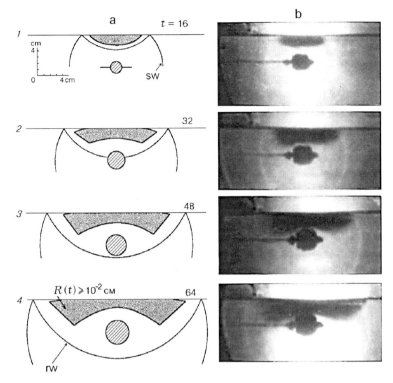

Fig. 7.5. Numerical (**a**) and experimental (**b**) results for the dynamics of the visible cavitation zone ($h/2 = 5.3$ cm)

the figure shows only the qualitative dynamics of the cavitation zone with decreasing explosion depth. As in the experiment (see below, for example, Fig.7.16), the periphery part of the cavitation zone in this case went down more sharply and its main density fell at the periphery zone with respect to the charge.

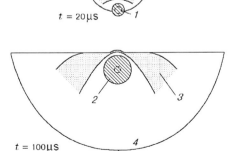

Fig. 7.6. Qualitative dynamics of the cavitation zone ($h/2 = 3$ cm): (1) initial size of charge (*dotted line*), (2) cavity with detonation products, (3) cavitation, (4) rarefaction wave front

7.2.3 Parameters of Rarefaction Wave in the Cavitation Zone

The system (7.7) enables calculations of the parameters and dynamics of the tensile stress field in the cavitation zone, which appeared to be essentially dependent upon the front steepness of rarefaction wave. We note that this concept in the given statement corresponds to the time of growing of absolute value of tensile stresses in the rarefaction wave front [32, 33] in the medium with a compressible liquid component. This parameter in the first equation of the system can be formally regulated by the multiplier σ_0. It can be taken to be equal to unit at $t = 0$ (the case of instantaneous application of the maximum negative pressure) or presented as a time function governing the law of increase of tensile stresses in the medium (the steepness of the rarefaction wave front). The function is taken from some additional considerations and is determined experimentally or numerically, for example, from the difference of the time of arrival of the characteristics of the rarefaction wave with zero and maximum amplitudes at a point under consideration [2].

An approximate estimates (based on the data of [2]) of the "steepness of the rarefaction wave front" for the case $h/2 = 5.3$ cm (curves 1 and 2) and $h/2 = 14$ cm (curves 3 and 4) are presented in Figure 7.7 for vertical (1, 3) and horizontal (2, 4) beams at different distances from the charge. With allowance for these data further studies were carried out for the time interval of the "front steepness," $\tau_{fr} = 0$–1 µs.

The calculation results for the parameters and dynamics of the field of tensile stresses are presented in Figs. 7.8–7.10. Figure 7.8 shows the time variation of the volume concentration k of gas phase (a) and pressure in the rarefaction wave p (b) at the point with the coordinates $z = 3.8$ cm and $r = 0$ (on the symmetry axis) during the explosion of a 1.2-g charge at a depth of 5.3 cm. The calculation was performed for $\tau_{fr} = 0$ µs (curves 1, instantaneous jump of the amplitude of tensile stresses), $\tau_{fr} = 0.1$ µs (curves 2), and $\tau_{fr} = 1$ µs (curves 3). For comparison we present the dependence $p(t)$ in the one-

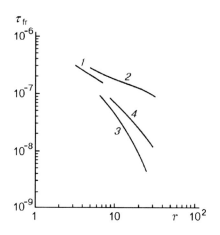

Fig. 7.7. Estimates of the front steepness in the rarefaction wave

7.2 Tensile Stress, Structure of Cavitation Region, Spalls

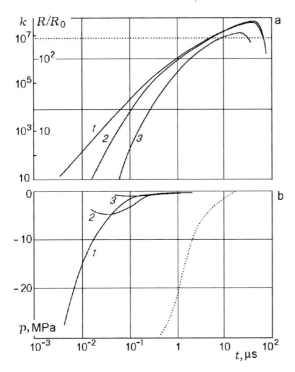

Fig. 7.8. The effect of the front steepness in the rarefaction wave on ultimate tensile stresses and the cavitation zone dynamics

phase liquid and the k value as the lower boundary of the visible zone (dotted lines). The initial parameters were $k_0 = 10^{-11}$, $R_0 = 5 \cdot 10^{-5}$ cm, $W \simeq 1.2$ g, $a_0 = 0.53$ cm, $h/2 = 5.3$ cm, and $p_{\text{det}} = 4 \cdot 10^4$ atm.

It is easy to see that the steep slope of the front for only 1 μs (curves 3) led to the decrease of the tensile stresses allowed by the cavitating medium by two orders of magnitude. The calculations show that the change of the volume gas concentration k was not so much affected by the steep slope.

For the same conditions, Fig. 7.9 presents the calculation results for the distribution of durations (Δt_{max}^-, Fig. 7.9a) and maximum amplitudes (Δp_{max}^-, Fig. 7.9b) for tensile stresses at vertical (H) and horizontal (r_1, Fig. 7.9c) axes. The figures at the curves 1, 2, and 3 corresponded to the times of occurrence of the steep slope $\tau_{\text{fr}} = 0$, 0.1, and 1 μs. One can see that the two-phase medium could maintain the negative pressures up to hundreds of atmospheres (curve 1, Fig. 7.9b and c) only within 0.1 μs (curve 1', Fig. 7.9a), over which the pressure decreased by an order of magnitude and more. This result complies with the known experimental data: a real liquid subject to instantaneous tension of 100 atm preserved the pressure only over 0.2 μs.

As indicated above, pressure in the front of a real rarefaction wave achieves its maximum over a finite time period. According to the calculations, over this time the size of cavitation nuclei increases considerably and the volume concentration increases by several orders of magnitude appreciably changing

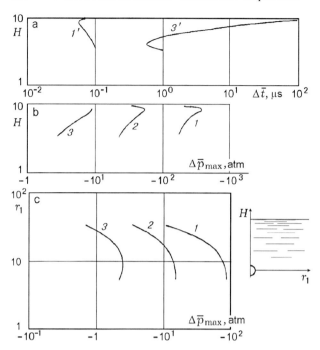

Fig. 7.9a–c. Distribution of the wave field parameters in the cavitation zone near the free surface for various steepnesses of the rarefaction wave front

the state of the medium. As a result, calculated negative pressures, the maximum values allowed by the medium, decrease by two orders of magnitude (Fig. 7.9b and c, curves 2 and 3) as τ_{fr} changes from 0 to 1 µs. In terms of the order of magnitude these values are comparable with the experimental data.

Figure 7.10 shows the angle distribution θ_1 (the distance $r_1 = 4.4$ cm from the charge center is fixed) for maximum tensile stresses $\Delta \bar{p}_{max}^-$ (Fig. 7.10a), maximum volume concentration k_{max} of gaseous phase (Fig. 7.10b), and the duration of the negative pressure phase Δt_{max}^- (Fig. 7.10c).

The calculations were performed for the above initial parameters and $\tau_{fr} = 0.1$ µs. Obviously, the abrupt change of k_{max} over the angle range $\cos \theta_1 = 0.4$–1.0 was accompanied by adequatly abrupt change of the duration of exposure to tensile stresses and relatively slight change of their amplitudes.

We would like to remind readers that the calculated and experimental rarefaction wave profiles at the point with coordinates $\theta_1 = 0$ and $r_1 = 14$ cm for explosion of a 1.2-g charge at a depth of 18.5 cm were compared in Chap. 6 (Fig. 6.8b). There the experimental curves lie between two calculated profiles of the rarefaction phase: curve 3 (initial parameters $R_0 = 5 \cdot 10^{-4}$ cm, $k_0 = 10^{-11}$, $\tau_{fr} = 5$ µs) and curve 4 (initial parameters $R_0 = 5 \cdot 10^{-4}$ cm, $k_0 = 10^{-10}$, $\tau_{fr} = 1$ µs). The parameters were selected on the basis of the above ranges of the gas content and real conditions of the experiment. Profile 3 was the closest to the experimental data in the order of magnitude and the shape.

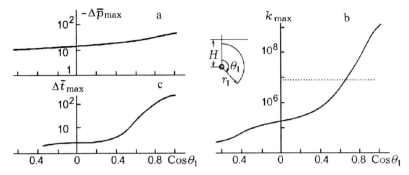

Fig. 7.10a–c. Angle distribution of tensile stresses and concentration

A principal distinction of the results obtained within the two-phase presentation of a real liquid with microinhomogeneities from the one-phase model is evident. We note that the cavitation nuclei in the rarefaction wave can fail to reach visible size near the free surface at a relatively large distance from the charge, nevertheless the amplitude of tensile stresses in this zone will noticeably decrease as well. This seeming paradox was mentioned by Cole [1]: an essentially underestimated amplitude (with respect to expected value) was recorded in the rarefaction phase in the absence of visible cavitation.

The results suggest that the proposed two-phase model of a real liquid enables calculation of the dynamics of rarefaction fields and cavitation zones close to real values using the data of actual gas content in liquids, without considering the problem on the formation of cavitation nuclei.

In conclusion we would like to mention the papers [34, 35] dealing with the one-dimensional problem on the development of cavitation and the rarefaction wave behind the nonuniformly moving piston in a channel. Assuming that the velocity and density in the medium are related by an unequivocal functional dependence, the authors obtained a solution from which it followed that *the profile of the cavitation wave was not distorted* and the change of the amplitude in the wave was proportional to the change of the volume concentration of gas. A discrepancy between this conclusion and the above results is obvious.

7.3 Formation of Jet Flows and Their Hydrodynamic Models

Since analysis of the structure of the cavitation zone in axisymmetrical statements is complicated, we ran a series of test experiments in plane statement (Fig. 7.1, right). A cylindrical trotyl charge was placed between two plane metal plates (15–20 mm thick) positioned in parallel at a distance of 10–15 mm from each other and rigidly fixed to one another. Both plates were

314   7 Jet Flows at Shallow Underwater Explosions

lengthened by Plexiglas plates, their inner surfaces were integrated so that a part of the walls between which the process occurred was transparent (see Fig. 7.1, right). The assemblage (slot cuvette with the charge) was placed upright in a tank filled with water (Fig. 7.1, left), the lower boundary of the transparent walls was slightly submerged under the free surface.

The structure dynamics of the cavitation zone for two depths (10 and 6 cm) of the charge of $R_{\mathrm{ch}} = 3.2$ mm is shown in Fig. 7.11a,b by two successive photos. As was shown above, for cylindrical charges $R_{\max}/R_{\mathrm{ch}}$ far exceeded

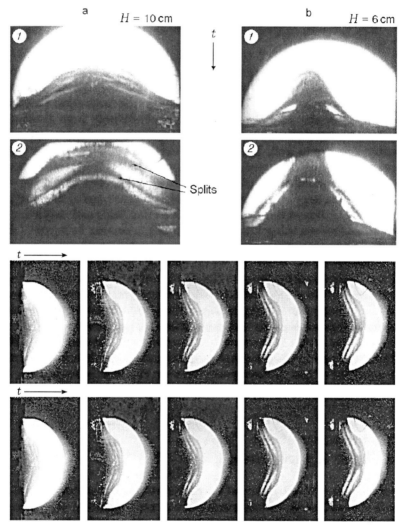

Fig. 7.11a,b. Development of spalls in the cavitation zone at underwater explosion (plane statement)

7.3 Formation of Jet Flows and Their Hydrodynamic Models

the ratio for spherical charges [42] and equaled about 130, i.e., $H/R_{\max} \simeq 0.25$ for $H = 10$ cm. The first photo (Fig. 7.11a, explosion at the depth of $\geq 30R_{\text{ch}}$) demonstrated two frames from the high-speed registration of the series formation (Fig. 7.11c) of plane spalls determining the stratified structure of the zone of intense development of cavitation characteristic of brittle failure of solids under loading by plane shock waves. *As the depth decreased (Fig. 7.11b), the spall zone concentrated near the symmetry axis, which indicated a tendency to the formation of a cumulative pit on the free surface.*

In the model of a sultan proposed by M.A. Lavrentyev, the availability of a cumulative pit on the free surface of a liquid was an important condition. The experimental data presented in Fig. 7.11b supported the idea of the formation of the pit. We have to check whether this condition is necessary for the formation of a sultan.

It was established experimentally that at depths of about 10 cm (Fig. 7.11a) spall layers had begun to cavitate already. This result is of principal importance, since current concepts of the formation mechanism of plane spalls was based on the assumption that the liquid disintegrates in the rarefaction wave field at some critical isobars, along the "split" line. The appearance of visible bubbles (≥ 0.1 mm) was identified with the beginning of spallation.

The fact of the development of cavitation throughout the region limited by the unloading wave front prejudices this concept. In addition, according to experimental data of different authors, the critical values of tensile stresses necessary to initiate spallation (as required by Lavrentyev's model) vary over a rather wide range from zero to hundreds of negative atmospheres and depend on the experimental arrangement. Therefore the necessity of their introduction is not justified and the character of load application appears to be more important than the load value.

Thus, the experimental studies of the development of the cavitation zone during underwater explosion of an HE charge near the free surface of a liquid proved that "the splash dome" arising at the initial stage of the explosion at the depth of about 30 charge radii (and more) "was a stratified structure of a series of cavitating spall layers". In particular, one can assume that the plane layered spall formed under explosion at a depth of 10 cm did not result in the formation of a cumulative pit on the free surface and the absence of a sultan jet during the explosion or considerable weakening of the effect would confirm the leading role of the shock wave. However, the experiment did not support the expected effect and called to question the governing role of the cumulative pit in the formation of a sultan.

The high-speed recording of the development of surface effects during explosion of a charge in the slot cuvette at a depth of 7.5 cm (Fig. 7.12) clearly illustrated the structure of the flow. The time between frames was 1 ms, the recording was done at a small angle from above the liquid surface. One can see that a spall and a vertical jet beneath the former and above the cavity

316   7 Jet Flows at Shallow Underwater Explosions

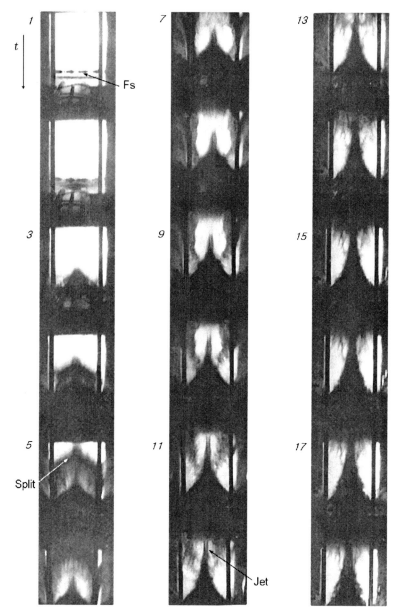

Fig. 7.12. Development of a jet under the separated dome

with detonation products (frame 5) were formed practically simultaneously. Then the jet penetrated a spall dome (Fig. 7.13) and formed a vertical jet flow (sultan) to the atmosphere. It should be noted that further expansion of the explosion cavity led to depressurizing and throwing out of detonation

7.3 Formation of Jet Flows and Their Hydrodynamic Models

Fig. 7.13. Development of the dome and rise of the vertical sultan jet

products into the atmosphere. As a consequence, a surface closure of the explosive cavity was observed, which tended to the formation of a secondary vertical jets.

When analyzing the phenomena produced on the free surface of a liquid by underwater explosions, researchers were mainly focused on describing the external parameters of sultans. The experiments, theoretical studies, and models discussed in this section made it possible to reveal the formation mechanism and conditions for such jets, determine main parameters of sultans, and understand their structure. It was found that depending on the explosion depth an entire class of cumulative jets was formed on the free surface, from a vertical jet tandem at the initial stage of the first expansion of the cavity with detonation products to the "eruption" of the vertical throwing out under the emergence of a giant cavity with detonation products of a nuclear charge on the surface.

Below we consider the flows under explosion at depths of $H = (0\text{--}2) \cdot R_{\max}$, where $R_{\max}$ is the maximum radius of the cavity with detonation products under explosion in an unbounded liquid. Mainly lightweight charges were used in the experiments. Based on the research results for the explosions of large charges, for example, from [1, 8–10], the laboratory data are reasonably extrapolated to large-scale experiments.

7.3.1 Formation of Vertical Jets on the Free Surface (Experiment, $H < R_{\max}$)

It is convenient to consider the flow features when geometrical parameters are reduced to the value of the first maximum of the explosive cavity, which is automatically related to the charge weight, if the density and the reaction heat are known. It is known that the maximum radii of the explosive cavity for different oscillations $R_{\max,i}$ and the oscillation periods T_i are determined by the equations

$$R_{\max,i} \simeq \left[\frac{3}{4\pi}\frac{\alpha_i EW}{p_0}\right]^{1/3} \quad \text{and} \quad T_i \simeq 1.83\sqrt{\frac{\rho_0}{p_0}}\left[\frac{3}{4\pi}\frac{\alpha_i EW}{p_0}\right]^{1/3},$$

where p_0 is the hydrostatic pressure in bar; E is the heat of explosive transformation of the HE in kcal/g; α_i is the coefficient determining the part of the energy remaining in the detonation products, i is the pulsation number, and W is the charge weight in g.

For 1-g charges one can readily obtain the values of maximum radii of the cavities with detonation products $R_{\max,i} \simeq 23\alpha_i^{1/3}$ (cm), as well as the oscillation periods $T_i \simeq 42\alpha_i^{1/3} 10^{-3}$ (s), where $\alpha_{1,2,3}$ is equal to 0.41, 0.14, and 0.076 [1], respectively, for spherical charges used in the experiments. In real numbers, we get $R_{\max,1,2,3} \simeq 17.1$, 11.9, and 9.7 cm and $T_{1,2,3} \simeq 31.3$, 21.9, and 17.8 ms.

Figures 7.14–7.16 present different stages of the development of the surface phenomena at depths of 15, 5, and 2 cm (or $\bar{H} = H/R_{\max} \simeq 0.88$, 0.29, and 0.12). Numbers under the frames correspond to ms.

Analysis of the experimental results showed that the distinct vertical jet appeared under the dome at the times reckoned from the explosion start (below $\tau = t\sqrt{p_0/\rho_0}/R_{\mathrm{ch}}$):

$$\tau(H) = \begin{cases} 30.1 \text{ (or 17 ms for } W \simeq 1 \text{ g) at } \bar{H} \simeq 0.9\,, \\ 8.8\text{--}10.6 \text{ (or 5--6 ms) at } \bar{H} \simeq 0.6\,, \\ 1.33 \text{ (or 0.75 ms) at } \bar{H} \simeq 0.3\,, \\ 0.44 \text{ (or 0.25 ms) at } \bar{H} \simeq 0.12\,. \end{cases}$$

The initiation of the jet with rather insignificant maximum velocity of the jet rise of about 15 m/s and height 1.3–1.5 m fell at the explosion depth $\bar{H} \simeq 1$–0.9. The main structure of throwing out on the free surface at these explosion depths was a cone-shaped splash dome. At the bottom of the photos in Fig. 7.14 one can see a "nonhermetic" cavity with detonation products at these depths (the time interval $t_2 = 30$–36 ms and later): there was a piped region produced by the surface closing of the explosive cavity after depressurizing and formation of upward and downward vertical jets.

With decreasing explosion depth the jet became more distinct: at $\bar{H} \simeq 0.6$ the jet velocity was about 30 m/s, its average diameter at a hight of 1.2 m was equal to 1–1.5 cm, while near the bottom (the dome vertex height was 0.4 m) the diameter was 5–7 cm. Further decrease in $\bar{H}$ led to a sharp growth of the jet velocity, which reached 100 m/s at $\bar{H} \simeq 0.3$. The velocity was about 300 m/s at $\bar{H} \simeq 0.15$ and the jet was clearly visible against the background of the dome of splashes (Fig. 7.15). The tendency to form a jet was retained at shallower depths (Fig. 7.16). Following [36, 37], the velocity of rise of the throwing out vertex (in this case, the dome vertex) on the free surface under explosion of a charge at depths twice the charge radius at the initial time could exceed 2 km/s, the Mach number of the incident flow could achieve 4 for small charges and 7 for charges of weight of about 100 kg.

The change of the maximum jet velocity v_{jet} (m/s) is shown in Fig. 7.17a. Here the data on the velocity of the dome vertex v_k (m/s) [36, 37] obtained for the range $\bar{H} \simeq 0.06$–0.1 for a 0.2-g TNT charge and a 100-kg trotyl charge are presented for comparison.

7.3 Formation of Jet Flows and Their Hydrodynamic Models 319

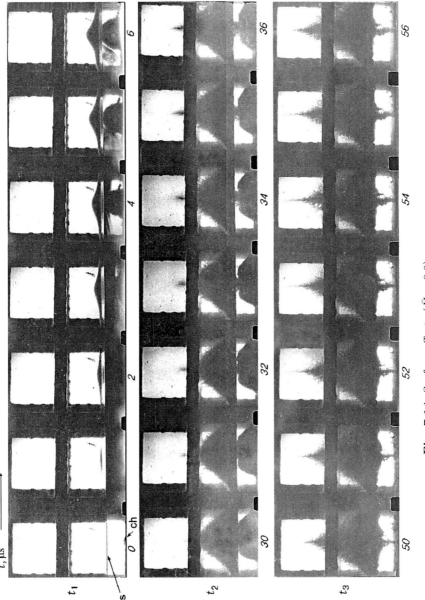

Fig. 7.14. Surface effects ($\bar{H} \simeq 0.9$)

320   7 Jet Flows at Shallow Underwater Explosions

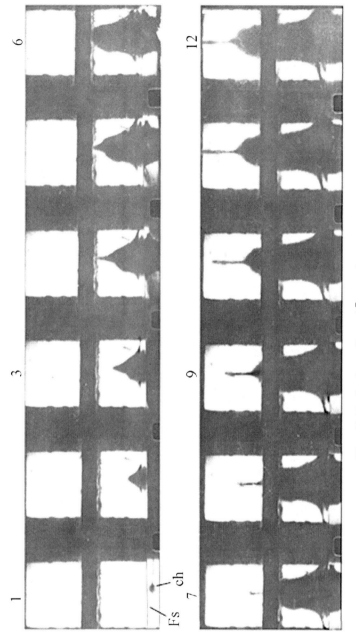

Fig. 7.15. Surface effects ($\bar{H} \simeq 0.3$)

7.3 Formation of Jet Flows and Their Hydrodynamic Models 321

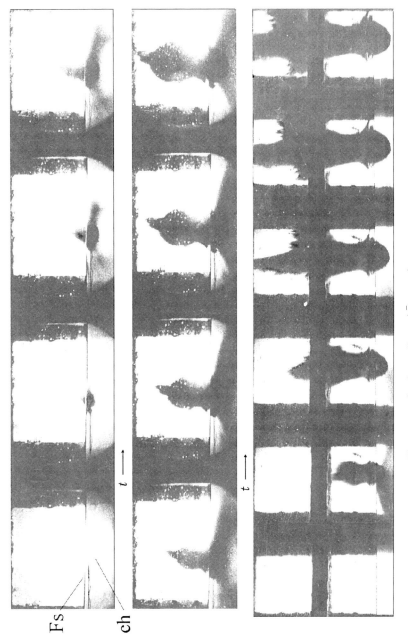

Fig. 7.16. Surface effects ($\bar{H} \simeq 0.1$)

By maximum jet velocity we mean the velocity recorded at the moment of exit of the jet from the dome. The dependence

$$v_{\text{jet}} \simeq \frac{16.5}{\bar{H}^{1.5}} \text{ m/s}$$

is a good approximation of the experimental data on the jet velocity over a fairly wide range of explosion depths $0.15 \leq \bar{H} \leq 1$ shown in Fig. 7.17a. Experimental data are shown by circles, numerical data are shown by a solid line. Experimental studies show that for the given range of explosion depths and the above times, the velocities of resulting jets are much higher than those of the vertex of the splash dome. Therefore, one can assume that the experimental data of [36, 37] for the velocity v_k of the "throwing out" peak obtained for the explosion depths $H/R_{\text{ch}} \simeq 2\text{--}3$ and $W = 0.2$ g (see figure) were probably associated with the jet perforating the dome rather than with the dome itself. In particular, this conclusion follows from the fact that the data for the "dome" v_k and for the jet v_{jet} shown in Fig. 7.17a and summed up as dots in Fig. 7.17b were approximated by one dependence over the explosion depth range $0.06 \leq \bar{H} \leq 1$ (Fig. 7.17b):

$$v_{\text{jet},k} \simeq \frac{16}{(\bar{H} - 0.03)^{1.5}} \text{ m/s} .$$

Another important characteristic of the vertical high-velocity flow is the amount of liquid in the sultan. The data of [8] and [10] diverge. The dependence $m_{\text{jet}} \simeq 540 \cdot W \cdot H/R_{\text{ch}}$ is presented in [8] and it is assumed that the mass of the sultan is determined by the weight of the liquid in the gas cavity at the moment of completion of the main stage of the process (apparently it is about an oscillation half-period). In [10], this dependence was defined

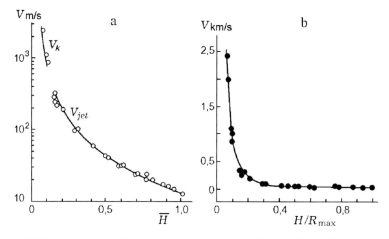

Fig. 7.17. Jet peak and dome velocity as a function of the explosion depth

7.3 Formation of Jet Flows and Their Hydrodynamic Models

as $m_{\text{jet}} \simeq 150 \cdot W$ for the explosion depth $H \simeq W^{1/3}$ m (W is the charge weight in kg). Moreover, at both sides of this depth the weight of the liquid decreased to $50W$.

To solve this problem, a special device was used in the experiments to catch the sultan jet at different depths from the liquid surface and to determine its mass. The device was a cylinder with a cone-shaped convergent upward entry and a bottom with a pulverizer that prevented the jet under impact from reflecting backward, and thus, kept the jet mass in the trap.

To eliminate the effect of splashes surrounding the sultan, the trap was placed at a height of several meters, the charge was centered perpendicularly under water relative to the narrow entry hole of the trap. The development of the jet and its infiltration into the trap were recorded using a high-speed camera. Figure 7.18a shows typical photos for the charge of weight of about 1 g at a depth $H = 10$ cm and a trap placed at a height of 2.4 m above the free surface.

The amount of liquid in the sultan jet appeared to depend essentially both on the explosion depth and on the trap height. The experiments showed that the liquid mass in the jet was 260 and 550 g for the trap height 2.4 m and $H = 5$ and 10 cm, respectively, and 440 g for $H = 10$ cm and the trap height 4.4 m. To ensure "pure" measurements of the amount of liquid in the jet,

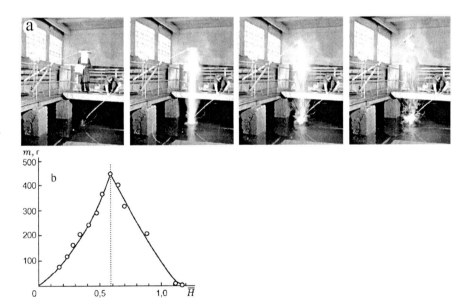

Fig. 7.18a,b. Measuring technique for the sultan mass (**a**). Mass distribution in the sultan as a function of the explosion depth (**b**)

the trap was placed at the highest level feasible in experimental conditions so that the splash dome could not get into it.

Figure 7.18b shows the change of the amount of liquid in the jet at a height of 4.4 m depending on the explosion depth. The dependence had a distinct maximum at the depth $\bar{H} \simeq 0.6$, which was close to the data of [10] on the optimum blasting depth ($H \simeq 0.1$ m $\propto W^{1/3}$ for $W \simeq 1$ g or $\bar{H} \simeq 0.1/0.17 \simeq 0.6$). These data differed in the maximum amount of the liquid in the throwing out $m_{\text{jet}}\big|_{L=4.4} \simeq 450W$ and $m_{\text{jet}}\big|_{L=2.4} \simeq 500W$ from those presented in [10].

Thus, the jet determined not only the maximum velocity of the sultan, but also its height and the amount of liquid in it. The experimental data obtained for the complicated flow structure near the free surface suggested the necessity of a detailed study of the mechanism of its formation. In this case, it was natural to try to separate the effect of the shock wave and the expanding cavity in the experimental and numerical models of the process. The role of the shock wave, following the above results, consisted at least in formation of spalls and of the cumulative pit (as predicted by M.A. Lavrentyev).

This point is of principal importance, since if the assumption on the mechanism of the classical cumulation in the formation of jets produced by underwater explosion is valid, the cumulative pit resulting from the spallation and the dynamics of the explosive cavity should be equally responsible for the nature of these flows. To prove the validity of this model, a "pure" statement was required, since one could assume that the great curvature radius of the "split" surface, as in the case of explosion at the depth of 7.5–10 cm, could play its role in the formation of a sultan.

The problem was solved for low explosion energies (on the order of 100 J) in model experiments with an exploding wire. The amount of the energy stored in the capacitor bank and transferred into the liquid as a compression wave was small due to the low-rate energy release (easily regulated by the

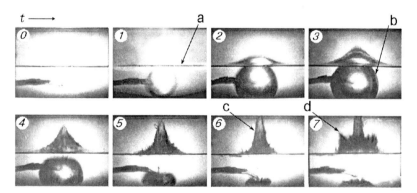

Fig. 7.19. Development of a sultan at underwater explosion of a wire near the free surface

parameters of the electric circuit), thus, the formation of the cavitation zone, and, therefore, the formation of spalls were excluded.

Figure 7.19 shows the development of the flow on the free surface during explosion of a wire (axial symmetry), which demonstrated the formation of a vertical jet in the absence of the cumulative pit on the surface. Three important conclusions followed from this result:

- The main role in the mechanism of jet formation was played by the explosive cavity
- We can neglect by shock wave effect
- The development of a sultan could be simulated within the framework of the model of an ideal incompressible liquid.

7.3.2 An Analog Model of a Sultan

Let us consider an unsteady liquid flow with the free surface at underwater explosion in the plane statement [19, 38]. Let an ideal incompressible imponderable liquid occupy the lower half-plane $z \leq 0$ (region $Q(t)$) containing a cavity of $R(t)$ with detonation products at a depth H from the free surface $\xi(t)$. The pressure at $\xi(t)$ is constant and equal to atmospheric pressure p_{at}, at $R(t)$ it changes following the adiabatic law $p(t) = p(0)[S(t)/S_0]^{-\gamma}$, where the initial pressure in the detonation products $p(0)$ and the adiabatic index γ are known.

The motion of the liquid is potential: $v = -\nabla \varphi$.

Statement of the problem:

$$Q(t): \Delta\varphi = 0, \quad \varphi \to 0 \quad \text{as} \quad |\mathbf{r}| \to \infty, \tag{7.8}$$

$$\xi(t): p = 1, \quad \varphi_t - \frac{1}{2}(\nabla\varphi)^2 = 0, \tag{7.9}$$

$$R(t): p(t) = p_{det}[S(t)/S_0]^{-\gamma}, \quad \varphi_t - \frac{1}{2}(\nabla\varphi)^2 = p(t) - 1. \tag{7.10}$$

The initial conditions $(t = 0)$ *are as follows:*

At $\xi(0)$ (horizontal surface): $\varphi = 0$, $p = 1$,

At $R(0)$ (a circle of radius R_{ch}): $\varphi = 1$, $p(0) = p_{det}$

In $Q(0)$: the section of the explosive cavity $S(0) = S_0 = \pi R_{ch}^2$.

Here,

$$\varphi' = \varphi\sqrt{p_{at}/\rho_0}\,R_{ch}, \; t' = t\sqrt{\rho_0/p_{at}}\,R_{ch}, \; \mathbf{r}' = \mathbf{r}R_{ch}, \; p' = pp_0 \text{ and } H' = HR_{ch}\,,$$

and $Q_b(\mathbf{r}, t) = 0$ is the equation of the boundary of the region $Q(t)$, $p_{det} = p(0)/p_{at}$, $\gamma = 3$, dimensional quantities are marked by primes.

The problem consists in the search for the function $\varphi = \varphi(\mathbf{r}, t)$ in the variable region $Q(t)$. The problem was solved by combining the method of electrohydrodynamic analogies (EHDA) and the method of finite-differences for the Cauchy–Lagrange integrals. The first was used to solve the Laplace equation (7.8), the second for Eqs. (7.9) and (7.10). The solution was reduced to the determination of the distribution of the values of the velocity potential $\varphi_{j,k,i}$ and its gradients $\nabla \varphi_{j,k,i}$ at the boundaries $\xi(t_i)$ and $R(t_i)$ at every moment of time t_i. The subscripts i, j, and k stand for the time increment and points at ξ and R, respectively.

The statement does not mention the kinematic condition

$$\frac{\partial Q_\mathrm{b}}{\partial t} - \nabla \varphi \nabla Q_\mathrm{b} = 0$$

at the boundary of the region $Q(t)$ that is automatically met in solving the Laplace equation using the EHDA method, since when it is implemented at the conducting paper to determine the potential and distribution, and constructing the current lines for all time intervals, the same "liquid particles" (points) are traced at both boundaries of the region $Q(t)$.

Using the definition $\mathrm{d}\varphi/\mathrm{d}t = \varphi_t + \mathbf{v}\nabla\varphi = \varphi_t - v^2$, the conditions (7.9) and (7.10) at the boundaries of the region for the potential can be written in the differential form

$$\xi(t_{i+1}): \quad \varphi_{j,i+1} = \varphi_{j,i} - \frac{1}{2} v_{j,i}^2 \Delta t_i , \qquad (7.11)$$

and

$$R(t_{i+1}): \quad \varphi_{k,i+1} = \varphi_{k,i} - \frac{1}{2} v_{k,i}^2 \Delta t_i + (p(t_i) - 1)\Delta t_i .$$

As was mentioned above, to solve the problems concerned with explosive processes, the so-called impulsive statement is often used. The statement suggests that when only instantaneous pressures affect the liquid (for a very short time over which the position of the boundaries of the region practically does not change) on the basis of the law of conservation of momentum,

$$\nabla \left(\frac{\mathrm{d}\varphi}{\mathrm{d}t} - p/\rho_0 \right) = 0 ,$$

the potential on the boundary can be determined as

$$\varphi = \rho_0^{-1} \int_0^{\tau_0} p\, \mathrm{d}t ,$$

where τ_0 sets the duration of exposure to the explosive load [39]. This value on the boundary $R(0)$ is taken as an analog unit.

The relation of the hydrodynamic potential φ_h to the electric potential φ_e is expressed by the scale coefficients different for each moment of time t_i.

Determining these coefficients, for the given time one can realize the known boundary values of the hydrodynamic potential through their electric analogs on the conducting paper. For this purpose, the boundary contours are covered by tires wound on insulation frames (shaped as the boundaries) [40] with leads at the points j and k. When the values of the potential, according to (7.11), are established at the boundaries, the distribution of φ in $Q(t)$ is practically obtained and we have only to mark it on the paper as equipotential lines. Then, selecting the increment of r, we determine the distribution of hydrodynamic velocities $v_{j,k,i}$ along the boundaries ξ and R as $\Delta\varphi/\Delta r$ (here Δ denotes the increment and Δr is taken along the normal to equipotential lines) and construct the flow lines. Thus, at the time t_i the flow pattern is completely determined and for the next time increment $t_{i+1} = t_i + \Delta t_i$ using the data on $v_{j,k,i}$, one can find a new position of the boundaries ξ_{i+1}, R_{i+1}, and from (7.11) the new distribution of $\varphi_{j,k,i+1}$ on them.

This distribution is realized again on the conducting paper upon determination of the scale factor and the calculation is continued for any time interval that is of interest for the process under study. Figure 7.20 shows the calculation results for the initial stage of unsteady liquid flow with the free boundary for different times at $p_{\text{det}} = 4 \cdot 10^4$ (determined on the assumption that the detonation throughout the HE charge volume is instantaneous), $\bar{H} = 4$, $R_{\text{ch}} = 1.5$ cm, $\rho_0 = 1$ g/cm^3, $p_{\text{a}} = 1$ bar, and $\tau_0 = 10^{-7}$ s (in the dimensionless form $6.66 \cdot 10^{-5}$). In view of the flow symmetry with respect to the vertical axis passing through the center of the explosive cavity, only halves of the half-plane (the symmetry axis on the conducting paper is made by the line of the cut) are considered in the calculation (and in the figure).

Figure 7.20 shows the equipotential lines $\varphi = $ const and the flow lines (the numbers 1–6 correspond to the times $t = 0, 2.7, 5.8, 87, 127$, and 208 µs). The entire pattern of the development of the jet (sultan) is shown in Fig. 7.21 for the times 0, 2.7, 5.8, 87, 127, 208, 308, and 408 µs. Here for the same times (Fig. 21b, 127 µs) the equipotential lines and the lines of flow supplement the pattern of the forming flow.

Figure 7.22 presents the diagrams for the fluctuation of the velocity v of liquid particles on the free surface (I, on the axis; II and III, along the boundary) and on the boundary of the explosive cavity (1 and 3 are the top and bottom particles on the axis; 2 represents particles on the horizontal diameter). One can see that for approximately 10 µs the flow character at all points is identical, 5–6 µs after the start the acceleration of liquid particles changes the sign due to the atmospheric pressure, their velocities start decreasing and about 10–20 µs after the decrease is several-fold. The velocities of liquid particles on the free surface first "follow" the behavior of the boundary of the gas cavity, but "stabilize" over the interval of about 20–40 µs and start growing again. For the points II and III, this growth stops at 100 µs and the character of their dynamics becomes identical again to that of the points on

328   7 Jet Flows at Shallow Underwater Explosions

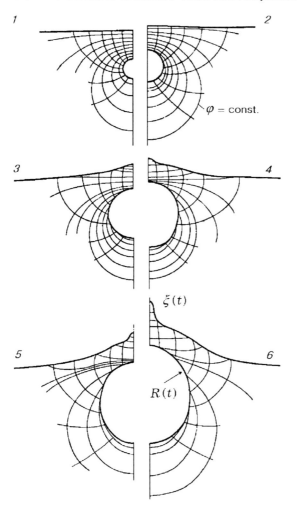

Fig. 7.20. Initial stage of the dynamics of the boundaries $\xi(t)$ and $R(t)$

the surface of the explosive cavity, while the velocity of the point I on the symmetry axis continues growing.

Comparing Figs. 7.20–7.22, one can find that the reason for this effect is the onset of formation of a vertical cumulative jet over the interval 20–40 μs. Analysis of the calculation results shows that the distinctive features of the mechanism of this effect are as follows: already over the interval $5 < t < 20$ μs (Fig. 7.20, 3) the lines of the flow bend over the symmetry axis. At almost all points shown in Fig. 7.22, a decrease in the velocity is observed, which after some rise (at the points II and III) continues except for the vicinity of the axis and the free surface (point I). Here it becomes apparent the inertial properties of the liquid layer over the explosive cavity that has acquired a strong impulse at the initial moments of time.

7.3 Formation of Jet Flows and Their Hydrodynamic Models   329

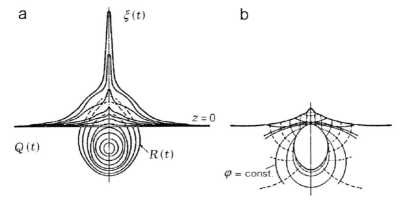

Fig. 7.21a,b. Development of a sultan under the expansion of an explosive cavity with initial pressure $p(0) = 4 \cdot 10^3$ MPa at the depth $H' = 4 \cdot R_{ch}$ under the free surface

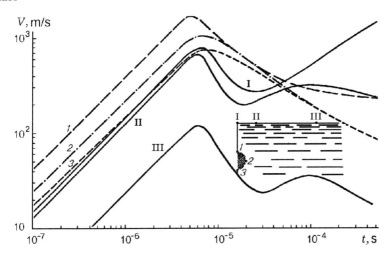

Fig. 7.22. Dynamics of mass velocities at characteristic points at the boundaries $\xi(t)$ and $R(t)$

Pressure in the detonation products and the expansion rate of the cavity continue decreasing, while the acceleration of the liquid near the point I increases. Tensile stresses arising in the region of the axis between the free surface and the explosive cavity lead to the local pressure decrease, and, as a result, to the turn and cumulation of the flow onto the symmetry axis: the growth of the particle velocity at the points I–III (over the time 20–90 μs) confirms this effect. Together with the vertical flow it results in the formation of a directional throwing out, a sultan. The velocity of liquid particles near the symmetry axis increases sharply, while that of the particles II and III remote from the axis decreases. Soon these particles start to "follow" the behavior

330   7 Jet Flows at Shallow Underwater Explosions

of the gas cavity whose effect on them is more tangible. This determines the structure of the resulting flow as a narrow vertical jet.

We note the effect of the cumulative flow on the dynamics of the upper half of the explosive cavity, whose velocities (at the point 1) decrease much slower than those of the lower half (points 2 and 3): the explosive cavity expands upward and transforms into an ellipsoid.

As was mentioned above, a necessary element of the sultan model proposed by M.A. Lavrentiev was a cumulative pit on the free surface of the liquid. The calculations made using the model for the free surface with a cumulative pit showed that the pit could amplify the effect of the formation of the directional jet under explosion at depths equal to several charge radii but does not effect on the mechanism of its formation.

7.3.3 Hydrodynamic Model of a Sultan: Pulsed Motion of a Solid from Beneath the Free Surface

Interpreting the formation mechanism of vertical jet flows produced by underwater explosions on the free surface allows a rather graphic model of a sultan to be constructed. Indeed, the sultan jet is shaped due to the momentum obtained by the liquid with a maximum velocity at the axis, abrupt decrease of the velocity of the cavity with detonation products, and occurrence of tensile stresses near the axis and flow cumulation on the axis. Obviously, these effects can be simulated assuming that the explosive cavity acts as a piston with the same feature of motion "acceleration–deceleration". An unsteady flow under impulsive motion of a solid submerged in a liquid from beneath the free surface can serve as the model [19, 38].

Let us consider the problem in the plane statement. Let a cylinder of radius R move upright with the velocity V_0 in an ideal incompressible liquid occupying the lower half-plane $z \leq 0$. At the time $t = 0$ the cylinder is at a distance $\bar{H} = H/R > 1$ under the horizontal free surface $\xi(0)$. As it occurs partially (or completely) over the surface at the time $t = t_0$, the cylinder stops instantaneously. There is a liquid layer on top of the cylinder, which is imparted an impulse and is "left on its own." How will the liquid behave then? This problem can be solved using the EHDA method. It is reduced to seeking the unknown boundary $\xi(t)$ and the potential $\varphi(\mathbf{r}, t)$ such that at $t \geq 0$

$$Q(t): \ \Delta\varphi = 0\,, \quad \varphi \to 0 \quad \text{at} \quad |\mathbf{r}| \to \infty\,;$$

$$\text{on} \quad \xi(t): \ p = 1\,, \quad \varphi_t - \frac{1}{2}(\nabla\varphi)^2 = 0\,;$$

and

$$\text{on} \quad R: \ \frac{\partial\varphi}{\partial r} = \begin{cases} -V_0 \cos\theta & \text{at } t_0 > t \geq 0\,, \\ 0 & \text{at } t \geq t_0\,. \end{cases}$$

At $t = 0: \xi(0)$ is the horizontal surface, at which $\varphi = 0$, $p = 1$. Here r and θ are the polar coordinates, the angle θ is reckoned from the vertical axis.

7.3 Formation of Jet Flows and Their Hydrodynamic Models

The kinematic condition is omitted here, since, as was noted before, it is satisfied automatically in finding the boundary $\xi(t)$. The condition on R at $t \geq t_0$ is met simply by cutting a circle of radius R on the conducting paper. The entire pattern of launching of the jet by a solid under impulsive motion from beneath the free surface is shown in Fig. 7.23. It is clear that in spite of the distinctions in the dynamics of the "pistons" (the explosive cavity expands, while the solid retains the size and moves only translationally) the flow character is amazingly identical (compare Figs. 7.19 and 7.23)!

Analysis of the dynamics of mass velocities in the calculations show that after the cylinder stops the velocity of particles on the free surface (near the axis) first noticeably decreases, then rapidly increases, which practically repeats the dynamics and velocity distribution for the case of explosion. The moment of stoppage of the body corresponds to the moment of emergence of negative acceleration of the wall of the cavity with detonation products. The calculation (Fig. 7.23) performed for $V_0 = 10$ m/s, $R = 5$ cm, and $\bar{H} = 1.4$, shows that some time after the stoppage the liquid layer on the solid surface concentrates to the axis of symmetry and forms a jet instead of flowing down. The pattern does not change qualitatively, if the cylinder R having the velocity V_0 at $t = 0$ moves with negative acceleration up to $V = 0$ or instantaneously brings an impulse to the liquid and remains at the same place. The calculation results are supported by experimental studies.

Experimental simulation of sultans was performed for the plane and axisymmetrical cases with metal spheres, hollow half-cylinders or half-spheres. The momentum was driven to the hollow bodies by an explosion of a distributed or concentrated charge (HE or exploding wire) placed immediately

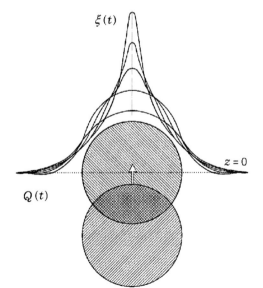

Fig. 7.23. Formation of a jet during impulsive motion of a solid from beneath the free surface (calculation)

under the surface of the body, whose internal part (facing the charge) was covered by special layers adsorbing shock waves in case of need. The waves that streamlined the thrown surface were already rather weak and did not distort the integrity of the liquid above the thrown surface. As suggested by the numerical estimates, the initial distribution of the particle velocities on the free surface was an important factor within these statements and should have relatively abrupt gradient with the maximum at the symmetry axis. The curvature radius of the thrown surface was selected of the same order as the maximum radius of the explosive cavity in an unbounded liquid.

Figure 7.24a and b illustrates the experimental simulation of the process of the jet formation and acceleration by impulsive motion of a solid. The figure shows the case of impulsive motion of a solid sphere from underwater and the formation of a jet after the sphere stoppage. The horizontal line in the lower part of frames is the free surface separating liquid from air. Figure 7.24b shows a half-sphere of diameter 8 cm (its peak point is at a depth of 2 cm) accelerated from under the liquid surface by a cavity resulting from the explosion of a wire and was naturally way decelerated above the liquid surface. The recording speed was 600 frames per second, the energy accumulated in the discharge gap was about 200 J. Figure 7.25 shows further development of the jet accelerated by a half-sphere. The recording speed was 25 frames

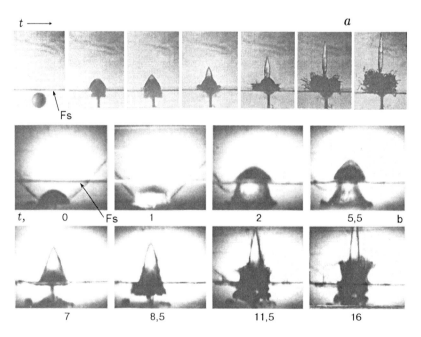

Fig. 7.24a,b. Jet formation under the acceleration of a solid sphere by mechanical pulse (**a**) and of a half-sphere thrown by explosion of a wire (**b**) from beneath the liquid surface (experiments)

7.3 Formation of Jet Flows and Their Hydrodynamic Models 333

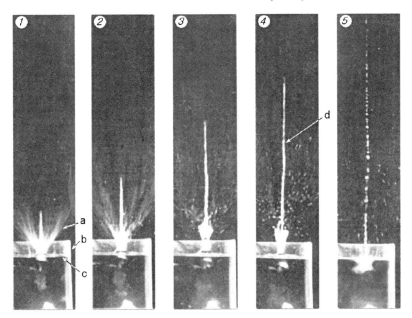

Fig. 7.25. Late stage of the development of a cumulative flow resulting from the pulse motion of a half-sphere

per second, the vertical scale of the frame was about 2 m (the depth of submergence of the half-sphere was 4 cm). Due to the velocity gradient the jet expanded and then disintegrated into drops. Varying the depth of initial submergence and the velocity of the accelerated body, one could obtain free flight of the jet.

Figure 7.26 shows the development of a sultan jet during launching of a half-cylinder by a 1.2-g HE charge from the depth of 2 cm (the time interval is 20 ms). The jet developed against a background of a vertical "scale rule" with the distance between the grades of 1 m. The high-speed photos showed that the thrown body imparted an impulse to the liquid layer above it and the deceleration of the body created conditions for modeling the formation of a jet during underwater explosion.

The above experimental results and the numerical analysis support a conclusion on the similarity of the formation mechanisms of jets at shallow underwater explosions and throwing of a liquid layer by a spherical or cylindrical solid from beneath the free surface.

7.3.4 Abnormal Intensification Mechanism of the First Pulsation

The second stage of underwater explosion at depths $H \geq R_{\max}$ starts after the cavity with detonation products reaches its maximum size and also contributes to the formation of a system of cumulative jets [19, 38, 41] at un-

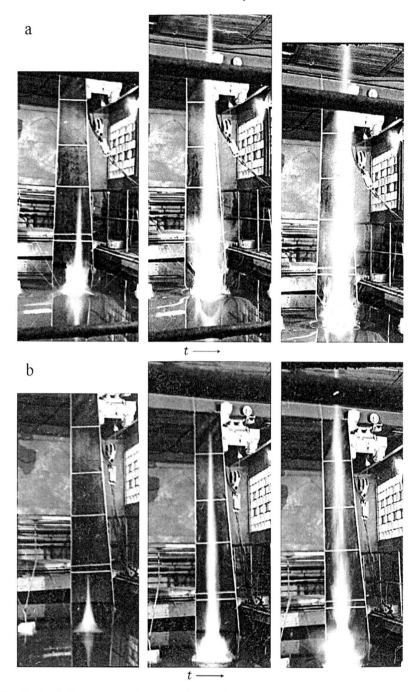

Fig. 7.26. Full-scale experiment on the acceleration of a hemicylindrical solid shell by HE charge and a cumulative jet formation, 5 m length

derwater explosions. Here the explosive cavity does not depressurize, a dome of fairly thin liquid layer is formed on the surface, while at the moment of stoppage of the cavity the pressure in it is almost zero and the cavity can be considered empty. Under the effect of hydrostatic pressure the cavity collapses and remaining "empty" over a considerable part of the half-period radiates a compression wave (the first pulsation) at the end of collapse.

A sharp and unobvious amplification of the maximum amplitude of the first pulsation at underwater explosions near the free surface (as compared with deep explosions) was established experimentally in [1]. The change of the amplitude (sometimes several-fold) is characteristic of a very narrow range of the submergence depth of the HE charge, which, as established in numerous studies for charges of various weight [1], are close to the maximum radius of the cavity with detonation products under the explosion in an unbounded liquid.

Figure 7.27 presents the data of [1] on the change of the relative value of the maximum amplitude of the first pulsation $\bar{p}_1 = p_1/p_{1,\infty}$ depending on the charge depth, expressed in the terms of maximum radius of the explosive cavity for the charges of weights 137 kg (circles) and 250 g (triangles). The following features of the experimental arrangement should be noted. The data on the explosion of a 137-kg charge were given with allowance for potential reflections from the free surface and the bottom, but the effect of the boundaries on the size and shape of the cavity during the first and the second half-periods were naturally. The maximum cavity diameter for 137-kg charge was equal to about 17 m (trotyl equivalent), while the depth of the explosion site was only 33 m. The data for the 250-g charge did not exclude the effect of the reflections from the boundaries for the magnitude and the shape of the pressure wave, while the explosion depth to the maximum cavity diameter ratio was more satisfactory and equaled six.

The accuracy of positioning the charge is very important. However, the data presented in Fig. 7.27 [1] indicated the existence of an abnormal increase in the amplitude of the first pulsation for charges of different weight in the region $\bar{H} = H/R_{\max} \simeq 1$. The results for 1.2-g charge (marked by *) demonstrated the same character of the dependence $\bar{p}_1(\bar{H})$ (but already in a practically "semi-infinite" region). It should be noted that the diagrams showed an abrupt disappearance of the secondary pressure field (of the first pulsation) under a slight decrease of the explosion depth as compared with the "abnormal" value.

Cole [1] cited the explanation proposed by Kirkwood for the abnormal effect, though he did not consider it convincing: at the stage of maximum expansion the cavity depressurized and atmospheric air mixed with detonation products. Then, due to some further reactions an additional energy releases in the mixture, this effect led to the amplification of the first pulsation under subsequent collapse of the cavity.

336   7 Jet Flows at Shallow Underwater Explosions

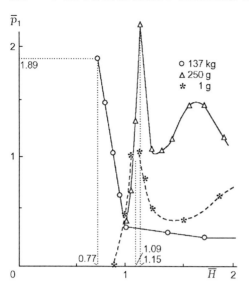

Fig. 7.27. Abnormal increase of the maximum amplitude of the first pulsation (experiment)

We would like to note that during the collapse some part of the liquid near the free surface in the region of the symmetry axis continued upward inertial motion due to the impulse received during the first half-period. As a result, a rarefaction wave and hence a cavitation zone could have occurred between the free surface and the explosive cavity. High-speed recording showed this region as thread-like clusters composed of cavitation cavities located along the particle trajectories. Naturally, at relatively long exposure of the recording the pattern would have a form of a continuous region, which might lead to the wrong conclusion on the depressurization of the cavity at abnormal explosion depths.

To clarify the nature of this abnormal effect we ran experiments with simultaneous recording of pressure (using pressure gauges), optical visualization of secondary compression waves at the shadow setup and the shape of the collapsing cavity for different explosion depths. The results of the experiments proved direct relation between the pressure field of the first oscillation to the change of the form of the explosive cavity collapsing near the free surface [19,41]. As the initial position of the charge approached the free surface the spherical (for $\bar{H} > 2$) shape of the gas cavity started distorting and for the depth range $\bar{H} = 1$–2 acquired the form of an oblate ellipsoid of revolution, as the cavity collapsed.

These depths were also characterized by the decrease of the amplitude of the first oscillation (Fig. 7.27,*), which was quite natural if the condition of spherical cumulation was violated. When $\bar{H} \simeq 1$, the collapse of the cavity was accompanied by the formation of a cumulative jet directed from the free surface inside the liquid (Fig. 7.28, plane statement, explosion of a wire). In this case, the free surface acquired a complex shape: it had a peak at the symmetry axis and two hollows symmetrical to the axis. The cumulative jet changed the flow topology: initially cylindrical explosive cavity acquired the form of two conjugate circles (torus for the case of spherical symmetry). Their expansion (frame 8, Fig. 7.28) formed two radial jets on the free surface. At $\bar{H} < 1$ the explosion products penetrated the atmosphere and the first oscillation vanished completely.

One can assume that the penetration of the high-velocity cumulative jet formed due to the first collapse of the explosive cavity near the free surface into the liquid was the most probable cause of the abnormal pressure gain.

Let us consider a mathematical model of the phenomenon in the plane statement. Let an empty cavity of radius $R(0)$ and with pressure $p = 0$ have a center was at a depth H from the free surface of the liquid at infinity and be placed under the free surface $\xi(0)$. Let us study the dynamics of the shape of the boundaries $\xi(t)$ and $R(t)$ by solving the problem on the velocity potential φ considering that

$$\text{in} \quad Q(t): \quad \Delta\varphi = 0, \quad \varphi \to 0 \quad \text{at} \quad |\mathbf{r}| \to \infty;$$

$$\text{at} \quad \xi(t): \quad p = 1, \quad \varphi_t - \frac{1}{2}(\nabla\varphi)^2 = 0;$$

and

$$\text{at} \quad R(t): \quad p = 0, \quad \varphi_t - \frac{1}{2}(\nabla\varphi)^2 = -1.$$

When $t = 0$:

At $\xi(0) - \varphi = 0$, $p = p_{\text{at}} = \text{const}$
At $R(0) - \varphi = 1$.

The initial forms of $\xi(0)$ and $R(0)$ are taken from the experimental data. The calculation was performed for $H = 19$ cm, $R_+ = 19$ cm, $R_- = 16$ cm, where

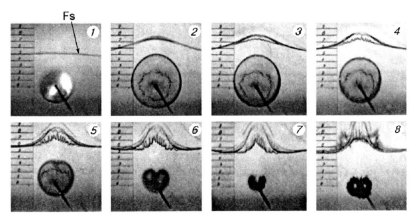

Fig. 7.28. The effect of the explosion depth on the shape of a cavity collapsing near the free surface (experiment)

R_+, R_- are the vertical and horizontal maximum cavity radii at the moment $t = 0$, respectively. The calculation results obtained using the combination of the EHDA method and the finite-difference method are shown in Fig. 7.29. One can see that the cavity collapse near the free surface was accompanied by the formation of an intense cumulative jet oriented from the free surface. By the moment when the jet reached the lower boundary of cavity the velocity of its head part V_{top} had achieved 300 m/s, while the average velocity of the cavity was about 20 m/s. One can estimate the order of the jet header at the initial moments of its penetration into the liquid was $p_{\text{pent}} \approx 2 \cdot 10^3$ atm. Following the stationary cumulation, the penetration velocity was of the order of $V_{\text{top}}/2$.

This result corresponded to $t \simeq 17.8$ ms, while the complete time of the collapse of an empty cylindrical cavity of radius 17.5 cm in the unbounded liquid was 26 ms [42]. Further collapse of the cavity (at $t > 18$ ms) resulted in a rapid increase of the velocity of its boundary and the cumulative flow, respectively. The estimate of the velocity of collapsing cylindrical cavity in an unbound compressible liquid in the acoustic approximation showed that at initial gas pressure $p_0 = 10^{-8}$ bar in the cavity and $\gamma = 1.25$ the velocity of the cavity wall in the region $R_{\min}$ could achieve values of the order of the speed of sound in unperturbed liquid. In this case, one can expect a considerable increase in the velocity of the cumulative jet during collapse. We note that, following the experimental results, the period of the first oscillation at abnormal depths remained practically the same and within the accuracy of experimental measurements was equal to the oscillation period under explosion in an unbound liquid.

In conclusion we note that the calculation of the pressure field for the abnormal effect is concerned with the problem on an unsteady cumulative jet penetrating a liquid. This is a rather complicated problem with unknown free boundary that includes the problems of formation and penetration of the jet, which should be solved simultaneously. Some qualitative estimates

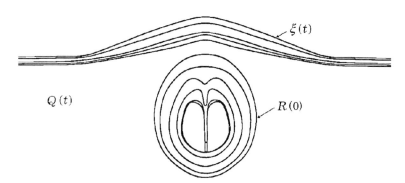

Fig. 7.29. Collapse of an empty cavity near the free surface at abnormal depths (calculation): $t = 0$, 12.95, 15.96, 17, 17.76, and 17.82 ms; $\bar{H} = H/R_+ = 1$

7.3.5 Two Models of Formation of Radial Sultans

As was shown above, as a result of the penetration of a cumulative jet into a liquid, the explosive cavity with detonation products is split into two cavities symmetrical with respect to the axis (in the plane case) or acquires the form of a torus (axisymmetrical case). Due to inertia of the liquid, these cavities collapse to a certain minimum size with pressure $p \gg p_{at}$, though much lower than during the first symmetrical oscillation. Then the cavities (torus) start expanding and in this sense the problem is equivalent to the first stage of the explosion considered above. But now the cavity with detonation products is of more complicated form and the free surface, as shown in Fig. 7.28, is far from being plane.

Within the model and the calculation method we shall consider a plane liquid flow arising under the expansion of two cavities symmetrical with respect to the axis. We take the initial forms of the cavities and the free surface from the solution of the problem on the collapse of an empty cylindrical cavity with the initial circular section [19] near the free surface and the depth of submergence of the center equal to 0.8 cavity radius and from the experimental simulation data for the case of exploding wires.

Let two gas cavities of $R(0)$ at the time $t = 0$ be placed at certain equal distance H from the free surface $\xi(0)$ symmetrically with respect to the axis (Fig. 7.30) in an ideal incompressible liquid. The pressure at $\xi(t)$ is constant and equal to p_{at}, at $R(t)$ it changes following the known adiabatic law. The solution of the problem consists of seeking for unsteady boundaries $\xi(t)$ and $R(t)$, and the velocity potential, for which the conditions stated in the problem on the sultan are valid. At $t = 0$, $\xi(0)$ and $R(0)$ have the form shown in Fig. 7.30. The situation was modeled for the explosion of a 1.2-g charge, the depth of the secondary "explosive" expansion of the cavities, $H = 13$ cm, was reckoned from the free surface at infinity to the upper point of the boundary $R(0)$, the initial pressure of the detonation products in the cavities was equal to $p(0) = 10^4$ bar.

The calculation results for the surfaces $\xi(t)$ and $R(t)$ are presented for the times 0, 110, 130, 160, and 190 μs. One can see that a radial jet (circular jet in the case of "explosive" expansion of a torus) developed symmetrically on the free surface of the liquid.

Analysis of the calculation results for the dynamics of mass velocities for the reference points marked in Fig. 7.30 showed how gradually a flow was formed with velocities much higher than those of peripheral and central points near the point with the fourth order number (the first is at the symmetry axis). This suggests that the flow on the free surface near this point became distinctly directional. Undoubtedly this flow was analogous to the first stage of the formation of sultans with the only distinction, the free surface here

340   7 Jet Flows at Shallow Underwater Explosions

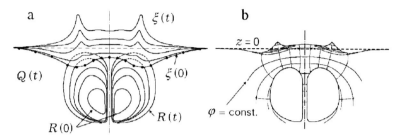

Fig. 7.30a,b. Calculation of the formation of radial throwing out: (a) general flow pattern, (b) the system of equipotentials and current-flow lines for $t = 130$ μs

had two cumulative pits with the axes inclined to the symmetry axis, and as the result the directional throwing out occurred at an angle to the symmetry axis.

The radial character of the sultan under explosion near the inclined free surface was also illustrated by the model experiments. Figure 7.31a shows the formation of a directional radial jet resulting from successive explosion of two wires. The first explosion I produced the inclined surface, the second II (frame 2) was performed with a delay determined from the preset deepening of the second "charge" $\bar{H} < 1$. Figure 7.31b illustrates the same effect of directional radial throwing out during subsequent explosion (with a delay for the formation of an explosive cavity) of two HE charges at the bottom of a water reservoir (the experiment was run at the Institute of Water Transport, Novosibirsk).

Radial sultans during the second expansion of the explosive cavity (Fig. 7.19, frame 7) also arose during modeling of the process by throwing a half-sphere (Fig. 7.25, frames 1 and 2, the result of the second oscillation of the explosive cavity under the thrown half-sphere). In this case, the role of the inclined surface was played by a curved surface near the sultan jet that was already formed at the first stage of throwing. Cole [1] pointed to the existence of radial fountains that occurred under shallow explosions of large HE charges (137 kg). As the explosion depth changed (downward from the surface) the radial fountains sometimes foreran vertical fountains. One of the mechanisms of their formation could be concerned with the cavitation effects. Indeed, strong tensile stresses occurring as a result of the reflection of powerful shock waves from the free surface resulted in the formation of a cavitation region with considerable concentration of cavitation bubbles, whose lifetime was comparable with the characteristic time of the process under consideration. The fast expansion of the cavity with detonation products occurred in the medium with essentially distorted continuousness. Then, by appearance the effect was equivalent to the explosions in soils near the free surface, at which, as a rule, radial throwing out unrelated to the above cumulative jets were observed. This reminds us of the so-called streamer structure (see Chap. 6)

7.3 Formation of Jet Flows and Their Hydrodynamic Models   341

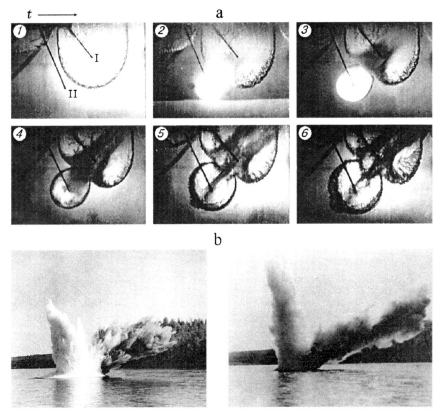

Fig. 7.31a,b. Experimental formation of radial jets: (a) modeling with exploding wires, (b) full-scale experiment

characterized by instabilities arising under the expansion of a cavitating liquid layer.

This model is illustrated qualitatively in Fig. 7.32 by the example of two experiments: explosion of a one-gram HE charge at a depth of 10 cm in a pure liquid (a) and in a liquid containing gas bubbles of diameter 3–5 mm and concentration of several tens of percents (b). It should be noted that underwater explosion in a bubbly medium can be used to model explosions in soils. Indeed, at certain concentration of the gas phase the bubbly liquid lost connectivity and explosive processes in it could develop following the accepted soil models, according to which:

- In the near zone, the soil behaved as an ideal liquid (in a model medium in the near zone, bubbles affected by powerful shock waves will fail and dissolve, thus the two-phase medium becomes uniform)
- In the middle zone, the wave strongly attenuated (modeled by the absorption of the shock wave by a bubbly medium)

342   7 Jet Flows at Shallow Underwater Explosions

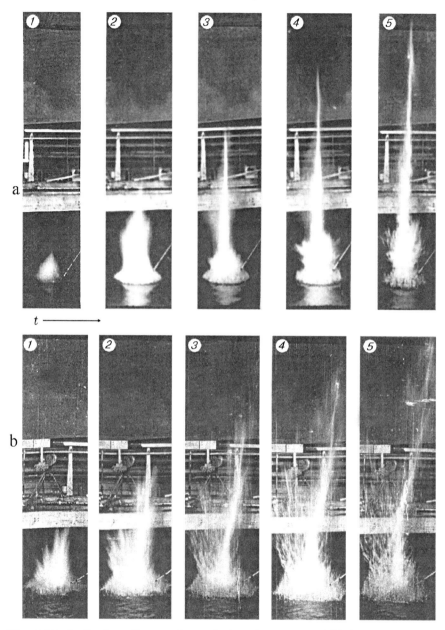

Fig. 7.32a,b. Surface effects under explosion of a one-gram HE charge at a depth of 10 cm (experiment): **(a)** uniform liquid, **(b)** bubbly medium

- In the far zone, the saturated bubbly medium behaved as loose soil (for example, the contents flew away from the free surface under the effect of expanding cavity with HE detonation products)

Above we considered the models of radial jet flows for the depths of the order of 30 charge radii and less. It is noteworthy that radial flows occurred during the explosion of large HE charges at a depth of about $(50\text{–}60)R_{ch}$ [44] at the moment of arrival of the explosive cavity at the surface.

7.3.6 Basic Parameters of Sultans

Figure 7.33 demonstrates the effect of the explosion depth by the example of underwater explosions of light charges: $\bar{H} > 0.9$ was the region of existence of the first oscillation, $\bar{H} \approx 1.1$ was the abnormal effect, $\bar{H} < 0.9$ was the depressurizing of the explosive cavity, and $\bar{H} \approx 0.6$ was the maximum liquid throwing out in the jet. Over the interval $\bar{H} = 0.5\text{–}0.15$, the velocity of the jet peak increased by a factor of 6–7, while the amount of liquid in it decreased by an order of magnitude.

Based on the generalized data of [7,8,10] and the results presented in this chapter, the main parameters of the vertical emission can be presented as follows:

$h_{\max} \simeq 60 \cdot W^{1/4} \sim (\frac{EW}{\rho g})^{1/4}$ is the maximum rise height;

$H|_{h_{\max}} \simeq 0.5 \cdot W^{1/4}$ is the appropriate explosion depth;

$T_{\max} \simeq 2.25 \cdot W^{1/8} \sim (\frac{EW}{\rho g^5})^{1/8}$ is the time of rise to $h_{\max}$;

$M_{\max} \simeq 300 \cdot W \sim (\frac{EW}{gH})$ is the maximum amount of liquid in the jet;

$H_{M_{\max}} \simeq W^{1/3}$ is the explosion depth corresponding to $M_{\max}$;

$v_{\text{jet}} \simeq \frac{30W^{0.5}}{H^{1.5}} \sim (\frac{EW}{\rho H^3})^{1/2}$ is the maximum velocity of the jet peak for the explosion depths $\bar{H} = H/R_{\max} \simeq 0.15\text{–}1$.

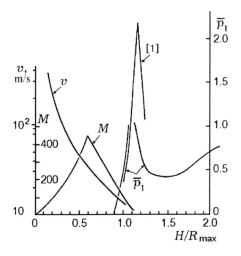

Fig. 7.33. Abnormal amplitudes of the first pulsation $\bar{p}_1$, dynamics of the liquid mass M, and the velocity of the sultan peak v (experiment)

Here: W and M are measured in kg; h and H in m; V_{jet} in m/s; T in s; E is the explosion heat, trotyl equivalent in kcal/kg; and the scatter is not more than 15%.

Figure 7.34 presents the general view of the development of a sultan under the explosion of a 10-kg charge at a depth of 1.1 m. One can see a distinct directional jet against the background of spallation effects.

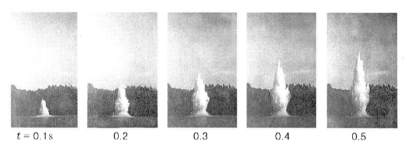

$t = 0.1\,\text{s}$ 0.2   0.3   0.4   0.5

Fig. 7.34. Development of a sultan (experiment, $W = 10$ kg, $H = 1, 1$ m)

7.3.7 Structure of Sultans: Jet Tandem, Analogy with a High-Velocity Penetration of a Body into Water

It is noteworthy that the structure of the first vertical throwing out at the initial stage of expansion of a cavity containing detonation products is not limited by the jet flow considered. The calculations show that if the explosion depth is only $4R_{\text{ch}}$, which is four times less than $R_{\max}$, the development of a jet and the expansion of an explosive cavity is accompanied by thickening of the circular layer at the jet bottom, which separates the detonation products from the atmosphere. The layer thickness decreases with time and at a certain moment the jet separates from the main liquid mass, the explosive cavity is depressurized, the explosion products are ejected through the circular rupture to the atmosphere forming a kind of a crown around the jet. The shape of the first separated jet resembles the structure of a cumulative jet under reverse cumulation [45, 46].

In [7,9], it was noted that for explosion depths of $\bar{H} \leq 0.9$, when the explosive cavity is depressurized by forming an open cavern, there are: (1) "a central column of apparently continuous liquid recorded against the background of a hollow cylindrical throwing out" at shallow depths; and (2) a "feather-like" surface throwing out [1, 7].

Further studies proved that all the flows are of the same nature and are governed by the fast surface closure of the depressurized cavern, which produces upward vertical and reverse jets, at $\bar{H} \approx 1$ radial jets supply the structure of throwing out at the stage of the first half-period of the expansion of the cavity with detonation products at shallow underwater explosions.

7.3 Formation of Jet Flows and Their Hydrodynamic Models

As mentioned in [47], the surface closure of the caverns was already described in the early 20th century when observing the process of submergence of a body in water. In more details the effect was studied in [4], where an unusual phenomenon was discovered. It appeared that at high-velocity penetration of a bullet into liquid (Fig. 7.35) the entire spectrum of jets is formed, which is practically coincident with the character of flows occurring at underwater explosion (though the order of events is reverse): the surface closure 1, the separation and collapse of the cavern 2, and its expansion 3 result, in accordance with the above mechanism, in the development of a vertical jet 4–7, in spite of the fact that the cavity collapses again and moves away from the free surface.

In conclusion, we would like to note that, according to [48], for large charges of ordinary HE and at nuclear explosions, the vertical throwing out at depths $\bar{H} > 1$ is developed as well. The time of its appearance is divisible by the oscillation period of the cavity with detonation products, while its appearance is conditioned by the availability of the free surface near the cavity at the moment of its maximum compression. This throwing out, following Ovsyannikov's model [6], also is of jet character; due to the large size of the explosive cavity and the effect of gravity it follows the cumulative mechanism, which is confirmed by the experimental and numerical data of Pritchett [48].

The summarizing picture of the surface effects corresponding to four characteristic depths (a–d), and the jet structure for each of them (for example, $a_{1,2,3}$) are shown schematically in Fig. 7.36. The dynamics of the cavity with detonation products with allowance for potential floating and relative deepening of the charge $\bar{H}$ [48] are shown on the right.

In conclusion, we summarize the basic features of the formation of a spectrum of cumulative jets at underwater explosion of a spherical HE charge at different depths:

– At shallow underwater explosions the basis of the throwing out is the jet tandem, the first jet is formed due to inertial motion of a liquid layer over the cavity with detonation products and the flow cumulation near

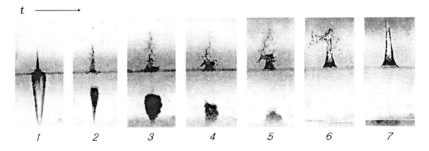

Fig. 7.35. Jet flows at high-velocity vertical penetration of a bullet into a liquid (experiment): the development of flows is inverse to the explosive process

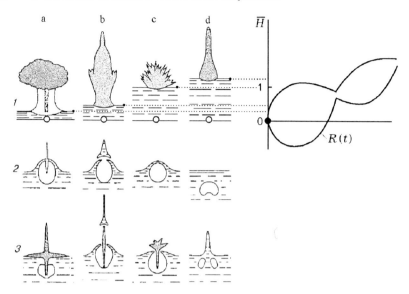

Fig. 7.36. Surface effects at underwater explosions and the schemes of their development (based on experimental data)

the symmetry axis, the second jet is a result of the surface closure of the cavern upon depressurizing of the cavity.
– At abnormal depths the explosion is accompanied by the formation of a downward vertical cumulative jet, change of the flow topology, and further development of radial cumulative jets under the secondary expansion of a toroidal cavity with detonation products.
– During large-scale explosions at depths corresponding to the region of existence of the first oscillation, there are vertical throwing out whose mechanism is determined by the development of the cumulative jet in the bottom part of the giant floating and collapsing explosive cavity (hundreds of meters in radius) in a heavy liquid, if in due time it appears near the free surface and the cumulative jet is capable of perforating the remaining liquid layer.

7.4 Jet Flows: Shallow Explosions of Circular Charges

This section deals with analysis of formation mechanisms of the sultans generated by powerful deep-water explosions. These processes are governed by the gravity forces that influence the rate of emergence of the explosive cavity during oscillations and the change of its topology. The latter is due to the development of the cumulative flow in the lower half of the spherical cavity during the second half-period of oscillations. The vertical cumulative jet perforates the upper boundary of the buoyant cavity, transforming the sphere

7.4 Jet Flows: Shallow Explosions of Circular Charges

into a torus. Furthermore, the oscillation process continues displaying all features of inertial motion. Several scenarios of the appearance of directional throwing out on the free surface are possible. One of them is associated with the estimate of the rate of emergence and the information on the pulsation period, which allows prediction of the probability of coincidence of two events. This is the concurrence of the time required for the formation of a cumulative jet and the time it takes the explosive cavity to reach the free surface. The scenario with an oscillating torus is also probable.

7.4.1 Gravity Effect

As was mentioned, gravity has a considerable effect on the dynamics of the explosive cavity at large-scale underwater explosions: a vertical motion (emergence) arises, whose impulse over the time t (assuming that the cavity remains spherically symmetrical) is determined by the integral

$$I = \frac{4}{3}\pi\rho g \int_0^t R^3(t)\,\mathrm{d}t\,.$$

With allowance for the added mass, which amounts to half the volume of the liquid displaced by the cavity, one can estimate the emergence velocity of the explosive cavity

$$V_* \simeq \frac{2g}{R^3}\int_0^t R^3(t)\,\mathrm{d}t\,.$$

Zamyshlyaev and Yakovlev [2] analyzed the development of the explosive cavity and proposed a simple approximation of its dynamics divided the expansion time equal to a half of the first oscillation period T into two intervals:

$$\frac{R}{R_{\max}} \simeq \frac{R_0}{R_{\max}}\left[1 + 3.73\left(\frac{R_{\max}}{R_0}\right)^{2.5}\frac{t}{T}\right]^{0.4} \quad \text{at} \quad t/T \leq 0.1$$

and

$$\frac{R}{R_{\max}} \simeq \left[\sin\left(\frac{\pi t}{T}\right)\right]^{0.36} \quad \text{at} \quad 0.1 \leq t/T \leq 0.5\,.$$

The formula for the impulse I can be expressed in terms of the relative variables $z = R/R_{\max}$ and $\tau = t/T$, and the limit of integration can be restricted by the expansion time $\tau = 1/2$. Then the impulse acquired by the system over the cycle is defined by the dependence

$$I = \frac{4}{3}\pi\rho g R_{\max}^3 T \int_0^{1/2} z^3\,\mathrm{d}\tau\,.$$

The multiplier under the integral can be expressed in terms of the explosion energy Q and the hydrostatic pressure p_0. One should remember that the potential energy of the system $U = p_0 V_{max}$ and Q are related by the dependence $U = \alpha_1 Q$, while $T = 1.14\alpha_1^{1/3} Q^{1/3} \rho^{1/2} p_0^{-5/6}$. Following [1], $\alpha_1 \simeq 0.41$ and $\alpha_2 \simeq 0.14$ for the first and the second oscillations, respectively. Using the data of [2] for $R(t)$, one can estimate the integral in the last determination of the impulse. It is equal to about 0.1. Thus, the impulse can be determined as

$$I \simeq 0.035 g \rho^{3/2} Q^{4/3} p_0^{-11/6} .$$

The emergence rate of the explosive cavity (in the vicinity of R_{min}) at $\tau = 1/2$ for a 1-kg charge exploded at a depth of 600 m estimated in accordance with the relation

$$V_* \simeq 0.228 g \rho^{1/2} (\alpha_1 EW)^{1/3} p_0^{-5/6}$$

yields a value of about 60 cm/s. For a 30-kg charge it is about 180 m/s at a specific explosion energy $E \simeq 1$ kcal/g.

One can assume that at the following motion stages the vertical impulse is retained, as well as the total energy of the system (as a sum of current values of the potential energy, the kinetic energies of the radial and forward motion, and the internal energy of explosion products with regard to the energy loss for radiation under the first oscillation). After the first oscillation the internal energy of compressed explosion products decreases considerably (to $0.14Q$), while the value of the cavity size at the first minimum corresponds to approximately three initial volumes of the HE charge. This result is easily obtained assuming that at an energy release of $0.14Q_1$ the same value of R_{max} can be reached only if the charge volume exceeds the initial value by a factor of α_1/α_2.

Within the framework of the Navier–Stokes equations under the assumption of incompressibility of a liquid, Pritchett [48] calculated the dynamics of the cavity with the detonation products of a 30-kiloton device blasted at a depth of 610 m. The calculations showed that after over 7.5 s the cavity came up for approximately 340 m, making three oscillations. The estimate of the emergence rate of the cavity (from the slope of the tangent to the trajectory of the initial stage of the second expansion) at the moment corresponding to the first minimum yielded a value of about 150 m/s. Thus, the dependence $V_*(H)$ provided a fairly acceptable estimate of this parameter.

Of particular interest are calculations of an explosive cavity, whose shape during the first collapse already transformed from a sphere into a toroid due to the development of an upward cumulative jet in the bottom part of cavity. The radius of the cavity at the first maximum was equal to 115 m: it corresponded to the pressure difference of 2.3 MPa between the upper and the lower points of the cavity [48]. Naturally, this suggested that spherical symmetry was not retained under these collapse conditions. Since the cumulative jet was directed upward, after it touched the upper boundary of the cavity a vortex-type flow started forming with a "core" as a rather stable toroid with detonation products and a vertical jet. It should be noted that in spite of this the collapse of the explosive cavity continued until the pressure in detonation products stopped it.

The data of Snay [49] on the effect of the reservoir walls as a semi-empirical dependence of the cavity oscillation period on the geometrical characteristics of the reservoir were cited in [48]. If T_f is the "free" oscillation period, η is the relation of the maximum cavity radius to the reservoir radius, and ζ is the distance to its bottom, the oscillation period is defined by the empirical dependence

$$T \simeq T_f(1 + 0.216\eta + 0.783\eta^2)(1 + 0.15\zeta),$$

which can be used not only for numerical analysis of the dynamics of the cavity in a naturally restricted calculation domain, but also for experimental studies both in laboratory conditions and in relatively shallow reservoirs.

We mentioned already that at relatively deep explosions of large HE charges (or nuclear explosions), an oscillating toroidal explosive cavity formed during the first oscillation in a heavy liquid continues to rise to the free surface. There is a question whether the "deep-water" sultans mentioned [1, 48] are a result of the dynamics of such a cavity near the free surface. This problem will be studied below, considering the experimental statements modeling the development of the surface effects by the example of shallow underwater explosions of circular charges [50].

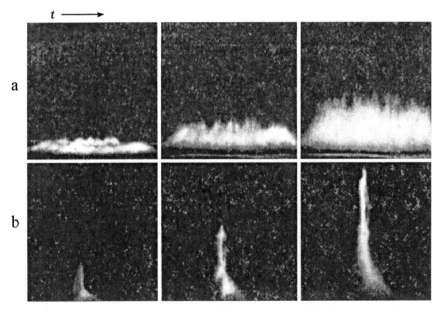

Fig. 7.37a,b. Development of circular (a) and central (b) sultans at underwater explosion of circular charges

7.4.2 Flow Structure Produced by Explosions of Circular Charges

We want to remind readers here that the features of the structure of the wave field arising under explosion of circular charges have been considered in detail in above sections. As was mentioned, if the circle radius is $a_r \geq 150\ R_{\mathrm{ch}}$, the final value of the detonation rate does not affect significantly the formation of the circular explosive cavity that keeps the form of a regular torus for long enough to get reliable modeling results.

The first experiments run in a natural reservoir, under the field conditions for circular charges with diameter of up to 2 m made of standard DC ($R_{\mathrm{ch}} = 1,5$ mm), confirmed the validity of the proposed model (Fig. 7.37). Figure 7.37a presents three successive moments of the high-speed recording of the development of a circular sultan under the explosion of DC charge, 2-m diameter, at a depth of 30 cm. We note that the maximum radius of the cavity with detonation products for a linear charge made of the same DC was approximately 20 cm. As the circle diameter decreased at fixed explosion depth, the character of the surface throwing out changed: Fig. 7.37b illustrates the development of a central sultan in the test with a circular charge 0.4 m in diameter, which confirmed the probability of appearance of a vertical throwing out on the free surface at deep large-scale explosions.

The structure of such flows was revealed in detailed laboratory studies with semi-ring charges whose plane was placed in parallel to the free surface, while the charge (or wire) ends were placed on a transparent vertical wall. In particular, such an arrangement made it possible to observe the flow development in the region of the circle axis. Figure 7.38 shows the high-speed photos of the explosion of a semi-ring 1 made of DC ($R_{\mathrm{ch}} = 0.0325$ cm) 18 cm in diameter at a depth of 9 cm below the free surface 2. The time interval between frames was 0.5 ms.

Two features should be emphasized: a cylindrical trace 3 (frames 3–7) in the cavitation zone (near the symmetry axis) and a thin vertical jet 4 on the surface in the dome center. The latter, as is known, consisted of split cavitating layers. It is obvious that the spall structure (with formation of a fountain-like throwing out 4) in the experiments with circular charges was determined by the form of a quasi-toroidal front of shock wave and by its

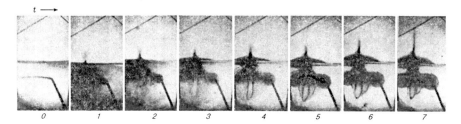

Fig. 7.38. Flow structure during explosion of a semi-ring HE charge

7.4 Jet Flows: Shallow Explosions of Circular Charges

focusing. To support this suggestion, the charge was placed in a dead-end circle of metal tube filled with water whose parameters were selected so that the leakproofness of the container and its size were preserved during the explosion. Naturally, only the shock wave "worked" under these conditions.

Thus, the experimental result was as follows: the fountain throwing out was observed in the experiment, while there was no central cylindrical trace in the cavitation zone. This result confirmed the fact that during the expansion of a toroidal cavity with detonation products, a jet was formed in the region of the symmetry axis of the circular charge due to the cumulation of the liquid flow on the axis. The details of the structure of the central jet were identified in similar arrangements with exploding wire. Figure 7.39 shows the sequence of the development of a flow on the free surface 2 and the cavity dynamics 1 during the explosion of a nichrome semi-ring wire (5 cm in diameter) at a depth of 4 cm. The time between frames was 4/3 ms. One can see that the energy per unit length of the conductor released under the discharge of a high-voltage capacitor bank was rather high in this experiment: the cavity was heavily deformed in the center and its maximum size was on the order of the explosion depth.

The peculiarities of this flow "section" in the central zone suggested that the vertical sultan forming on the free surface was a hollow central circular jet 3, whose nature was determined by the initial shape of the charge (rather than by the depressurizing of the cavity) and was apparently associated with the closure of the circular jet on the surface, as one of the structures of the vertical sultan in the geometry of this kind of flows. At later times (frames 5–7) the structure of the throwing out on the surface was similar to that in experiments with spherical charges.

The experimental results for the dependence of the height of the central jet $h(\tau)$ at underwater explosions of circular charges of diameter 40–80 cm (standard DC) at depths of $H = 20$–80 cm are summarized in Fig. 7.40. Despite the noticeable scatter in experimental data for the initial stage of rise of the sultan that might be due to the overlapping of the "fountain" throwing out of the split dome, we can make a conclusion on the general similarity of the development of central jets for different linear parameters of circular charges and explosion depths. Over the main time interval ($\tau \simeq 4 \cdot 10^1$–10^4),

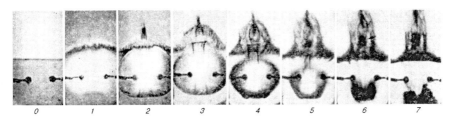

Fig. 7.39. The structure of the vertical jet during explosion of a semi-ring wire

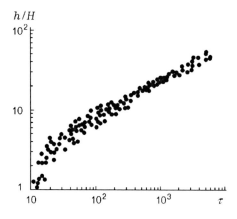

Fig. 7.40. The height of the central sultan jet during underwater explosion of circular charges

the dependence $h(\tau)$ was of power character and was approximated by the simple expression

$$\frac{h}{H} \simeq \tau^{1/2} \quad \text{where} \quad \tau = \left(\frac{EW}{\rho_{\text{ch}} H^5}\right)^{1/2} \cdot t$$

or

$$h \simeq \left(\frac{EW}{\rho_{\text{ch}} H}\right)^{1/4} t^{1/2}.$$

Here E is the explosion energy per HE unit mass, W is the HE mass, ρ_{ch} is the HE charge density, and t is the dimensional time.

7.4.2.1 Hydrodynamic model

It is logical to apply the above hydrodynamic model of a sultan to throwing of a liquid layer by a solid. However, in the case of circular charge this problem is further complicated by the necessity of modeling of the convergent flow, whose dynamics changes considerably during the expansion of a toroidal cavity. Nevertheless, it appears that if the dimensions of a solid tore correspond to the maximum dimensions of the cavity with detonation products, the expected effect can be achieved.

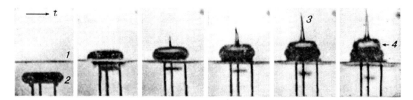

Fig. 7.41. Formation of circular and central jets during impulsive motion of a solid tore from under the liquid surface

Figure 7.41 shows the development of circular 4 and central 3 jets during the impulsive motion from beneath the liquid surface 1 of a solid toroid 2 that was imparted an upright impulse (high-speed recording). The impulse was imparted to the structure shown in the figure by the explosion of an HE microcharge using a device rigidly connected with the toroid and placed at a depth of several toroid diameters. The interval between the first frames was 4 ms, all subsequent intervals were 4.3 ms. One can see that, as in the field experiments, two jet systems were formed after the toroid stop: a circular (full analogy with throwing of cylindrical and spherical shells) and axial due to cumulation of flow to the axis that acquired radial mass velocities in the upward motion of a solid toroidal surface.

The results show that one of the possible formation mechanisms of the jet flow on the free surface formed at underwater explosion of large charges is determined by the flow cumulation to the axis at the stage of expansion of the toroid formed during the emergence of an initial spherical cavity with detonation products in a heavy liquid and the change of the flow structure with the formation of a stable toroidal cavity.

7.5 Shallow Underwater Explosions, Surface Water Waves

As the conclusion to this section I would like to mention a few results, some of which were practically inaccessible to Russian readers for a long time. Nevertheless, even a little information would be useful to have for a more complete representation different aspects of problems of shallow underwater explosions.

A numerical implementation of a generalized hydrodynamics has been used by Rogers and Szymczak to compute a number of violent surface motions (characterized by the collision of different portions of the free surface) [51]. In particular in a test of the predictions of the generalized hydrodynamics, the evolution of a plume generated by underwater explosions was compared with a computed plume history. An important diagnostic tool for studying violent surface motions has been the analysis of the sum of the kinetic and internal energies of the flow. Accordingly, one includes some results of an investigation into mechanisms for energy dissipation, as well as a description of some relations between energy loss and modes of cavity collapse.

A theory on the generation and collapse of water craters created by explosions in shallow water was developed in [52] by Khangaonkar and Lemehaute. A leading wave generated by a crater lip radiates outward while the water crater collapses towards the center. This phenomenon is similar to the dam break problem on a dry bed. The water edge reaches the center and forms a cylindrical dissipative bore the height of which decays rapidly with distance from the center. The method is applied to wave records obtained in the laboratory, ponds, and lakes with TNT explosions and also tested with the wave records from nuclear explosions, with yields ranging over 11 orders of magnitude. It is found consistently that the characteristic size of water crater created by underwater explosions in shallow water can be determined as $R(c) = W0.25$, with $R(c)$ in feet, and W in pounds of TNT. The relative hydrodynamic energy dissipation is nearly a constant 40% of the potential energy of the initial crater with lip.

Mechanisms in the generation of airblasts by underwater explosions were studied by Malme, Carbonell, and Dyer in [53]. Small-scale experimental tests were conducted to explore the formation and propagation of airblasts resulting from underwater explosions. The objects of the work were to describe the mechanisms of the explosion energy transferring through the interface and to relate the pressure field and water motion to the observed airblast pressures. The resulting airblast signature was found to be produced by acoustic radiation from a large surface area following the passage of the transmitted shock. When the interface velocity above the charge is supersonic, the character of the airblast signature becomes more shock-like and the explosion is said to be "shallow". While the airblast generating mechanisms of deep explosions are still present, their effect is generally overshadowed by the shock wave associated with the supersonically rising dome and plume. The intensity of this shock is closely related to the Mach number of the plume. A method of calculating the airblast from shallow explosions was developed. This method considers the driving mechanism of the plume to be the impulse transmitted by the explosion to the water above the charge. The plume motion is strongly influenced by the reaction of the air and is predicted analytically using supported-shock theory. This permits an approximate calculation of the peak airblast pressure using existing theory on off-axis shock-wave generation by supersonic blunt bodies.

Craig [54] has carried out a series of experiments on the water wave amplitude vs depth of charge detonation at underwater explosions. The position of the center of charges (25.4 mm in diameter) was varied from +6.4 to −610 mm relative to the water surface (0 mm). Craig noted that the shape, size, and the plume created in the result of the momentum of the collapsing water are apparently influenced by details of the explosive bubble collapse, by the remnant of the water encasing the steam of the mushroom cloud, and by water back from the jet. It's known that the wave amplitude vs. depth of a given charge of explosive has two maxima. One maximum corresponds to so called the upper critical depth (UCD) and occurs when the charge is between one-half and just fully submerged. The work [54] was done to confirm the UCD phenomenon as well as to provide quantitative data useful for calibrating a computer model of the phenomena. Increasing the charge radius by a factor of two (50.8 mm in diameter) did not result in any clear qualitative change.

As shown in [54], the explosive bubble had a general downward motion contrary to expectations from consideration of buoyancy alone. However, this motion is not in qualitative disagreement with the "image-source" model [55]. A point of concern in the experiments performed in a smaller tank by Ch. Mader [56] was the cause of the root extending from the bottom of the explosive bubble. It was assumed that the root can be due to either the small tank size, the method of support of the charge, or the MDF. The analogous studies were carried out by Hendricks and Smith [57] on above and below surface effects of one-pound underwater explosion. One needs to take into account that in this work a root was called a stem.

One can note that the root does not seem to have been previously recognized as separate from the bubble. Tracings of the bubble outline according to [58] could be interpreted as an approximate hemispherical bubble and a root. According to the experimental data of Kriebel [59] the tracing of the bubble generated by exploding wires can be explained in the same way. The root in these experiments was identified as a jet caused by inward collapse of the walls of the water column just before the cavity becomes fully expanded.

Hirt and Rivard [60] have investigated some aspects of explosively generated water cavities near the ocean surface to reproduce (in particular, upward and downward moving jets) in the numerical simulations the results of experiments and calculations received in [4, 54]. The authors [60] have supported the experimental observation of [4] regarding the source depth that produce the maximum upward jet mass. They also shown that the combination of following conditions can create the ideal situation for maximum effect: (1) it's desirable that the explosive cavity pressure acts on the water surface until it expands, and (2) one would like the cavity pressure to equilibrate with the atmosphere. It was noted, in particular, that a significant question arises with regard to the scaling that can be expected between nuclear and HE explosive sources at or near the UCD. Their opinion is unlikely that HE and nuclear sources can be scaled simply in terms of yield because, for instance, nuclear devices is producing a significant amount of vaporized water, a feature not associated with HE tests.

The survey of experimental and theoretical efforts devoted to obtaining an understanding of explosion-generated water waves was presented by Hirt in [61]. The subject of this report was using the appropriate computational model to reproduce experimental data for high explosive tests, which would be a useful tool both for exploring scaling relationships and assessing the applicability of similar scaling rules for nuclear sources. The author has noted [61] that although several investigations have used hydrodynamic computer codes to simulate nuclear bursts in water, for example, [62], and hydrodynamic phenomena such as jets, roots, stems and plumes [4, 54] these studies were qualitative and no detailed attempt has been made to verify computational models with experimental data for water wave parameters. In [61] a state-of-the-art computer program for free-surface hydrodynamics [63] was applied to model near-surface explosions and to investigate scaling processes. Based on these studies, modified scaling procedure including gravity and atmospheric pressure effects were developed.

Theoretical problems for explosion-generated wave propagation were also studied by Pace et al. [64], LeNehaute [65] and Houston, and Chou [66].

References

1. R. Cole: *Underwater Explosions* (Dover, New York, 1965)
2. B.V. Zamyshlyaev, Yu.S. Yakovlev: *Dynamic Loads at Underwater Explosion* (Sudostroenie, Leningrad, 1967)
3. V.P. Korobeinikov, B.D. Khristoforov: *Underwater Explosion*, Itogi Nauki Tekh. Gidromekh, 9, pp. 54–119 (1976)
4. V.K. Kedrinskii: *Surface Effects at Underwater Explosions (Review)*, Zh. Prikl. Mekh. i Tekh. Fiz. **19**, 4, pp. 66–87 (1978)
5. M.A. Lavrentiev, B.D. Shabat *Hydrodynamics Problems and Mathematical Models* (Nauka, Moscow, 1973)
6. L.V. Ovsyannikov: On floating of a bubble. In: *Some Mathematics and Mechanics Problems* (Nauka, Leningrad, 1970)
7. H. Kolsky, I.I. Lewis et al.: *Splashes from Underwater Explosion*, Proc. Roy. Soc. A **196**, pp. 379–402 (1949)
8. V.L. Zaonegin, L.S. Kozachenko, V.N. Kostyuchenko: *Experimental Study of Gas Bubble and Sultan Growth at Underwater Explosion*, Zh. Prikl. Mekh. i Tekh. Fiz. **1**, 2 (1960)
9. L.S. Kozachenko, B.D. Khristoforov: *Surface Effects at Underwater Explosions*, Fiz. Goren. Vzryva **8**, 3, pp. 433–439 (1972)
10. V.G. Stepanov, Yu.S. Navagin et al.: *Hydroexplosive Pressing of Ship Units* (Sudostroenie, Leningrad, 1966)
11. B.D. Khristoforov: *Interaction of Shock Wave in Water with Free Surface*, Zh. Prikl. Mekh. i Tekh. Fiz. **2**, 1, pp. 30–38 (1961)
12. A.G. Ryabinin, S.A. Khristianovich, A.A. Grib: *On the Reflection of Plane Shock Wave in Water from Free Surface*, Prikl. Mat. Mekh. **20**, 4 (1956)
13. A.G. Ryabinin: Reflection of a shock wave from the free surface in water. Ph.D. thesis, Leningrad (1957)
14. B.I. Zaslavskii: *On Reflection of Spherical Shock Wave in Water from Free Surface*, Zh. Prikl. Mekh. i Tekh. Fiz. **4**, 6, pp. 50–59 (1963)
15. M. Dubesset, M. Lavergne: *Calcul de Cavtation due aux Explosions Sous-Marines a Faible Profoundeur*, Intern. J. Acoustica **20**, 5 (1968)
16. R.A. Wentzell, H.D. Scott, R.P. Chapman: *Cavitation due Shock Pulses Reflected from the Sea Surface*, J. Acoust. Soc. Am. **46**, 2 (1969)
17. L. Briggs: *Limiting Negative Pressure of Water*, J. Appl. Phys. **21**, 7 (1950)
18. Ch.L. Mader: Detonations near the water surface. In: *Los Alamos Scientific Laboratory of the University of California*, report LA-4958, US-34 (Los Alamos, New Mexico, 87544, June 1972)
19. V.K. Kedrinskii: *The Experimental Research and Hydrodynamical Models of a "Sultan"*, Arch. Mechanics **26**, 3, pp. 535–540 (1974)
20. G. Flinn: Physics of acoustic cavitation in liquids. In: W. Mason (ed.) *Physical Acoustics*, vol. 1 B, (Academic Press, 1964) pp. 7–138
21. L.R. Gavrilov: Content of free gas in liquids and methods for measuring it. In: L. Rozenberg (ed.) *Physic Base of Ultrasound Technology*, Part 2 (M.: Nauka, 1970) pp. 395–426
22. V.K. Kedrinskii: On relaxation of tensile stresses in cavitating liquid. In: P. Pravica (ed.) *Proc. 13th Intern. Congress on Acoustics*, vol. 1 (Dragan Srnic Press, Sabac, 1989) pp. 327–330

23. V.K. Kedrinskii: *Dynamics of Cavitation Zone at Underwater Explosion Near Free Surface*, Zh. Prikl. Mekh. i Tekh. Fiz. **16**, 5 (1975)
24. B.S. Kogarko: *On One Model of Cavitative Liquid*, Dokl. Akad. Nauk SSSR **137**, 6, pp. 1331–1333 (1961)
25. S.V. Iordanskii: *On the Equations of Liquid Motion with Gas Bubbles*, Zh. Prikl. Mekh. i Tekh. Fiz. **1**, 3, pp. 102–110 (1960)
26. L. van Wijngaarden: *O the Equations of Motion for Mixtures of Liquid and Gas Bubbles*, J. Fluid Mechanics **33**, 3, pp. 465–474 (1968)
27. V.K. Kedrinskii: Dynamics of a cavity and a wave. In: V. Monakhov (ed.) *Dynamics of Continuum*, vol. 38 (Novosibirsk, 1979) pp. 48–70
28. I.S. Gradshtein, I.M. Ryzhik: *Tables of Integrals, Sums, Series, and Products* (M.: Fiz.-Mat. Lit., 1962)
29. M.M. Hasan, K.S. Jyengar: *Size and Growth of Cavitation Bubble Nuclei*, Nature **199**, 4897 (1963) p. 995
30. M. Strasberg: *Onset of Ultrasonic Cavitation in Tap Water*, J. Acoust. Soc. Am. **31**, 2, p. 163 (1959)
31. Se Din-Yu: *Bubble Growth in a Viscous Liquid Under Short-Time Pulse*, Theor. Osnovy Inzh. Rasch., 4 (1970) pp. 121–124
32. V.K. Kedrinskii: *Negative Pressure Profile in a Cavitation Zone at Underwater Explosion Near Free Surface*, Acta Astronaut. **3**, 7–8, pp. 623–632 (1976)
33. V.K. Kedrinskii: Influence of the developing cavitation zone on the rarefaction wave parameters. In: S. Kutateladze (ed.) *Nonlinear Wave Processes in Two-Phase Media*, Proc. of the 20th Siberian Thermophys. Workshop (Novosibirsk 1976) pp. 90–99
34. B.S. Kogarko: *One Dimensional Unsteady Motion of Liquid with Arising and Development of Cavitation*, Dokl. Akad. Nauk SSSR **155**, 5, pp. 779–782 (1964)
35. Yu.A. Boguslavskii: *To the Question About Arising and Development of Cavitative Rarefaction Wave*, Acoust. Journ. **13**, 4, pp. 538–540 (1967)
36. M.A. Tsykulin: *On Arising of Shock Wave in Air at Underwater Explosion*, Zh. Prikl. Mekh. i Tekh. Fiz. **2**, 1 (1961)
37. V.N. Kostyuchenko, N.N. Simonov: *Experimental Study of Shock Wave in Air at Underwater Explosion in Shallow Water*, Zh. Prikl. Mekh. i Tekh. Fiz. **1**, 1 (1960)
38. V.K. Kedrinskii: *About Underwater Explosion Near Free Surface*, Dokl. Akad. Nauk SSSR **212**, 2 (1973)
39. N.E. Kochin, I.A. Kibel', N.V. Rose: *Theoretical Hydromechanics*, Part 1, (M.:Fiz.-Mat. Lit., 1963) Chap. 4
40. F.P. Fil'chakov, V.I. Panchishin: *Modeling of Potential Fields on Conducting Paper* (Naukova Dumka, Kiev, 1961)
41. V.K. Kedrinskii: *Sultan Formation at Underwater Explosion*, Izv. Akad. Nauk SSSR, Mekh. Zhidk. Gaza, 4 (1971)
42. V.K. Kedrinskii: On oscillation of a cylindrical gas cavity in unbounded liquid. In: M. Lavrentyev (ed.) *Dynamics of Continuum*, vol. 8 (Novosibirsk, Institute of Hydrodynamics, 1971) pp. 163–168
43. D.F. Hopkins, J.M. Robertson: *Penetration of 2D Jet of Incompressible Liquid*, Mechanics, 1 (1969)
44. H.G. Snay: Underwasser-Explosionen Hydromechanische Vorgange und Wirkungen. In: *Naval Hydrodynamics*, Publ. 515. Chap. XII. (Nat. Acad. Sci., Nat. Research Council, 1957)

45. V.M. Titov: *About Regimes of Hydrodynamic Cumulation at Shell Collapse*, Dokl. Akad. Nauk SSSR **247**, 5 (1979)
46. V.K. Kedrinskii: *Hydrodynamics of Explosion (Review)*, Zh. Prikl. Mekh. i Tekh. Fiz. **28**, 4, pp. 23–48 (1987)
47. G. Birkhoff: *Hydrodynamics* (Inostr. Lit., Moscow, 1963)
48. J.W. Pritchett: Incompressible calculation of underwater explosion phenomena. In: *Proc. 2nd Int. Conf. on Numerical Methods in Fluid Dynamics* (Springer, Berlin Heidelberg New York, 1971) pp. 422–428
49. H.G. Snay: Model tests and scaling. In: *Naval Ordnance Lab. Rep.*, NOLTR-63-257 (1964)
50. V.K. Kedrinskii, V.T. Kuzavov: *Underwater Explosion of Coil Charge Near Free Surface*, Zh. Prikl. Mekh. i Tekh. Fiz. **24**, 4, pp. 124–130 (1983)
51. J.C.W. Rogers, W.G. Szymczak: Philosophical Transactions of the Royal Society of London, Series A-Mathematical Physical and Engineering Sciences **355**, 1724 (1997) pp. 649–663
52. T. Khangaonkar, B. Lemehaute: Applied Ocean Research **14**, 2 (1992) pp. 141–154
53. Ch.I. Malme, J.R. Carbonell, I. Dyer: Mechanisms in the generation of airblasts by underwater explosions. In: *US National Ordnance Laboratory, White Oak, Matyland*, report NOLTR 66-88, Sept 23 1966
54. B.G. Craig Experimental observations of underwater detonations near the water surface. In: *Los Alamos Scientific Laboratory of the University of California*, Los Alamos, New Mexico 87544, report LA-5548-MS, UC-34, Apr 1974
55. A.R. Bryant: The Gas Globe: Underwater Explosion Research 2 (Published by ONR), (1950)
56. Ch.L. Mader: Detonation near the water surface. In: *Los Alamos Scientific Laboratory of the University of California*, Los Alamos, New Mexico 87544, report LA-4958, US-34, June 1972
57. J.W. Hendricks, D.L. Smith: Above and below surface effects of one-pound underwater explosion. In: *Hydra I*, (US), report NRDL-TR-480 (1960)
58. F.H. Young, R.R. Hammond: A non-spherical model describing the motion of a shallow underwater explosion bubble. In: (US), report NRDL-TR-771 (1964)
59. A.R. Kriebel: Simulation of underwater nuclear bursts at shallow depth with exploding wires. In: report URS7028-1 (1971)
60. C.W. Hirt, W.C. Rivard: Wave generation from explosively formed cavities. In: *Defense Nuclear Agency*, report DNA-TR-82-131 (1983)
61. C.W. Hirt: Phenomena and scaling of shallow explosions in water. In: *Flow Science, Inc.*, report FSI-85-3-1 (1985)
62. M.B. Fogel et al.: Water waves from underwater nuclear explosions. In: *Pacifica Technology*, report PT-U82-0572 (1983)
63. B.D. Nichols, C.W. Hirt, R.S. Hotchkiss: SOLA-VOF: a solution algorithm for transient fluid flow with multiple free boundaries. In: *Los Alamos Scientific Laboratory*, report LA-8355 (1980)
64. C.E. Pace et al.: Effect of charge depth of submergence on wave height and energy coupling. In: report WES-TR-1-647-4 (1968)
65. B. LeMehaute: Theory of explosion-generated water waves: Advances in Hydroscience, **7**, 1 (1971)
66. J.R. Houston, L. Chou: Numerical modeling of explosion waves. In: *US Army Engineer Waterways Experiment Station*, Technical report (1982)

8 Conclusion: Comments on the Models

Analysis of different approaches to studying high-velocity hydrodynamic flows proves that "heuristic" models (i.e. ones guessed on the basis of some physic principles) often allow one to describe the phenomena when no exact solutions can be obtained or an approximate mathematical model cannot be adequately justified. In these cases, experiments play the role of "the existence theorem" and set the limits of reliability of the approaches.

Underwater Explosion

Indeed, formally, the transition from the exact acoustic model of the invariance of the function $G = r \cdot (h + u^2/2)$ on the characteristics c_0 to its invariance on the characteristics $(c + u)$:

$$\frac{\partial G}{\partial t} + c_0 \frac{\partial G}{\partial r} = 0, \quad \Rightarrow \quad \frac{\partial G}{\partial t} + (c+u) \frac{\partial G}{\partial r} = 0.$$

is unproved. But this approach made it possible to solve the problem of underwater explosion, i.e. to calculate the dynamics of an explosive cavity and the wave field parameters from the near zone to the asymptotic. Rather reliable correlation between the calculation results and experimental data proved the validity of the idea.

A solution for cylindrical symmetry, the most complicated case in modeling the explosion in an incompressible liquid, was obtained owing to a kind of "inverse problem" that involves:

- Deducing an equation for the dynamics of a cavity in a compressible liquid

$$R\left(1 - \frac{\dot{R}}{c}\right)\ddot{R} + \beta\left(1 - \frac{\dot{R}}{3c}\right)\dot{R}^2 = \frac{1}{2}\left(1 + \frac{\dot{R}}{c}\right)H + \frac{R}{c}\left(1 - \frac{\dot{R}}{c}\right)\frac{dH}{dt} \quad (8.1)$$

- Experimental determination of the coefficient β ($\beta = 1$)
- The limiting transition in the speed of sound ($c \to \infty$)
- The transformation of (8.1) into

$$R\ddot{R} + \dot{R}^2 = \frac{1}{2}H \quad (8.2)$$

The first integral of the equation multiplied by 2π,

$$\frac{\pi p(0) R(0)^2}{\gamma - 1} \left[1 - \left(\frac{R(0)}{R}\right)^{2\gamma - 2} \right] = 2\pi \rho_0 R^2 \dot{R}^2 + \pi p_\infty (R^2 - R(0)^2),$$

is the law of energy conservation, according to which the decrease in the internal energy of detonation products (in the left-hand part of the equation) should be equal to the change of the potential energy of the liquid–cavity system (the second summand in the right-hand part) and the kinetic energy increase. Hence it follows that the expression

$$E = 2\pi \rho_0 R^2 \dot{R}^2$$

determines the kinetic energy E of the *incompressible* liquid surrounding a cylindrical cavity. We keep in mind that this is eventually a result of the asymptotic approximation and the "empirical correction," but since the result is favorable, it has the right to exist.

We note that Eq. (8.2) allows for an analytical solution for an empty cavity, which in the dimensionless form is rewritten as

$$\tau^2 = 2 \cdot (1 - y^2),$$

where $y = R/R_0$ and $\tau = t \cdot \sqrt{p_\infty/\rho}/R_0$. Hence one can readily determine the first oscillation period of the cylindrical explosive cavity $\tau_* = 2\tau_{\max} = 2\sqrt{2}$, if R_0 is the radius of the explosive cavity with detonation products at the moment of maximum expansion. A comparison shows that the analytical estimate of this parameter differs from the experimental data only by several percent.

Bubble Cavitation and Cavitation Fracture of a Liquid under Shock-Wave Loading

The definition of the state of a real liquid with natural microinhomogeneities as that of a two-phase medium has radically modified the approaches to the description of cavitation phenomena. The main point is that it enables logical combination of the two problems: the evolution dynamics of cavitation phenomena and the strength of a liquid. The problems are at bottom reduced to one problem on *wave processes in a cavitating liquid*, in which the structure and the parameters of the wave field both depend on and determine the cavitation processes.

The processes can be described within a two-phase pk-model suggested for nonreacting bubbly media:

$$\Delta p - c_l^{-2} \frac{\partial^2 p}{\partial t^2} = -\rho_0 k_0 \frac{\partial^2 k}{\partial t^2}, \tag{8.3}$$

$$\frac{\partial^2 k}{\partial t^2} = \frac{3k^{1/3}}{\rho_0 R_0^2} (p_0 k^{-\gamma} - p) + \frac{1}{6k} \left(\frac{\partial k}{\partial t}\right)^2. \tag{8.4}$$

8 Conclusion: Comments on the Models   361

The model contains only two sought functions: the mean pressure p in the bubbly medium and the volume concentration k of the gas phase. Within the two approximations considering physical features of any bubbly systems, Eq. (8.3) can be rewritten as

$$\Delta p \simeq p \;.$$

The first approximation, the incompressibility of a liquid component, is quite justified, since the compressibility of a cavitating liquid is mainly determined by the compressibility of the gas–vapor phase. The second approximation is the allowance for the microscale of the heterogeneous structure of the medium through the new spatial variable $\zeta = \alpha r k^{1/6}$, and is rather "heuristic" because of the absence of a rigorous proof of validity of the required inequalities (Chap. 6). However, practical coincidence of the structure of a visible cavitation zone calculated using this model with the experimental data makes this approach quite justified.

As a result, the pressure in a cavitating liquid is determined from the Helmholtz equation, while the concentration k in the system of Eqs. (8.3–8.4) is solved using the spatial coordinates r and θ as parameters (in the axisymmetrical statement). Experiments proved that this approach is correct. Moreover, in describing the development of the cavitation zone near the bottom of a vertical tube filled with water that obtains a downward acceleration pulse the approach yields an analytical solution of Eq. (8.4) and an explicit estimate for the relaxation of the wave field in a cavitating liquid.

It is believed that the modeling of cavitation is complicated by the natural size distribution of nuclei. How should one consider the dynamics of their summary effect on the process, structure, and parameters of the wave field? An example of the calculation of the influence of dispersion effects in a bubbly medium on the its "acoustic transparency" is illustrated in Chap. 5. In this example, the governing system contains a certain discrete set of the Rayleigh equations for each kind of bubbles. Application of the sophisticated two-phase model makes it possible to explain the experimental data that might seem strange at first sight. Later on it is shown that this sophistication is often redundant. For example, in the problems on impulsive loading of liquids with rarefaction wave amplitudes of tens or even hundreds of negative atmospheres it suffices to use the "integral" Eq. (8.4) describing the dynamics of an average volumetric concentration. It was established in ad hoc studies that in rather strong rarefaction fields, the originally disperse systems acquire at early stages a monodisperse structure that is retained throughout the cavitation process.

The idea of *instantaneous relaxation of tensile stresses* in the zone of intense cavitation is justified only "physically." However, it enables a considerable simplification of the mathematical model and removes the restriction on the development of cavitation up to "dense packing" of bubbles. Analysis of the flow parameters in the cavitation zone within the model of instantaneous relaxation yields an unexpected conclusion: the mass velocity profile is "frozen" in the cavitation zone. Experiments on measuring the mass velocity by means of special traces and flash X-radiography in a hydrodynamic shock tube confirm this result. Thus, the model of instantaneous relaxation is justified. We note that this model enables solving the problem on spallation in cavitating liquids during shock-wave loading.

We mentioned that the cavitation processes are the initial stage of the disintegration of liquid media with microinhomogeneities by impulsive loading. The final stage of the disintegration is the formation of separate cavitating fragments (spalls), their further decay into smaller parts, and, as a result, formation of a gas–droplet medium. The entire process is defined as the inversion of the two-phase state of the medium. It seems that its complexity does not allow for an intelligible mathematical modeling even for the development of a physical analog of the transition process. Nevertheless, an adequate physical model is formulated assuming that for $k \approx 0.6\text{–}0.7$ the dense packing of bubbles immediately transforms into a dense package of elastic nonconfluent liquid drops. This model is called "sandy" model.

Calculations of the dispersion of such a shell with drops and comparison with the data of high-speed recording of the dynamic failure of liquid shells and shells composed of sand having diverse particle sizes shows that the model works and explains obscure facts of particle size distribution over the section of the disintegrating shell and the effect of the "fog" in the central part of a gas–vapor cloud formed by the explosion.

If we carry the "heuristic" idea of restricting the size of a liquid sample destroyed during shock-wave loading to seeming absurdity, it would be reduced to the size of a droplet. But this idea turned out to be fruitful for presenting a physical disintegration model and allowed one to reveal the fine structure of the inversion transition of one two-phase state (cavitating liquid) into another, the gas-droplet system.

Sultans

In solving the problems of the structure and mechanisms of the formation of directed throwing out on the free surface at shallow underwater explosions, the key role was played by the experiment–mathematical model–physical model–experiment sequence. Experimental results were used to simplify the mathematical model and the statement of the appropriate problem. The solution of the problem was a step towards constructing a physical model and verifying the understanding of the process mechanisms in model experiments. Thus, all problems of the complicated structure of vertical and radial sultans were solved successively and a new class of cumulative jets formed at various explosion depths was discovered. These studies were completed by the demonstration of hydrodynamic models of sultans formed under impulsive motion of a solid body from under the free surface and high-velocity penetration of the body into water, and the development of a direct method of measuring the change of the liquid mass in the sultan jet.

CPSIA information can be obtained at www.ICGtesting.com
Printed in the USA
LVOW070752221212

312896LV00004B/248/P